DOMESTICATION, CONSERVATION AND USE OF ANIMAL RESOURCES

World Animal Science

Editors-in-Chief:
Prof. A. Neimann-Sørensen,
Royal Veterinary and Agricultural University,
Institute of Animal Science,
23 Rolighedsvej,
1958 Copenhagen,
Denmark.

Prof. D.E. Tribe,
Australian Universities' International
Development Program (AUIDP)
P.O. Box 1142,
Canberra City,
A.C.T. 2601,
Australia.

Volumes in the Series

Subseries A: Basic information
Domestication, conservation and use of animal resources (ISBN 0-444-42068-1) — A 1
Development of animal production systems (ISBN 0-444-42050-9) — A 2
Anatomical basis of animal production
Dynamic biochemistry of animal production (ISBN 0-444-42052-5) — A 3
Basic physiological systems
Physiology of production
Microbiology of animals and animal products
Ethology
General and quantitative genetics (ISBN 0-444-42203-X) — A 4

Subseries B: Disciplinary approach
Genetic appraisal, selection and breeding of animals
Cattle genetic resources
Genetic resources of pig, sheep and goat
Animal reproduction
Animal feeding and nutrition
Grazing animals (ISBN 0-444-41835-0) — B 1
Feed science
Animal housing and accommodation
Bioclimatology and adaptation
Health management and preventive veterinary medicine
Parasites, pests and predators (ISBN 0-444-42175-0) — B 2
Animal production and environmental-health management
Meat science, milk science and technology

Subseries C: Production-system approach
Beef-cattle production
Dairy-cattle production
Sheep and goat production (ISBN 0-444-41989-6) — C 1
Pig production
Horse breeding and management
Buffalo production
Production of other traditional domesticated mammals and of newly or partially
    domesticated mammals
Furred-animal production
Poultry production
Production of animals other than mammals and birds
Laboratory animals
Concluding volume (and cumulative index)

World Animal Science, A 1

DOMESTICATION, CONSERVATION AND USE OF ANIMAL RESOURCES

Edited by

L. PEEL

Department of Agriculture and Horticulture, University of Reading, Earley Gate, Reading RG6 2AT, Great Britain

and

D.E. TRIBE

AUIDP, P.O. Box 1142, Canberra City, A.C.T. 2601, Australia

ELSEVIER
Amsterdam — Oxford — New York — Tokyo 1983

ELSEVIER SCIENCE PUBLISHERS B.V.
1, Molenwerf
P.O. Box 211, 1000 AE Amsterdam, The Netherlands

Distributors for the United States and Canada:

ELSEVIER SCIENCE PUBLISHING COMPANY INC.
52, Vanderbilt Avenue
New York, NY, 10017

Library of Congress Cataloging in Publication Data
Main entry under title:

Domestication, conservation, and use of animal resources.

  (World animal science ; A1)
  Includes bibliographical references and index.
  1. Domestic animals. 2. Domestication. 3. Germ-
plasm resources, Animal. I. Peel, Lynnette J. (Lynnette
Jean), 1938-   . II. Tribe, D. E. (Derek Edward),
1926-   . III. Series: World animal science ; v. A1.
SF61.D65 1983   636   83-1498
ISBN 0-444-42068-1 (U.S.)

ISBN 0-444-42068-1 (Vol. A 1)

ISBN 0-444-41836-9 (Series)

Printed in The Netherlands

General Preface

Several factors make it desirable at this time to collect and integrate existing knowledge in Animal Science in its widest sense.

Millions of people in the world are today suffering from starvation or malnutrition and the number will increase with the inevitable rise in the world population. This poses an inexorable challenge to all scientists involved in problems underlying the production of food for man. Yet the development of livestock industries does not only aim to improve the nutritional standards of the human population, important though that is. From man's point of view animals are multipurpose and their use can have different objectives: economic, social and ecological. In addition to being important as sources of food, clothing and certain forms of power, animals can also represent forms of wealth, recreation, means of employing labour, aesthetic enjoyment, and determinants of landscape.

Animal production must increasingly compete with other forms of production for resources, especially energy, but also for land, water, finance and labour. This creates a greater need than ever to develop systems which maximize efficiency. At the same time, these systems need also to meet other requirements. They must be environmentally beneficial, ethically defensible, socially acceptable, and relevant to the particular aims, needs and resources of the community they are designed to serve.

Rapid advances in knowledge within practically all areas of the animal sciences are now being made. Solutions to many of the problems which face the livestock producer, whether he is working in, say, a cattle feed-lot in the U.S.A. or in a traditional system of village goat production in West Africa, are now resulting from the research being carried out in the various disciplines of Animal Sciences. However, too often the results of this research remain confined to the specialized journals of different scientific disciplines when, increasingly, the approach of those working in animal production needs to be interdisciplinary and global. Furthermore, Animal Science has attained a new dimension in recent years. Whatever form it takes, animal production constantly influences and interacts with the other components of the total ecosystem within which it operates. New disciplines, like ecology, ethology and conservation have become important; new forms of production, such as aquaculture and the use of non-conventional feed resources are receiving increasing attention.

The scientists and planners in animal production have to work within the framework of these developments. Extensive inquiries among such specialists in many parts of the world have revealed the need for a comprehensive and up-to-date review of the Animal Science literature covering the entire range of technical knowledge that is now required in animal production and development. Therefore, Elsevier Scientific Publishing Company has initiated this major work of reference under the title "World Animal Science". Inevitably, such a task must cover many volumes and involve the collaboration of a great number of editors and authors. Through an elaborate preparatory phase and with continuous editorial guidance from the Chief Editors and the Editorial Advisory Board, including scientists from all parts of the world, the aim has been to produce an integrated series of volumes which, although not encyclopaedic and not intended to be exhaustive in each branch of knowledge, does give appropriate emphasis to coherence and applicability. With this in mind the series has been divided into three parts: the first set of volumes provides information on the anatomical, genetical, biochemical, physiological and microbiological bases of animal production; the next fourteen volumes are each devoted to a particular discipline important to animal production, e.g. reproduction, breeding, feed science, bioclimatology and adaptation; and finally, in the last thirteen volumes, production systems are described on a species basis, covering beef and dairy cattle, sheep, pigs, horses, buffaloes, poultry and some newly or partially domesticated mammals.

Emphasis is laid throughout on the careful reviewing and integration of significant knowledge in the total field of animal production with the aim of reporting not only what is known but also of drawing attention to important gaps in our knowledge. Account is taken of current trends in thinking and development and controversial topics are also dealt with, e.g. ethical aspects of animal use. Traditional farm animals, i.e. cattle, horses, sheep, goats and pigs, are given major emphasis and their production systems are treated in special volumes. There are also separate volumes on the conservation and use of their genetic resources. Other forms of animal production, such as poultry and fur production, are given less attention in the early volumes but still have their special volumes dealing with production systems. Other topics of less over-all importance or about which less is known, e.g. aquaculture, are treated in separate chapters within volumes, while some highly specialized types of animal production, e.g. silkworms, are described in only a single chapter. Because of the increasing importance of animal production in less developed areas of the world, attention is paid to domesticated mammals such as buffalo, camel, lama, alpaca, yak, reindeer and elephant, and to such newly or partially domesticated mammals as the eland, oryx, red deer and musk-ox. The series deals with both intensive and extensive animal production systems. In this way an attempt has been made to set the whole of world animal production into an appropriate and contemporary perspective.

Although, by editorial concept and by cross-referencing between volumes, the series functions as a single entity, each volume is nevertheless constituted as an independent unit, suitable for separate use. To achieve this and still avoid undue overlap, each volume only summarizes those essential elements of topics which are treated in detail in other volumes. Thus each volume aims to approach the breadth of World Animal Science from its own particular point of view, with supporting references to details in other volumes.

The series is written primarily for use by people who have a specialist interest in animal production as students, teachers, extension officers and consultants, policy-makers, and research scientists. The volumes are planned for world-wide use, which implies that the information presented covers systems and principles of more than local or national interest.

Contents

Chapter 1

Domestication, Dispersal and Use of Animals in Europe

S. BÖKÖNYI

1. THE SOURCES OF EUROPEAN ANIMAL HUSBANDRY

The origins of European animal husbandry go back to four main sources:

(1) The Near Eastern centres of animal domestication from where (a) sheep and goat, the most important species of the earliest animal husbandry of Europe, had their origin; (b) at least two other species had been domesticated earlier than in Europe and might have been carried over to Europe.

(2) Southeast Europe where (a) the first European domestication of the economically most important domestic animals of the temperate belt occurred; and (b) where the first real animal husbandry of Europe with five domestic species took shape.

(3) Individual locations on the European mainland where Mesolithic or even late Paleolithic communities carried out isolated attempts at carnivore (dog) or omnivore domestication which never resulted in a real animal husbandry, keeping several species.

(4) Domestication of the locally available wild forms of domesticated animals already known (after the introduction of the animal husbandry from the southeast) in order to increase the number of domestic stock all over Europe.

In order to determine what was contributed by each of these sources to the development of animal husbandry in Europe it seems to be useful to carefully check the four sources from the viewpoints of zoology, zoogeography, prehistory and also human cultural dispersal.

In the development of European animal husbandry the Near East undoubtedly played a role. The question is, how essential was this role? If nowadays prehistorians or archaeozoologists discuss the origin of the Neolithic domestic fauna of Europe, two opinions will clash. According to the group holding the first opinion, the domestic fauna consisting of five species with an overwhelming majority of sheep and goats, which can be found already at the earliest beginning of the Neolithic in Southeast Europe, had its origin in Southwest Asia. The species of this domestic fauna were domesticated there and arrived already in domesticated form in Southeast Europe. The other group of specialists holds the opinion that the animal husbandry of Southeast Europe had a local origin which was contemporaneous with that of Southwest Asia or even preceded it, or at least certain species were domesticated earlier in Southeast Europe than in Southwest Asia.

A well-known principle of archaeozoology is that an animal species would be domesticated only where its domesticable wild form lived. Starting out with this principle, one can divide the five Neolithic domestic species into two groups: the wild forms of sheep and goat did not live on the East Euro-

Chapter 1 references, p. 19

pean mainland in post-Pleistocene times; on the other hand, those of cattle, pig and dog did. At the same time, the wild forms of all five Neolithic domestic species existed in Southwest Asia in the early Holocene, the time of earliest domestication.

This means that the small ruminants could not have been domesticated in Southeast Europe; for their place of origin one has to look to Southwest Asia from where they reached Europe already in domesticated form. In this respect the animal remains of the Franchthi Cave at Nauplion, Greece, gave useful evidence. The late Pleistocene and early Holocene layers of this cave yielded no bones of the wild forms of sheep and goat, and these small ruminants appeared already in evidently domesticated form in the pre-pottery Neolithic.

The three other species could have been and undoubtedly were domesticated in Europe too; it is another question whether their domestication happened there earlier or later than in Southwest Asia.

The case of dog is rather clearer since Turnbull and Reed (1974) proved the presence of the domestic dog in the Zarzian (terminal Pleistocene) level of the Palegawra Cave, Northeast Iraq, from around 12 000 B.P. This is the earliest date for domestic dog at present; the earliest finds in Europe go back only to the 8th millenium B.C. and those of the Balkans to the middle of the 7th millenium B.C. Probably also the pig was earlier domesticated in Southwest Asia than in Southeast Europe since the pigs of Qalat Jarmo, Northeast Iraq, were c. 500 years earlier than those of Argissa-Magula, Greece, and even they were not locally domesticated, which means that somewhere else in Northeast Iraq or West Iran pig domestication had taken place earlier.

The case is not so simple with cattle. The earliest finds of domestic cattle, from Argissa-Magula, Greece, and Çatal Hüyük, Anatolia, are roughly contemporaneous and come from the middle of the 7th millenium B.C. In my opinion one has to suppose the existence of one single large centre of cattle domestication in the northeastern basin of the Mediterranean, including Anatolia, in the 7th to 6th milleniums B.C.

The next question is — and this leads to the second point from above — did the special type of domestic fauna consisting of the above-mentioned five species, with an overwhelming caprovine majority, develop in this form in Southwest Asia and get from there to Europe or were only some of its species domesticated there, with the others joining them in Southeast Europe?

Faunas containing all five domestic species of the Neolithic first appeared in Southwest Asia around 6000 B.C. In the whole area, in every archaeological site, the caprovines — sheep and goat — represented the overwhelming majority among domestic animals; the ratios of other domestic species were rather variable. The importance of animal husbandry in comparison to that of hunting was also rather variable. In most of the sites the ratio of domestic animals was far higher than that of the hunted ones; however, there were some settlements where the importance of hunting as a meat and raw material source still exceeded that of animal keeping.

While the five Neolithic domestic species appeared together in Southwest Asia first around 6000 B.C., this animal husbandry was present in Southeast Europe as early as the middle of the 7th millenium B.C. This means that although four or even five species of the earliest Neolithic animal husbandry had been domesticated earlier in Southwest Asia than in Southeast Europe and these species occurred alone or in different combinations in sites earlier than anywhere in Southeast Europe, the typical Neolithic animal husbandry with five domestic species did not develop in the Near East.

At present, one thing can be stated with certainty: a domestic fauna based on caprovines and most probably containing also dog and pig coming from Southwest Asia reached Southeast Europe around 6500 B.C. Whether the domestic cattle had originally been a member of this fauna, or joined it in Europe, remains an open question.

In the southern part of the Balkan Peninsula not much was added to the domestic stock of this animal husbandry. Cattle domestication was never very extensive in the Greek Neolithic for the simple reason that aurochs never lived there in great quantity. Aurochs like large plains with forested spots, and such are very rare in Greece. The greatest number of aurochs could have lived in Thessaly and Eastern Macedonia, however, they were much scarcer everywhere than for instance in the Carpathian Basin.

The domestication of the pig was of greater importance to the Greek Neolithic settlements. Wild swine were abundant on the Greek mainland during the Neolithic, and man was eager to try his hand at this domesticable stock. Freshly domesticated animals are found quite frequently in the bone samples of the settlements, and domesticated pigs often show quite close relations to the local wild ones. In the Greek early Neolithic sites there was no trace of local dog domestication. Some domestic dogs and wild wolves occurred but no transitional animals could be found which would have proved a local dog domestication. The difference of the relative importance of cattle and pig domestication shows at every site; in the earliest Greek domestic fauna there were more pigs than cattle. At the same time the number of dogs was extremely low.

For a long time it was an axiom in archaeology and archaeozoology that domestication had begun at the dawn of the Neolithic. Recently, however, more and more data show that the beginnings of domestication go back to the Mesolithic or even the late Paleolithic. Nevertheless, these early attempts at domestication exclusively concerned carnivores (dog) and omnivores (pig), animals which could live on human food remnants. The earliest European domestication of this kind happened in the Knie Cave in Thuringia, Germany, where primitive dogs (domestic wolves) were found in the Magdalénien culture of the late Paleolithic c. 8300 B.C. Just 200 years later were the Mesolithic dogs of Star Carr, Great Britain, the dog found in the Senckenberg Moor near Frankfurt, Germany, and the dogs of 10 late Paleolithic and Mesolithic sites of the Crimean peninsula. The Maglemose dogs of Denmark are again somewhat later, their age being 6810 B.C.

The dog domestication of Vlasac, in the Iron Gate gorge of the Danube in Yugoslavia, is around 1000 years later, and its special interest is that it was done by the surviving Mesolithic population in a remote valley at a time when the whole area was already invaded by Neolithic animal keepers coming from the south.

From Crimean late Paleolithic and Neolithic caves, also, 'domestic' pigs were described; nevertheless, their domestic nature still needs confirmation. Icoana, an Epipaleolithic site on the Rumanian side of the Iron Gate gorge, also yielded domesticated-like pigs but one needs more studies to settle the question of their status.

Connected to the success of these attempts at dog and pig domestication was certainly the fact that both species could survive on human food. At the same time, however, this fact also hindered the large-scale keeping of these species because they would be food competitors with man. The grazing and browsing ruminants, on the other hand, exploited and converted into meat a type of food which could not be used by man, dog or pig either — this was the cellulose of straw and hay. At the same time the sheep domestication supposed by some authors to have occurred (Radulesco and Samson, 1962; Zeuner, 1963) did not happen in Europe due to the lack of wild sheep on

Chapter 1 references, p. 19

the European mainland as mentioned above. The allegedly wild sheep of the above authors are in fact primitive domestic sheep coming with the first wave of domestic animals from Southwest Asia via Southeast Europe.

Although it was often debated before, it is rather obvious now (and also the above examples showed) that domestication did not begin in one single centre. Just the opposite: as man arrived at a particular stage of economic and cultural development at the same time in many areas, he began to control and domesticate the locally available animal species. Thus each species was domesticated at about the same time in different places. And what is designated the earliest site of domestication of a given species can be a matter of sheer chance, depending on where archaeological research has been most successful. According to our present knowledge, the earliest animal domestication undoubtedly took place in Southwest Asia, but after the acquisition of the first domestic animals and being familiar with the use of domestic animals and the technique of domestication, man started domestication in Europe, too. Here, aurochs and wild swine were extremely good objects for such domestication, and also a third, though economically less important, species, the wolf, was present. Man really domesticated them, particularly the aurochs, in order to increase his domestic stock. The large size of the aurochs populations of certain areas resulted in the development of some large cattle domestication centres in Europe. The most important one was undoubtedly on the Hungarian Plain where immediately after the end of the early Neolithic the auroch became the most commonly hunted animal; this hunting was closely connected with the capture and domestication of its calves (Bökönyi, 1974). Another large cattle domestication centre also existed on the Wallachian Plain, and others on the smaller and larger plains of Central and Western Europe. On the other hand, there is hardly any trace of cattle domestication in the mountains, the aurochs being an animal of the plains.

The large-scale domestication of aurochs ceased with the end of the Neolithic primarily because the greater part of the domesticable wild stock was exterminated. At that time man switched over to the wild swine whose domestication was on rather a small scale in the Neolithic, with the exception of the early Neolithic of Greece (Bökönyi, 1973) and some sites of the Late Linear Pottery culture in Central Europe (Müller, 1964; Murray, 1970), but began to flourish in the Copper and Bronze Ages.

2. THE DISPERSAL OF ANIMAL HUSBANDRY AND DOMESTIC ANIMALS IN EUROPE

As the preceding section shows, not counting isolated experiments with dog and occasionally pig domestication, the animal husbandry with five domestic species — cattle, sheep, goat, pig and dog — first appeared in Southeast Europe, more exactly in Greece, around the middle of the 7th millenium B.C. This was a seemingly perfect animal husbandry for the European Neolithic, as demonstrated by the fact that during the Neolithic there were changes in the ratios of the different domestic species but in the course of the whole period no new species were domesticated by Neolithic man, and it took 3000 years before a sixth species, the horse, was acquired to broaden the circle of domestic animals.

The leading element of this earliest animal husbandry of Southeast Europe was the caprovines, and this was so in every early Neolithic site of Greece, independent of the geographic setting, in Argissa-Magula and Achilleion in Thessaly, Nea Nikomedeia in Greek Macedonia, Lerna and the Franchthi

Cave on the Peloponnese and Knossos on Crete. Pigs were far less common, followed by cattle and then dog. This domestic fauna thrived in Greece as the geographic and climatic conditions were very similar to those of Southwest Asia, the homeland of the most important species. An indication of how successful this animal husbandry was in Greece is that the inhabitants of the sites were hardly forced to hunt and fish in order to complete their animal protein needs.

During the early Neolithic this domestic fauna started to advance to the north. Although the route is not yet clear in detail, the main directions can already be determined. The advance reached the southern part of Yugoslav Macedonia within the 7th millenium B.C. and other parts of Yugoslavia and also Bulgaria in the 6th millenium B.C.

In Yugoslav Macedonia, environment and climate were very similar to those of Central and Northern Greece. It is not surprising therefore that in the early Neolithic of the region (Starčevo Culture) a domestic fauna very similar to that of early Neolithic Greece could be found. The only difference was that there the cattle—pig ratio was opposite to that in the Greek sites, thus cattle were more frequent than pigs. Even the very low wild ratio resembled that of the Greek sites.

In the course of the 6th millenium B.C. the early Neolithic animal keepers invaded the whole Balkan peninsula and advanced partly in the river valleys and partly along the sea coast to the north in three directions. On a straight northern route they arrived in the Carpathian Basin by the end of the 6th millenium B.C. On a northeastern route they reached the south Ukraine, around the middle of the 5th millenium B.C. (in the southern part of East Europe one must also reckon with waves of animal keepers coming from the Near East through the Caucasus and the Transcaspian region), on a third route they reached the Appenine peninsula, went along the northern coastal region of the Mediterranean and inundated the southern part of West Europe in the course of the 5th millenium B.C.

During the times when the Neolithic domestic fauna introduced by the Starčevo Culture extended and gained a firm foothold over the whole Balkans, the local Mesolithic population still made a living out of hunting and first of all fishing in inaccessible places, and made isolated attempts at dog domestication. In this respect the best data are yielded by sites of the Iron Gate gorge of the Danube, Lepenski Vir and Vlasac on the Yugoslav side and Icoana, Razvrata and Veterani on the Rumanian side. In the Yugoslav sites the dog is the only domestic species and in Vlasac its local domestication can be proved. The occurrence of domesticated dogs is very probable also in the Rumanian sites; in Icoana there were found traces of a so-called 'suid-economy', an economy based on the exploitation of pigs.

The northern advance of animal husbandry was probably connected with the so-called 'climatic optimum', a warm wave in the early Holocene, which also took wild animal species of Mediterranean or even Southern Asiatic origin up to the Carpathian Basin. Interestingly enough the end station of this advance was in the Carpathian Basin, and neither the southern wild species nor the southern-type animal husbandry based on the caprovines went any further in their original forms. Nevertheless, domestic animals of this animal husbandry reached regions lying to the north through trade or quite often as booty. The situation was similar in the South Ukraine and in South France. In this way the domestic animals reached practically all of temperate Europe by the end of the 4th millenium B.C.

The earliest animal husbandry of the southern parts of Europe was rather uniform. In most of the sites the caprovines represented the overwhelming majority (between the two small ruminants the sheep were always more numerous than the goats and sometimes accounted for around 90% of the caprovines), followed by cattle; the pig was far less prevalent and the dog was always very rare. In a few sites, particularly in well-forested areas, cattle surpassed the frequency of caprovines but that was all.

This caprovine-based domestic fauna did well in the hot and dry environment of the South Balkans and along the Mediterranean, and its leading species bred very well. The result was that this form of animal husbandry did not have to undergo any essential changes in these areas for a long time. Man was also not forced to hunt or fish very much because he could get enough food and raw material of animal origin from animal husbandry.

It was not the same in the northern half of the Balkans, the South Ukraine and the Carpathian Basin. In the latter regions this animal husbandry of the dry climate reached the forests and swamps. Sheep and goat are certainly not the species of wet environments and did not breed at all well there. Therefore man was forced to hunt and fish more in order to complete his food needs. In early Neolithic settlements of these regions a very high ratio of the animal remains come from wild mammals, birds and fishes, and huge shell heaps reveal that collecting also played an essential part in securing food for the human population. Thus this animal husbandry had to change and in fact did so.

After becoming deeply rooted in Southeast Europe, the animal husbandry began to adapt itself to the local conditions. The caprovine-based husbandry was alien and unable to develop further in the temperate belt of Europe. It was alien because its most important species could find suitable living conditions only in the Southern Balkans at best, and it was unable to develop further because its leading species had no domesticable wild forms locally and their number could therefore not be increased through local domestication. In this respect one must not forget that Neolithic people were peasants, and for peasants two real values have existed, land and livestock. Land for cultivation was enough in those times but the primitive agricultural instruments and the lack of draught power strongly limited the size of cultivated land. On the other hand, the care of a larger amount of livestock, up to a certain number, was not an inexecutable additional work. Also one cannot exclude the possibility that the larger amount of livestock, particularly that of cattle, meant also some kind of social prestige — as with nomadic peoples even today.

As already mentioned, in temperate Europe there were two domestic species which were well-adapted to the local environmental conditions and had domesticable local wild forms at the same time: the cattle and the pig. (A third one was the dog but economically it was unimportant). The Neolithic peasants had therefore to switch from caprovines to cattle and pig, and this they did. The whole process began in the early Neolithic; however, it reached its climax only in the middle Neolithic. This change over can be studied best in the Carpathian Basin and its neighbouring territories where from the middle Neolithic onwards cattle became the most frequent domestic animal in practically every site of each culture, generally followed by pig.

In this period hunting also entered into the service of domestication. Since domestication could start only with young animals, in order to be able to capture them man had to kill the adult individuals of the domesticable wild forms because they tried to protect their young. Therefore the animal bone samples of such sites were full of adult aurochs and wild swine bones. In every middle and late Neolithic site of Hungary the aurochs was the main animal hunted and this, together with the high ratio of domestic cattle and pig, is so characteristic for these sites that their age can be determined through it — even without having any artefacts.

In these sites also hunting itself was more important than in those of the early Neolithic. The reason was not only that it was connected with a large-scale domestication but also that it had to replace the losses caused by the troubles of the change-over to the new type of animal husbandry.

In the Northern Balkans, Rumania and the Carpathian Basin cattle became the most frequent species in the animal husbandry. Behind them pig and caprovines alternatively occupied the second place in the Balkans and in Rumania, while in the Carpathian Basin pig firmly kept the second place before the caprovines. There caprovines were in second place only in one culture, the Linear Pottery culture (Bökönyi, 1974). In fact the Linear Pottery culture was the main propagator of the cattle-based animal husbandry in Europe. Between c. 4600 and 3700 B.C. this culture had a wide distribution, probably the widest among early and middle Neolithic cultures, extending from Western Russia and Northeast Rumania, to Eastern France and Holland. In its early phase caprovines occupied the second and pig the third place, in the late phase they changed places, probably as a result of the switch-over to local resources. Dog bones were remarkably rare (Müller, 1964). In the Linear Pottery culture this new type of animal husbandry seems to have been quite productive. The best evidence is that the ratio of hunted animals' bones is never higher than 10—20% in its sites.

Without wanting to go into the small details of Neolithic animal husbandry in Europe one should still touch upon the development found in the late Neolithic (Herpály culture) of Hungary because animal keeping and hunting showed very special features there. The importance of animal keeping sharply decreased there without any changes in the ratios of the domestic species. The frequency of domestic animals fell to about 30% and the amazingly increased hunting almost exclusively served the domestication. Primarily aurochs was hunted, and the cattle samples of the sites were full of freshly domesticated individuals connecting the wild and domestic populations.

This phase is often called the period of 'domestication fever' when man, recognizing the secondary uses (milk, wool, draught power, etc.) of his domestic animals, tried to increase the number of his domestic animals by all means. By the end of this period man had almost completely eradicated the largest wild aurochs population of Europe and put an end to large-scale cattle domestication for ever.

With the end of the Neolithic an important phase of development came to an end in animal husbandry in Europe. In the following Copper and Bronze Ages processes of a completely different nature took place. In these latter periods knowledge about animal keeping reached such a level that animal husbandry was able to supply the human population with enough meat, milk, wool, draught power, etc., for its needs and at the same time not only keep the domestic stock population at its original level but even increase it, and all this essentially without any local domestication. As mentioned above, the domestication of cattle practically ceased, that of the pig still played a role for a certain time; however, it did not contribute considerably to the increase in the domestic pig stock. As a new species of domestic fauna, the domestic horse occurred around the middle of the 4th millenium B.C. The importance of hunting strongly decreased not only because man did not need to hunt, having a successful animal keeping, but also because land used for agriculture considerably increased and a great part of the habitat of the wild animals was destroyed.

European animal husbandry was influenced by eastern and southeastern elements at the beginning of the Copper Age, around the mid-4th millenium B.C., and by a strong southeastern wave at the end of the period, by the second half of the 3rd millenium B.C. The eastern elements caused only

Chapter 1 references, p. 19

unimportant changes in animal husbandry. At the beginning the domestic horses were introduced one by one or in small groups to the Carpathian Basin and the Northeastern Balkans from the Ukraine, without developing any real horse keeping there. The result of the intrusion of southeastern elements of the early Copper Age was also only a few more caprovines in the domestic fauna. But the strong southeastern wave of the late Copper Age completely transformed the domestic fauna of a large part of Central Europe into one similar to that of the early Neolithic. (This is not surprising because the origins of the two waves were similar.) At once the caprovines became the leading species of the domestic fauna again, and only the rare occurrence of the domestic horse and the low ratio of wild animals distinguished this fauna from that of the early Neolithic.

It is true that this domestic fauna subsequently dissolved into that of the indigenous European population; however, its effect was long lasting. In the first third of the Bronze Age caprovines were still nearly as frequent or even more frequent than cattle in the domestic fauna, and the large sheep, coming from the Mesopotamian centre of the earliest conscious sheep breeding (Kraus, 1966), having better wool than that of the Neolithic sheep, transformed not only the sheep population but also human clothing, making possible a switch-over from leather clothes to woollen ones.

Towards the end of the Copper Age the domestic horse became widely known in Europe; however, it reached a high frequency as the result of real horse keeping only at the beginning of the Bronze Age. The people of the Bell Beaker culture seem to have been the best propagators of the keeping of this new domestic species; the frequency of horses sometimes amounted to as much as 40% in their settlements in the Carpathian Basin. (Interestingly enough these were horses of eastern origin very much resembling those of the Ukrainian domestication centre.) In the Bell Beaker settlements of Western Europe horses were not as numerous as in the Carpathian Basin, but generally speaking there was no Bell Beaker settlement in Western Europe without domestic horses; even Ireland is no exception in this respect.

The earliest horse domestication took place in the South Ukraine around the middle of the 4th millenium B.C., and the first settlement where bones of domestic horses were uncovered was the Copper Age (Srednij Stog culture) site of Dereivka near the town of Kremenchug, on the right bank of the Dnieper River (Bibikova, 1967).

The domestication and dispersal of horse meant much more than the addition of a new species to the list of domestic animals. Its importance was not the result of its direct economic use; in this respect it was surpassed by several other species, first of all the cattle. Its real importance was that it made revolutionary changes in transport and warfare. The swift horse and the spoked-wheeled cart connected with it considerably accelerated travel and transportation and in this way also the range of trade. In warfare, the effect of chariots first used in the Bronze Age and the cavalry that appeared in the Iron Age can be compared to the tanks of the 20th century.

In the course of the Bronze Age the strongest effect on animal husbandry was produced by a climatic change that took place in the 2nd millenium B.C. This climatic change was of much smaller scale than those of the Pleistocene; nevertheless, it exerted a strong influence on the flora, fauna, and economy. The origins of this climatic change reached back to the Neolithic but it was mainly felt in the 2nd millenium B.C. Its essence was that the hot and dry climate of the Neolithic gradually changed into a cold and wet one. As a result the forestation increased and forest species preferring a wet environment and surviving under a cold climate became frequent in the wild fauna. Red deer, in particular, an animal of the dense forest, became common, and arctic species (e.g. wolverine) migrated from the north to Central Europe.

The domestic animals cannot leave an area because of a climatic change but since the species preferring the new situation breed better, man will keep them first of all. In other words, nature forces the animal keepers to change their domestic stock according to the new circumstances. In the course of the Bronze Age the caprovines and the horse, animals of the dry environment, gradually decreased in number in favour of the pig, a typical animal of the wet environment. With the horse something else also happened: the narrow-hoofed steppe horses of the early Bronze Age transformed into a wide-hoofed, forest type, a clear adaptation to the soft, swampy soil of the new environment, by the late Bronze Age.

Thus at the end of the Bronze Age an animal husbandry based on cattle and pigs with fewer caprovines, horses and dogs existed and grew into the Iron Age type. The deterioration of the climate, but also other factors, caused the animal husbandry of the latter period to sink to a low level: the domestic animals were of extremely primitive type and of unbelievably small size.

This animal husbandry was not able to regenerate unaided. The conscious breeding of sheep and cattle reached Europe from Southwest Asia and that of the horse came over from Persia and Middle Asia in the Iron Age. The first affected the Greeks, the second the Scythians, and immediately started to flourish in both places. The Greeks took over the knowledge of breeding from the Scythians and imported Scythian horses from the Carpathian Basin as well, and the tens of thousands of breeding mares from the Scythian and Persian booties of Philip of Macedonia and Alexander the Great also helped considerably in improving Greek horses.

Besides the dispersal of good eastern horses, the Scythians also played an essential part in the propagation of the domestic hen, taken over from the Persians. From the Greeks they acquired new animal species, the cat and the ass; the first one became the luxury animal of persons of rank. The ass also could not have been very common and well known, as shown by the fact, mentioned by Herodotus, that the horses of the Scythian cavalry were often disturbed by the Persians' asses and mules. In the propagation of the domestic hen the Celts also had an important role.

The Greeks' knowledge of conscious animal breeding, together with their improved breeds, was taken over and developed further by the Romans. The Roman animal breeding was based on a nearly scientific agricultural literature and due to this fact they developed improved breeds of high productivity which spread all over the Roman Empire, and their influence could be noticed even outside the boundaries of the Empire.

The Romans introduced new domestic species, originating mostly in Africa or Asia, such as the ass, the camel, the cat, the peacock, the guinea hen, and also completed the dispersal of the domestic hen in Europe. In fact they laid the foundations of the modern European domestic fauna. Although this fauna certainly suffered heavy losses after the fall of the Roman Empire, and most of its improved breeds disappeared, the germs of the conscious animal breeding which had been developed by the Romans survived. Roman animal husbandry remained alive in some towns miraculously surviving the whole migration period, and single Roman luxury breeds, e.g. greyhounds, and heavy horses, also remained in the possession of chieftains or noblemen. When conscious animal breeding came to life again with the Renaissance it began on the classical basis.

Chapter 1 references, p. 19

3. EARLY DOMESTIC ANIMALS OF EUROPE

The emergence and development of domestic animals were the result of human activity. Thus the quality of domestic animals in every period was the function of the quality, success or failure of man's activity in domesticating and keeping animals. It is true, that when man began domesticating animals he had acquired a lot of knowledge about their anatomy, biology, physiology and behaviour — information gathered by thousands of generations of hunters. But all this concerned wild animals, at the beginning man did not know well how to handle animals in captivity. He learnt this later, in the first phase of animal breeding which grew out of domestication. I call this phase *animal keeping*, that is the primitive form of animal breeding without conscious breeding selection and the control of feeding. Its most important characteristics are: (a) in a given population there is only one breed, and (b) the domestic animals are of a primitive type and — in the case of mammals — are smaller in size than in their wild form. The origins of breeding selection, however, do go back to prehistoric times, for with the introduction of castration of some males, from the late Neolithic onwards, man was practising a certain selection, i.e., he selected some males for breeding and excluded others. Nevertheless, this process was not conscious selection for high productivity, e.g., for size, strength, speed, meat production, milk yield, wool, etc.

The basis of the second phase, the real *animal breeding*, is conscious breeding selection and the control of both quantity and quality of feeding. It begins when man treats his domestic animals as individuals and not only as a herd. The best criteria of this advanced phase are: (a) different breeds live together in a population, (b) the size of animals increases but dwarf breeds also appear, and (c) the productivity of animals increases.

Thus the lack of breeding knowledge was the reason man did not feed his domestic animals properly in the period immediately following domestication. This resulted in early domesticated animals falling far behind their wild ancestors in size and showing characteristics of types of slow development.

3.1. Cattle

Neolithic cattle were 25—35 cm smaller in the withers than wild aurochs, their average withers height being 120—130 cm (Bökönyi, 1974). In the Neolithic in some regions even dwarf cattle occurred with c. 110 cm withers height. The withers height of the cattle continued to diminish by approximately 6—7 cm by the Bronze Age and by an additional 8—9 cm by the Iron Age. At that time the average withers height of domestic cattle hardly surpassed 1 m. With the first introduction of conscious animal breeding in the Roman Imperial Period the size of cattle abruptly increased. The average was around 125 cm but some individuals reached 140 cm or even more. With the fall of the Roman Empire the good cattle population was destroyed and the withers height began to decrease, reaching a new low of just above 1 m in the early Middle Ages, the 10th—13th centuries A.D. Thereafter only the newly introduced, conscious animal breeding of the Renaissance brought about an increase again.

From the Neolithic to the Iron Age one cattle breed existed in Europe; nevertheless it was rather variable in its skull form. Craniologically, most of the cattle were very close to the wild form (*primigenius* type), having a straight intercornual ridge, flat forehead and long, heavy horns. This is no surprise because with local domestication this was always the type that found its way into the stock of the domestic animals. Also the *brachyceros* (short-horned) type, representing a more advanced domesticated state, had appeared already in the early Neolithic, while hornless cattle were first found only at the end of the Neolithic. In the Copper Age, another representative of the advanced domesticated state, the *frontosus* type, occurred.

During the Iron Age the *brachyceros* type became more frequent, and among Scythian cattle hornless individuals occurred in large numbers. With the Romans a large longhorn cattle breed, a result of conscious breeding, was introduced in the provinces of the Empire all over Europe. It certainly exerted a strong influence on the aboriginal cattle populations, and scattered individuals, mostly draught oxen, reached territories outside the Empire, including those of the Celts and the Sarmatians. With the end of the Roman Empire this Roman cattle breed vanished.

3.2. Sheep

The Neolithic sheep were smaller than their wild form, their average withers height being around 60 cm. The large sheep of the southeastern migration wave of the late Copper Age substantially raised the withers height to nearly 70 cm by the early Bronze Age. The effect of this wave can be noticed all over Europe. From then the size curve of sheep was identical to that of cattle. In the Iron Age one can witness the same size decrease as in the case of cattle, the withers height fell back below 60 cm to the Neolithic and Copper Age level. In the Roman Imperial Period, also, the size of sheep increased by 8—9 cm on average, only to give way to another decrease in the early Middle Ages to a height even shorter than in the Iron Age. A slow increase began in the Renaissance which carried over into the scientific sheep breeding of the Modern Age.

Early sheep also belonged to one single breed whose males had heavy three-edged spirally twisted horns, while the females had short and untwisted horns or were hornless. Hornlessness occurred rather early among domestic sheep in Europe. The earliest ones were found in the early Neolithic (middle of the 7th millenium B.C.) of Achilleion, Greece. The sheep introduced at the end of the Copper Age did not differ much from the earlier ones. Besides their larger size only their horns were somewhat heavier and the ratio of hornless individuals seems to have been a little lower. Their essential characteristic that put them above the earlier sheep was that they had wool, a rather coarse wool mixed with kemps.

A considerable change took place in sheep breeding when by the middle of the 1st millenium B.C. the Greeks brought the methods of sheep breeding, along with a high level of wool processing, from their colonies in Asia Minor to Europe. As a result, in Greece a systematic sheep breeding was pursued as early as the Classical Period. A whole series of breeds came into being which differed from one another not only in their outer appearance but also in their use.

Conscious sheep breeding spread out of Greece in a wide radius. In the north it reached Scythia, resulting in the first occurrence of really fine wool in a tumulus grave in the vicinity of Nymphaeum, a Greek colonial town on the Crimean Peninsula in the 5th century B.C. (Ryder and Hedges, 1973). The place and date of this fine wool are very interesting because the legend of the golden fleece is also associated with the same area and time period.

The Romans took the basis of their conscious sheep breeding from the Greeks, through Magna Graecia. They developed it further and introduced their fine wool sheep to all their provinces. The Roman sheep did not differ craniologically from the aboriginal ones, but through their fine wool had clear superiority over them.

After the fall of the Roman Empire this breed disappeared, and the peoples of the Migration Period brought eastern breeds to Europe but these latter did not exert very much influence on the European sheep population except in Southeast Europe.

3.3. Goats

Due to the scarcity of finds the size changes of goat are not well known. Among prehistoric goats of the Balkans there were some males with extremely heavy horns, probably individuals which reached their full maturity. More large goats occurred in the Roman Imperial Period, probably from some Italian improved breed. As for their horn form, in prehistoric times many scimitar horned goats (*aegagrus* type) — resembling the wild ancestor — could be found, though the majority of the goats had twisted horns (*prisca* type). From the Roman Imperial Period onwards, the number of scimitar horned goats gradually decreased. The earliest hornless goats appeared in Europe quite late, in the Roman Imperial Period, and even at that time they were extremely rare: from the whole Empire three specimens are known.

3.4. Pigs

The Neolithic domestic pigs were mostly small animals; nevertheless, the average withers height was comparatively high, around 77 cm, because of the occurrence of freshly domesticated individuals. It was about the same in the Copper Age, too, but in the Bronze Age the withers height suddenly rose to nearly 80 cm due to increasing pig domestication resulting in more large, freshly domesticated animals. As with other domestic species the size of pig also sharply decreased in the Iron Age to about 71 cm. The upswing of the Roman Imperial Period was less in the case of pig, showing that the Romans did not care particularly much for the improvement of pig breeds. In the early Middle Ages another sharp decline occurred, below 67 cm, and with the end of the Middle Ages the size of pig began to rise very slowly.

Most of the Neolithic pigs were craniologically, and in their body shape, very close to the wild swine. Their skulls were long, narrow and wedge-shaped. This skull form was very suitable for grubbing, an essential method used by early domestic pigs for acquiring food. Also their trunks were narrow with a convex back. Clay figurines often show that these pigs still had a hairy crest along their back similar to that of the wild form.

From the end of the Neolithic onwards, pigs on a higher stage of domestication also appeared. They had a shortened skull with a concave profile and tooth abnormalities (crowded, overlapping teeth). In Classical Greece and also in the Roman Empire there were traces of conscious pig breeding, however, its importance was much smaller than in the case of other domestic species, and osteologically the existence of different breeds cannot be confirmed.

3.5. Dogs

The earliest domestic dogs were certainly smaller than the wild wolves but their skull form still resembled that of the wild ancestor in several traits: the forehead was flat, the brain-case was only moderately arched, the median crest was high, giving enough room to strong muscles. On the other hand, the facial part and the lower jaw were considerably shorter than in the wolf, and the large teeth, whose size decrease always lags one step behind the jaw shortening, stood crowded in them. The extremity bones were long and straight, pointing to a medium-sized running-dog type.

The somewhat smaller, so-called *turbary* dogs appeared somewhat later, representing a more developed stage and shortly became widespread in the Neolithic dog population. Their main craniological characteristics were the well-arched brain-case, flat, median crest and pointed facial part.

In the Roman Imperial Period dog breeding also had an upswing. In Roman towns four to six different breeds appeared, ranging from tiny lap dogs and dachshunds to greyhounds and mastiffs.

3.6. Horses

Probably the similar way of life ensured that the early domestic horses were very near to their wild ancestors. The similarity was so great both in skull and body shape that purely on an osteological basis wild and domestic horses could rarely be distinguished, and other distinguishing features have to be taken into account such as age, sex and bone distribution, pieces of horse gear, etc.

Among the early domestic horses of Europe two groups can be distinguished (Bökönyi, 1978). The size of the horses of Eastern Europe and the Carpathian Basin was very close to that of the horses domesticated in Dereivka, the first horse domestication centre. Their withers height was between 126 and 144 cm, on average about 138—139 cm. In the western part of Europe the prehistoric horses were not only rarer but also considerably smaller, with an average withers height of 126—127 cm.

In the Early Iron Age the Scythians brought large numbers of excellent horses to Europe. From the breeder's viewpoint they were much better than the contemporaneous western horses whose typical representative was the Celtic horse. Therefore it is not surprising that it was prestigious for chieftains and other distinguished persons living in the area of western horses to acquire animals from the eastern breed. Such horses — with Scythian gear — were found in the graves of their western masters (Bökönyi, 1968).

As already mentioned, Persian and Scythian horses played an important role in developing the Roman military horse, a comparatively heavy breed but not identical with the 'cold blood' occidental horses. The latter were the result of conscious breeding at the end of the Migration Period and in the early Middle Ages for war purposes, for in this period heavy knightly armour appeared and large, strong horses were needed for carrying riders in armour.

3.7. Other species

The history and development of the other European domestic species have not yet been studied in such detail and have therefore not been elucidated as well. We certainly know the time of their earliest appearance and sometimes that of their subsequent appearance or appearances (e.g. in the case of the guinea hen), we know the route of their introduction, we have some ideas about the effects exerted by conscious breeding and the development of their breeds; however, far less is known about other domestic species than is known about the six species discussed above.

4. THE USE OF EARLY DOMESTIC ANIMALS

It is not very easy to determine what the domestic animals were used for

---

Chapter 1 references, p. 19

in prehistoric times. There are three sources of information: (a) the kill-off-patterns or the age frequencies of the different species; (b) animal representations; (c) remains of tools or gear connected with different types of use.

The kill-off-patterns or frequencies of the different age groups in the animal bone sample of a Neolithic site can reveal a lot about the exploitation of the different species. It is quite obvious that animals kept for their flesh were killed at a young age not only because their meat was tastier but also because of the winter fodder shortage which meant that man could feed only the valuable breeding stock. The best proof of this is that in Neolithic sites of Southeast Europe 70—90% of the pigs were slaughtered in their juvenile or subadult age, thus before their first or second winter. On the other hand, animals with other uses (wool, milk, draught power, etc.) were killed mainly in their adult or mature age because they had to reach their full maturity in order to become productive with their secondary uses. When man needed the meat of such animals, he killed primarily young males. The frequencies of age and sex groups of sheep in Anzabegovo clearly demonstrate this: man killed c. 2/3 of the males before their first winter and more than 1/4 before their second winter, keeping only c. 8% of them for breeding. At the same time, he did not kill any females before their first winter but killed c. 80% of them in their second year of life (when some of them could certainly have produced young), and kept c. 20% of them for breeding (see Table 1.1). Looking at the division of age groups of the main domestic animals of Sitagroi-Photolivos, as shown in Table 1.2, exploitation types discussed above can be clearly distinguished.

As for animal representations, lucky finds can prove the occurrence of certain primary or secondary uses of domestic animals; for example, a figurine of a woolly sheep or a relief showing a cow being milked. Finally, tools or gear such as bits, wheels and shearing scissors are also good proofs of certain types of exploitation.

4.1. Meat

The main aim of domestication, whether for securing meat reserves or sacrificial animals, can be debated. One thing is certain, however, and that is that in the Neolithic man ate all five of his domestic species. It is not accidental that man had hunted the wild ancestors of four out of five of his Neolithic domestic animal species for their meat. Man killed most of his domestic animals at an immature age. In this respect the dog was no exception, as is proved by the occurrence of skulls with open brain-cases (brain was considered a delicacy even in those times) and of extremity bones broken up for the marrow. Man ceased eating dog flesh only at the end of the Bronze Age, and this was probably the first food avoidance in Europe.

During the whole Neolithic, cattle were the main meat supplier animal species. One must not forget that the quantity of meat from one cow was equivalent to that from seven caprovines or four to five pigs. In this way c. 40% of all domestic animal meat came from cattle even in caprovine dominated animal husbandries. In husbandries which were based on cattle this ratio could be as high as 75—90%. In the early Neolithic, caprovine meat was also important but from the middle Neolithic onwards pig meat became the second most important meat after that of cattle.

4.2. Secondary uses

The secondary uses of domestic animals, the essence of which is man's exploitation of the animals without killing them, had already appeared in

TABLE 1.1

Frequencies of age and sex groups of sheep at Anzabegovo

		Female	Male	Undetermined	Total
Juvenile	Specimen	0	33	4	37
	%	0	64.7	66.7	44.1
Subadult	Specimen	22	14	1	37
	%	81.5	27.5	16.7	44.1
Adult	Specimen	5	4	1	10
	%	18.5	7.8	16.7	11.9
Undetermined age	Specimen	43	19	4	66
Total	Specimen	70	70	10	150

After Bökönyi (1976b).

TABLE 1.2

Division of age groups of the main domestic animals of Sitagroi-Photolivos

	Cattle (%)	Sheep (%)	Pigs (%)
Mature	35.3	39.0	10.0
Immature	64.7	61.0	90.0

After Bökönyi (1973).

the Neolithic. Nevertheless they first became important in the final phase of the Neolithic and in the Copper Age. This is apparent from the kill-off-patterns of cattle and caprovines in these periods when the number of adult and mature individuals increased considerably.

4.2.1. Cattle

The production of milk and draught power can be taken into consideration firstly as the secondary uses of cattle. But they also served as sacrificial animals (Behrens, 1964) and were used in bull-fights originating in Southwest Asia and introduced to Europe through Crete to Greece and Italy. In Rome bulls were sacrificed as part of the *suovetaurilia*, and as bone finds excavated at the *Niger Lapis*, where these rituals took place, prove, all the bulls were from the ancient, possibly prehistoric breed, demonstrating the deep roots of this custom (Blanc and Blanc, 1958—1959).

It is obvious that the earliest secondary use of cattle or of domestic animals in general, was milking, and the utilization by man of the superfluous quantity of milk would have been quite natural (Thévenin, 1947). Cattle on a low level of domestication would have had hardly any milk besides the quantity required for feeding their calves but in the case of the death or killing of the calves man could use the milk of the cows. The first representation of cattle milking and milk processing is quite late but I am convinced that milking goes back far more in time. The representation originates from the temple of Nin-Hursag in Ur, after 2400 B.C. (Zeuner, 1963), in which it is interesting to note that cows are being milked from the rear, as is usual with goats. From this fact one may infer that the milking of cows may have emerged after the milking of the goats. Milking was a particularly important

Chapter 1 references, p. 19

use of cattle in Classical Greece and Rome. Aristotle mentions that cows of the large Molossos breed gave one full amphora of milk a day. In Rome the colostrum milk and whipped cream were considered delicacies, and the breakfast of the average Roman was cow milk (Keller, 1909, 1913).

Nevertheless, the draught power of cattle was also used both in Greece and the Roman Empire, and the large Roman oxen were highly appreciated even outside the Empire. Concerning the use of draught power of cattle, Murray (1970) suggests, from the high ratio of adult oxen in Linear Pottery sites, some kind of unwheeled vehicle, like the travois, during the Neolithic of Europe, and Gimbutas (1970) is of the opinion that the wagon was introduced to Europe as early as the 4th millenium B.C.

4.2.2. Sheep

The secondary uses of sheep were first of all wool and then milk. The earliest representation of wool sheep is from Tepé Sarab in Eastern Iran (Bökönyi, 1974) from about the 6th millenium B.C. Woolly sheep may have reached Europe already in the course of the Neolithic; however, the first big wave of sheep with better wool arrived there at the end of the Copper Age, in the second half of the 2nd millenium B.C. Besides fleece finds in the Bronze Age of Denmark, this is also shown by the changed kill-off-patterns and the sheep statuettes of the Bronze Age representing woolly sheep. As already mentioned, sheep with really fine wool first occurred in Greece and were dispersed to the whole of Europe in the Roman Imperial Period. A good proof of the quality of Roman wool in comparison with the Bronze Age Danish wool is that while the latter wool had 31 hair fibres and only five wool fibres out of a total of 36 fibres, in the former the ratio of hair to wool fibres was 2:10. In Rome sheep milk was also consumed, although mostly in the form of cheese. As sacrificial animals sheep were not often used in pre-historic times but in Rome sheep were one of the three species used at the *suovetaurilia*. Peoples of the Migration Period often put whole sheep or sheep skulls and feet into the graves of the deceased.

4.2.3. Goats

The only secondary use of goats was milk production since woolly goats are not known from prehistoric and early historic periods. In classical times goat milk was highly appreciated and the cheese made from it was the most frequently mentioned cheese type. In Greece goats took the place of sheep in the local form of *suovetaurilia*.

4.2.4. Pigs

The pig is the only domestic animal species of Europe which has always been a meat animal without secondary uses, except for the role of sacrificial animal, but this was not a typical secondary use because it involved the killing of the animal. The real importance of the pig as a meat animal was first realized in the late Neolithic and the Bronze Age when it had to provide meat to replace the meat of those cattle and caprovines which had secondary uses and were therefore not killed in their immature age. Neolithic and Bronze Age clay figurines suggest that the fattening of pigs began in early prehistoric times. Classical sources from Greece mention that pigs were fattened on figs and peas. Both in Classical Greece and in Imperial Rome pork was the favourite meat type. When killed, pigs were first cut into pieces, then sprinkled with white flour and baked on a spit. The Romans knew several ways of conserving pork, ham, etc., and the manufacture of sausages was also widespread.

The pig was a kind of totemic animal of the Celts as shown by the extraordinary frequency of pig bones, skulls or skeletons cut in two in their whole length in Celtic graves and also by the statuettes found at settlements.

In Greece and Rome the pig was a rather common sacrificial animal, and Indo-Europeans killed pigs when making agreements, taking oaths, etc.

4.2.5. Dogs

Dogs could have served as hunting companions and herd-dogs very early, for recent observations show that wolves can shepherd groups of ungulates very well in order to separate their prey from the group. As watch dogs they helped man from the beginning of settled human life. With the beginning of conscious animal breeding a whole series of dog breeds developed for special purposes, e.g. lap dogs, dachshunds, mastiffs, greyhounds, etc.

The role of the dog as a sacrificial animal must also not be underestimated. In Lepenski Vir, the famous site of the surviving Mesolithic population in the Iron Gate gorge of the Danube, dog sacrifices were found in almost all of the buildings. From the Neolithic onwards, the dog was one of the domestic animals most frequently buried in the grave of its master.

4.2.6. Horses

The domestication of the horse differs in one essential point from that of the five domestic species mentioned above: whereas the domestication of the latter species was without precedent and for the purpose of securing meat reserves, the domestication of the horse occurred when man was aware of the so-called secondary uses of domesticated animals.

It is plausible, therefore, that in addition to domesticating the horse for its meat (the consumption of which was given up by man only after the conversion to Christianity) another motivation was to utilize it as a draught animal. Proof that the horse was used as a riding or draught animal immediately after its domestication comes from horse bits dated to the early Copper Age (Telegin, 1973). This theory is also supported by kill-off-patterns determined at Copper and Bronze Age settlements.

The horse achieved a close relationship with man. The chief reason for this was that the horse as a riding animal was a comrade-in-arms of its master whose life often depended on the horse. Probably this is why the horse was paramount in man's view of the afterlife. The practice of killing a horse and placing it in the grave of its dead master began immediately after its domestication. Horse sacrifice became even more frequent during the Bronze Age. The horse outnumbers all other wild and domestic species found in ritual graves up to the Christian era (Behrens, 1964). Horse skulls were believed to ward off evil spirits. Up to the most recent times, horse skulls were buried beneath the threshold or floors or were placed on poles in front of dwellings or boundaries throughout Christian Europe.

Horse milk was consumed only by East European nomadic and semi-nomadic peoples, also in the form of yoghurt and cheese.

5. SUMMARY

European animal husbandry goes back to four main sources: (1) the early domestication centres in Southwest Asia; (2) primary domestication in Southeast Europe; (3) isolated attempts with dog and/or pig domestication in the latest Paleolithic and the Mesolithic in Europe; (4) secondary domestication centres all over Europe.

The first domestic fauna of Europe consisting of cattle, sheep, goat, pig and dog, with a clear caprovine predominance, first occurred in the southern

Chapter 1 references, p. 19

half of the Balkan peninsula. Of these species the caprovines surely, the pig
and the dog probably, and the cattle possibly came from Southwest Asia.
This was a seemingly perfect animal husbandry for the Neolithic of Europe
as is proven by the facts that (a) through 3000 years man did not find it
necessary to broaden the circle of his domestic animals by adding new
species to it, and (b) in the southern half of Europe where this husbandry
found a favourable environment its species bred so well that man hardly
needed to hunt or fish to meet his needs for animal protein and raw materials.

In the 7th millenium B.C. this animal husbandry began to advance to
other parts of Europe. During the 6th millenium B.C. it inundated the whole
of the Balkans and entered the Carpathian Basin just before 5000 B.C. In the
course of the 6th millenium B.C. it also occupied the Appennine peninsula
and going along the northern coast of the Mediterranean it arrived in the
southern part of Western Europe in the 5th millenium B.C. On the east it
reached the South Ukraine around the middle of the 5th millenium B.C.

These were the northernmost points reached by the earliest European
animal husbandry in its original form. From there on only individual animals
or animal groups were moved from one settlement to the other by trade or
as spoils of war.

In the newly occupied parts of the continent the early animal husbandry
type had to adjust itself to the local environmental conditions. The essence
of this adjustment was a switch-over from caprovines that had no domesticable wild forms in Europe to cattle and pig whose wild forms were at hand
and whose number could therefore be increased not only by breeding but also
through local domestication. In this way new centres of domestication
developed throughout Europe; some of them, for example that of cattle in
the Carpathian Basin, became of decisive importance particularly in the final
phase of the Neolithic.

After the end of the Neolithic, large-scale domestication also came to an
abrupt end because man's knowledge about animal keeping reached such a
high level that animal husbandry alone was able to supply the human population with enough food and raw material of animal origin and to increase the
domestic stock as well. Among the domestic species a new and very important one appeared: the horse.

At the end of the Copper Age a new southeastern wave of domestic fauna
arrived in Europe, bringing a large number of domestic sheep. This domestic
fauna subsequently dissolved in that of the indigenous European population;
however, its effects lasted through several centuries.

Also a climatic change of the 2nd millenium B.C., when the earlier hot
and dry climate became cold and wet, had its influence on the animal husbandry, increasing the numbers of pig and decreasing those of the caprovines
and the horse. As a result of this climatic deterioration the animal husbandry
sank to the first low point of its history from where only the conscious
animal breeding of the Greeks and Romans helped it out. In the Roman
Imperial Period, based on a nearly scientific agricultural literature, animal
husbandry had an enormous upswing. Unfortunately this did not last long
for it disappeared with the fall of the Roman Empire. In the early Middle
Ages another decline came, and only the Renaissance lifted animal husbandry up to the former Roman level.

As for the early domestic animals of Europe, they represented a primitive
type characterized by small size and slow development. In the course of the
prehistoric times most domestic species underwent a size decrease process
whose deepest point was in the Iron Age. The conscious breeding selection
and the improved breeds of the Roman Imperial Period considerably raised

the size of the domestic animals, but after the fall of the Roman Empire not only most of the improved breeds disappeared but a general decline in size also followed. New breeds and another size increase occurred again only in the Renaissance.

Since the main aim of domestication was to secure meat reserves, the first domestic animals were nothing but living food conserves. Neolithic man ate all of his five domestic species; the first food prohibition came by the end of the Bronze Age in Europe when man gave up dog meat consumption. Nevertheless, he soon discovered the secondary uses of his domesticates, uses he could exploit without killing the animals. The use of milk, wool, hunting, companionship, house and herd watching appeared quite early, some of them already in the early Neolithic. Others, first of all draught power, riding or carrying of burden, came later but still in the Neolithic and became important in the Copper Age, with the domestication and dispersal of the horse, this typical beast of burden. The use of bulls in bull-fights also goes back to the Neolithic; however, it reached its first climax in Europe during the Bronze Age. One should not underestimate the sacrificial role of the domestic animals either. It began more or less simultaneously with the earliest domestication and lasted up to the Middle Ages, in some places even surviving the conversion of peoples to Christianity.

6. REFERENCES AND LITERATURE CONSULTED

Amschler, J.W., 1949. Ur- und frühgeschichtliche Haustierfunde aus Österreich. Arch. Aust., 3:1—100.

Behrens, H., 1964. Die Neolithisch-frühmetallzeitlichen Tierskelettfunde der Alten Welt. Deutscher Verlag der Wissenschaft, Berlin, 135 pp.

Bibikova, V.I., 1967. Kizučeniu drevneishykh domashnikh loshadei Vostočnoi Evropy (Studies of ancient domestic horses in Eastern Europe). Byull. Mosk. Obšč. Ispyt. Prir. Otd. Biol., 72:106—118.

Blanc, G.A. and Blanc, A.C., 1958—1959. Il bove della Stipe votiva del Niger Lapis nel Foro Romano. (The cattle of the 'stipe votiva' of Niger Lapis in the Forum Romanum.) Boll. Paletn. Ital., N.S., 12:7—57.

Boessneck, J., 1958. Zur Entwicklung vor- und frühgeschichtlicher Haus- und Wildtiere Bayerns im Rahmen der gleichzeitigen Tierwelt Mitteleuropas. Stud. an vor- u. frühgesch. Tierrest. Bayerns, II, München, 170 pp.

Bökönyi, S., 1968. Data on Iron Age horses of Central and Eastern Europe. Bull. Am. Sch. Prehist. Res. (Cambridge, MA), 25, 71 pp.

Bökönyi, S., 1969. Archaeological problems and methods of recognizing animal domestication. In: P.J. Ucko and G.W. Dimbleby (Editors), The Domestication and Exploitation of Plants and Animals. Duckworth, London, pp. 219—229.

Bökönyi, S., 1973. Stock breeding. In: D.R. Theocharis (Editor), Neolithic Greece. National Bank of Greece, Athens, pp. 165—178.

Bökönyi, S., 1974. History of Domestic Mammals in Central and Eastern Europe. Akadémiai Kiadó, Budapest, 597 pp.

Bökönyi, S., 1976a. Development of early stock rearing in the Near East. Nature (London), 264: 19—23.

Bökönyi, S., 1976b. The vertebrate fauna of Anza. In: M. Gimbutas (Editor), Neolithic Macedonia. Monumenta Archaeologica, I, University of California, Los Angeles, pp. 313—363.

Bökönyi, S., 1978. The earliest waves of domestic horses in East Europe. J. Indo-Eur. Stud., 6:17—76.

Bökönyi, S. and Kubasiewicz, M., 1961. Neolithische Tiere Polens und Ungarns in Ausgrabungen. I. Das Hausrind. Soc. Sci. Stetin, Szczecin, 94 pp.

Brentjes, B., 1965. Die Haustierwerdung im Orient. Die Neue Brehm-Bücherei, 344, Ziemsen Verlag, Wittenberg, 112 pp.

Clason, A.T., 1967. Animal and Man in Holland's Past, I—II Palaeohistoria, XII A & B, Wolters, Groningen, 436 pp.

Ducos, P., 1958. Le gisement de Châteauneuf-les-Martigues — les mammifères et les problèmes de domestication. Bull. Mus. Anthropol. Préhist., Monaco, V.

Gimbutas, M., 1970. Proto-Indo-European culture: the Kurgan culture during the fifth, fourth, and third millennia B.C. In: G. Cardona, H.M. Hoenigswald and A. Senn (Editors), Indo-European and Indo-Europeans. University of Pennsylvania Press, Philadelphia, pp. 155—197.

Hančar, F., 1955—56. Das Pferd in Prähistorischer und früher historischer Zeit. Wien. Beitr. Kulturgesch. Linguist., XI, Wien, Munchen, 650 pp.

Herre, W. and Röhrs, M., 1973. Haustiere — Zoologisch Gesehen. Gustav Fischer Verlag, Stuttgart, 240 pp.

Keller, C., 1909. Die Stammesgeschichte unserer Haustier. B.G. Teubner, Leipzig, 114 pp.

Keller, C., 1913. Uber die Huastierfunde von La Tène. Mitt. Thurg. Naturforsch. Ges., 20:140—153.

Kraus, F.R., 1966. Staatliche Viehhaltung im altbabylonischen Lande Larsa. Meded. K. Ned. Akad. Wetensch., Afd. Leterkd., Amsterdam, 29 (5), 65 pp.

Krysiak, K., 1951. Szczatki zwierzece z osady Neolitycznej w Cmielowie (Animal remains from the Cmielow Neolithic settlement). Wiad. Arch., 17: 165—228.

Krysiak, K., 1952. Szczatki zwierzece z osady Neolitycznej w Cmielowie (Animal remains from the Cmielow Neolithic settlement). Wiad. Arch., 18:251—290.

Müller, H.-H., 1964. Haustiere der mitteldeutschen Bandkeramiker. Naturwiss. Beitr. Vor-Frühgesch., 1. Akademie Verlag, Berlin, 181 pp.

Murray, J., 1970. The First European Agriculture. Edinburgh University Press, Edinburgh, 380 pp.

Radulesco, C. and Samson, P., 1962. Sur un centre de domestication du mouton dans le Mésolithique del la grotte 'La Adam' en Dobrogea. Z. Tierz. Züchtungsbiol., 76: 282—320.

Ryder, M.L. and Hedges, J.W., 1973. Ancient Scythian wool from the Crimea. Nature (London), 242: 480.

Telegin, D.J., 1973. Über einen ältesten Pferdezuchtherde in Europa. Actes VIIIe Congr. Int. Sci. Préhist. Protohist., II, Beograd, pp. 324—327.

Thévenin, R., 1947. Origine des Animax Domestiques, Presses Universitaires de France, Paris, 126 pp.

Turnbull, P.F. and Reed, C.A., 1974. The fauna from the terminal Pleistocene of Palegawra Cave, a Zarzian occupation site in northeastern Iraq. Fieldiana, Anthrop., 63: 81—146.

Zalkin, V.I., 1962. K istorii životnovodstva i okhoty v Vostočnoi Evrope. (The history of animal husbandry and hunting in Eastern Europe.) Mater. Issled. Arkheol. S.S.S.R., Vol. 107, 107 pp.

Zeuner, F.E., 1963. A History of Domesticated Animals. Hutchinson, London, 560 pp.

Chapter 2

Domestication and Use of Animals in the Americas

ELIZABETH S. WING

1. INTRODUCTION

Animals domesticated in the Western Hemisphere have contributed relatively little to the 20th century food economy. In contrast, a number of plants, such as corn, beans, peanuts, and potatoes, originating in the American tropics, are at the foundation of modern forage and subsistence crops. Although most of the animals domesticated in the New World are not now used primarily for food, they are used for a variety of other purposes. Turkeys are, for example, primarily a feast food, a delicacy reserved for Thanksgiving, Christmas, or weddings. Guinea pigs, originally used for food, are now only used as such in a circumscribed area in South America, primarily Ecuador and Peru. They have a much wider use now as experimental animals — to the extent, in fact, that experimental animals are called guinea pigs. Alpaca, raised only in the central highlands of South America, are farmed for their luxuriant wool. The name alpaca has become so synonymous with the warmth of alpaca wool that synthetic 'alpaca' garments are manufactured. Parrots and tropical fish, which may only be considered tame rather than fully domestic, are basic to the pet trade. Parrots were also kept in Precolumbian times perhaps as a ready source of colourful feathers. In Precolumbian times the New World domesticated and tamed animals also had a wide variety of roles although these were of course not necessarily the same as they are today.

Detailed information about these roles and the precise sequence of events leading to the domestication and subsequent spread of domestic forms is not well known. The centre of much of the New World animal domestication is in the South American Andes, and although much archaeological research has been conducted in this region, identification and interpretation of South American faunal remains, upon which studies of the origin of animal domestication are based, has only just begun. At the time of writing, the basic outline of South American animal domestication has begun to emerge but, clearly, continued research will modify and augment the picture of domestication as we see it now.

2. DEER

The single most important animal in the Americas during Precolumbian times was the deer. This relationship appears to be similar to the intensive use of red deer in Europe before the Neolithic (Jarman, 1972). Its primary use was for subsistence but hides and bone, a raw material for tool manufacture, were also vitally important. The species most widely used was the white-tailed deer, *Odocoileus virginianus.* More regional use was made of the mule deer, *Odocoileus hemionus*, in western North America, brocket deer, *Mazama* spp., in Middle and South America, and huemal deer, *Hippocamelus* spp., and dwarf deer, *Pudu* spp., in the Andes. Generally, use of deer became increasingly more intensive from Archaic times to European contact. Archaeological sites in the Western Hemisphere are only rarely completely devoid of deer remains. These exceptions are occupation sites such as those in the extreme north, the West Indies, or along the coast where subsistence was based entirely on marine resources. In sites with evidences of broad subsistence bases deer may not be an abundant element of the fauna, but since each individual deer would provide a relatively large quantity of meat, the importance of deer was usually significant. During Colonial times deer were exploited for the hide trade which reached considerable proportions. As many as 200 000 hides were shipped annually during the mid seventeen-hundreds from the Charles Town port alone.

Deer were not domesticated. However, some of the hunting practices resulted in fortuitous game management. These hunting methods, described from ethnographic studies, probably have antiquity in view of the long exploitation of deer. One of these is the concentrated hunting of deer during their rutting season when the bucks tend to be less timid (Smith, 1975). A number of different tribes are reported to have timed their hunt to this season, from September through to December. In these hunts, Indians have used the disguise of wearing a deer hide with antlers to stalk deer. In those few archaeological sites for which the sex and ratio of the deer remains have been determined, bucks and does are represented approximately equally (Smith, 1975). This suggests only that this type of hunting method was not exclusively practised.

More difficult to trace in the prehistoric record is the maintenance of buffer zones of prime deer land between lands occupied by two different tribes. Such a buffer zone or debatable area, described for a stretch of woodland between the Chippewa and Sioux in Wisconsin, is similar to strips of contested territory between many North American Indian Tribes (Hickerson, 1965). Hickerson (1965, p. 61) believes that 'The maintenance of the buffer, that is, the warfare which kept a large portion of the best deer habitat a buffer, was a function of the subsistence requirements of the Chippewa and Sioux.'

A third hunting practice also relying on habitats preferred by deer is based on the attraction of garden plots to deer. Subsistence farmers hunt deer and other garden pests, such as raccoon, as a part of their agricultural activities removing garden competitors (Linares, 1976). As Linares (1976) points out, the composition of the mammal faunas used by prehistoric horticulturalists in Panama was largely garden pests.

Quite a few reports refer to communal deer drives, some using natural barricades combined with constructed impoundments. The deer were driven by fire or simply by the advance of the hunters. Invariably the reports indicate that the impounded deer were killed (Larson, 1970). Even where guanaco, vicuña and deer were driven, the deer were killed for food while some of the guanaco might be added to the llama herds, and the vicuña were shorn and released (Browman, 1974). A few indications suggest that deer were kept in corrals, implying that they were tame and under human control (M. Pohl, personal communication, 1976; Swanton, 1922, p. 42). The next step toward control of the animals' reproduction and true domestication was apparently never taken.

The reasons why they were never domesticated, in spite of their wide and intensive use, have never been satisfactorily answered. The relationship between the Indians and deer had many of the characteristics that in other situations have led to domestication. The long period of close association between man and deer has allowed ample opportunity for man to become thoroughly familiar with the behaviour of deer. Such knowledge can be applied to developing ways to control the animal. Problems to be faced in the control of these animals as domestic stock are establishing subservience of the animals to the herder and supplying sufficient browse to maintain the herd. Behaviourally, deer are woodland species and nervous, high-strung animals. They feed mainly at night and twilight. Deer are not gregarious but, rather, form small groups during part of the year, then break up into solitary animals or does with their young of the year. The aspects of this behaviour pattern which lend themselves least well to development of a domestic herd are those related to social organization. Which of these characteristics is of prime importance is a matter for speculation; the continued maintenance of the hunter-prey relationship for millenia by Indians of many cultures is testimony to the difficulty and cost in terms of energy of domesticating deer.

3. DOGS

The oldest and most malleable domestic animal is the dog, *Canis familiaris*. Much evidence supports the hypothesis that this important domesticated animal is descended from Eurasian wolves. The earliest remains of this domesticate date to about 12 000 years ago and were found in material excavated from Palegawra Cave in Northeastern Iraq (Turnbull and Reed, 1974). Dogs accompanied people early in their migration into the New World. Evidence for this early spread into North America comes from dog remains found in Jaguar Cave, Idaho, and dated at 8400 B.C. (Lawrence, 1967). The subsequent dispersal of dogs throughout the Americas was also quite rapid. They presumably accompanied people in their early migration throughout the New World. By 5500 B.C., dogs already occurred in such different contexts as in association with now extinct species in Florida (Martin et al., 1974) and in a llama hunters' site in the Andes of Central Peru (Pires-Ferreira et al., 1976).

At the time of first contact, Europeans found a variety of different types of dogs in the Americas. We know, from illustrations of Indian dogs and from mummified prehistoric dogs, that American Indians had dogs with different coat colours and hair lengths and different body sizes and proportions of the body parts. We cannot know such characteristics as the colour and type of coat that most prehistoric dogs had or the form of their ears or how they carried their tails, but we can make some estimations of their body size and proportions based on their skeletal remains.

The range in prehistoric dog types may be more easily appreciated by a comparison of estimates of their size and proportion based on their skeletal remains rather than solely on a comparison of skeletal measurements. The measures of size used here are estimates of body weight based on correlations between body weight and condylo-basal skull length, height of the mandible, or length of the lower molar tooth row. Another measure of size is the height at the shoulders. Estimates of shoulder height are based on the relationship that the length of the humerus is on the average 26% of this height. The third measure used here is of skull proportion, more precisely, the relative length of the snout. This is expressed as a ratio between skull length (condylo-basal length) and snout length (orbit to alveolus l′). In this ratio the snout is shorter as the ratio increases toward 3 and, conversely,

Chapter 2 references, p. 38

TABLE 2.1

Measurements of dogs associated with sites dating 1000 B.C. or earlier

Site name	Koster	Indian Knoll	Perry Site	Devil's Den	S. Indian Field	Real Alta
Location (country, state)	U.S., IL	U.S., KY	U.S., AL	U.S., FL	U.S., FL	Ecuador, Chanduy
Time period	5000 B.C.	± 3000 B.C.		± 5500 B.C.	2000–1000 B.C.	3000 B.C.
Ref.	Hill, 1972	Haag, 1948, p. 122	Haag, 1948, p. 140	Martin and Webb, 1974	Gross, ms	Lathrap and Marcos, 1975
Condylo-basal length of skull[a] (mm)						
Number		18	10		5	1
Mean		142.4	143.8		154.0	160
Range		127.5–165	134–153		138.8–165.4	
Snout length (mm)						
Number	1	19	14		6	
Mean	66	63.2	65.7		68.9	
Range		54.2–73	59–73		61.8–72.5	
Ratio $\frac{\text{Condylo-basal}}{\text{Snout length}}$		2.25	2.19		2.24	
Length of lower molar row (mm)						
Number	1	21	28	2	21	2
Mean	31.1	30.6	30.0	29.6	32.1	32.6
Range		27.4–32.6	27.7–32.2	± 29–30.2	29.8–35.1	31–34.2
Height of mandible at M_1 (mm)	20.2			19.9		22.1
Estimated body wt. of average (g)[b]	11 567	8453	8590	11 179	10 505	11 682
Humerus length (mm)						
Number	1	21	33			2
Mean	116.5	115.3	122.5			127.8
Range		98.5–140	108–133			119.5–136
Estimated ht. at shoulder (humerus 26% of height) (mm)	448	443	471			491.5

[a] Measurements follow Von den Driesch (1976) except measurement of humerus which follows Haag (1948).

[b] (1) $\text{Log } y = 2.7755 (\log x) - 2.05$ y = body weight in g
    x = skull length in mm
    r = 0.98

(2) $\text{Log } y = 2.2574 (\log x) + 1.1164$ y = body weight in g
    x = height of mandible at M_1 in mm
    r = 0.98

(3) $\text{Log } y = 3.2735 (\log x) - 0.8873$ y = body weight in g
    x = length of lower molar row in mm
    r = 0.93

longer as it decreases to 2.

Paleo-Indian and Archaic dogs have been recovered from a number of sites in both North and South America. Some of these, such as the dogs from Jaguar Cave, Itasca Bison site, and Devil's Den, are associated with extinct animals. As well as isolated fragmentary remains, early during this time period, remains have been found of dogs that had been intentionally buried. The earliest dog burial was recovered from Rodgers Shelter in Missouri from a zone dated approximately 5500 B.C. (McMillan, 1970). Numerous other dog burials have been reported from archaic sites in Illinois, Kentucky, Alabama, Florida, and Ecuador (Haag, 1948; Hill, 1972; Gross, ms).

These early remains include both small and large dogs. The burials and most of the early finds are of dogs that had medium length snouts with a snout to skull length ratio of from 2.19 to 2.25, a range in size from an estimated 8 kg (18 lb) to 12 kg (26 lb) and which stood from 44.3 cm (17 in.) to 49.2 cm. (19 in.) high at the shoulder (Table 2.1). At this same time period, a few remains of a large type of dog have been found in sites in Minnesota and Idaho (Lawrence, 1968). These large dogs are comparable in size to an Eskimo dog which would weigh in the neighbourhood of 23 kg (50 lb) and if normally proportioned would stand approximately 58 cm (23 in.) high at the shoulders (Table 2.2). These finds indicate that already at this early time there existed at least two different types of dogs which presumably played two different roles in the human culture of which they were a part.

TABLE 2.2

Paleo-Indian and Archaic dogs

Site name	Jaguar Cave MC 51769	Jaguar Cave MC 51770	Basket Maker dogs	Itasca Bison	Birch Creek ISUM 19635	Eskimo dog USNM 6032
Location	U.S., ID	U.S., ID	U.S., SW	U.S., MN	U.S., ID	U.S., AK
Time period	8400 B.C.	8400 B.C.	A.D. 500–1700	5500 B.C.	2500 B.C.	20th century
Ref.	Lawrence, 1967	Lawrence, 1967	Haag, 1948, p. 141	Lawrence, 1948	Lawrence, 1968	Haag, 1948, p. 175
Condylo-basal length (mm)			147.8			204
Alveolar length P_1 —M_3 (mm)	67.9	±63.2	65.8			
Alveolar length C—M^2 (mm)				±86	87.5	87
Estimated body wt. (g)			9376			22 925

Remains of a great many dogs are associated with sites more recent than the 1st millenium B.C. The skeletal characteristics of these dogs and their archaeological contexts are extremely varied (Table 2.3). The snout lengths range from long-nosed dogs with a snout to skull length ratio of 2.09, to pug-nosed dogs with undershot jaws with a ratio of 2.71. The estimates of body weight also include extremes ranging from very small dogs weighing about 5 kg (11 lb) to large dogs weighing up to 21 kg (46 lb). The range in shoulder height is equally extreme, ranging from 34 cm (13 in.) to 63 cm (25 in.). Not only were a wide variety of dogs in existence during the two and a half millenia before European contact, but there is some evidence to

Chapter 2 references, p. 38

indicate that some Indians may have had two distinctively different types of dogs at the same time. Two dogs from the Palmer site in Florida are, for example, very different in size although they are representatives of two breeds that may have lived at the site contemporaneously (Table 2.3). Allen (1920, p. 486) also reports the large Indian dog and small dog or Techichi occurring together in the village site in Madisonville, Ohio.

In the classic study of aboriginal dogs by William Haag the generalization is made that 'similar cultural manifestations include similar-sized dogs' and that 'the older the (archaeological) horizon the smaller the dog' (Haag, 1948, p. 258). This is still generally true although discoveries of additional dog remains in the years subsequent to this study have increased the numbers of exceptions to this generalization. These have revealed the great complexity

TABLE 2.3

Measurements of dogs associated with archaeological contexts more recent than 1000 B.C. but before mixture with European strains of dogs

Site name Location (country, state) Time period Ref.	Lighthouse Peru, Lima 900 B.C. Haag, 1954, p. 139	Rosamachay Peru, Ayacucho 450 B.C.— A.D. 400	Pachacamac Peru, Lima 200 B.C.— A.D. 1476 Allen, 1920, p. 501	OGCH 20 Ecuador, Chanduy 500 B.C.— A.D. 300	Santa Luisa Mexico, Vera Cruz 300 B.C.— A.D. 300	Chalpa Mexico, Sinaloa A.D. 700— 1200
Condylo-basal length (mm)						
Number		1	5	2	1	1
Mean		128	137.4	145	±145	±140
Range			124—145	140—±150		
Snout length (mm)						
Number	1	1	5	1	1	1
Mean	70.4	55.3	50.8	66.3	66.1	64.3
Range			47—53	65.5—67.1		
Ratio $\frac{\text{Condylo-basal}}{\text{Snout length}}$		2.32	2.71	2.19	±2.19	±2.18
Length of lower molar row (mm)						
Number	1	1		2		4
Mean	32.8	28.8		31.2		30.4
Range				30.5—31.9		27.8—32
Height of mandible at M_1 (mm)						
Number		1		2	1	4
Mean		16.7		19	21.0	18.5
Range				18—20		17.6—19.1
Estimated body wt. of average (g)	11 885	6288	7656	8892	8892	9484
Humerus length (mm)						
Number		1	1	2	1	2
Mean		114	97	113.1	±115	106.8
Range				106.1—120		99.6—113.9
Estimated height at shoulder (mm)		438.5	373	435	442	411

TABLE 2.3 (Extension a)

Site name	N12/1	Palmer UF 95580	Palmer UF 95579	Chilca	Peru	Cancery Lane	Pecos Village Site	Governador
Location	Mexico, Sinaloa	U.S., FL	U.S., FL	Peru	Peru	Barbados	U.S., NM	U.S., NM
Time period	A.D. 900–1200	A.D. 600	A.D. 600	A.D. 1000–1400	A.D. 1400	A.D. 380–1500		A.D. 500–900
Ref.		Bullen and Adelaide, 1976	Bullen and Adelaide, 1976	Bullen and Adelaide, 1976	Allen, 1920, p. 473		Allen, 1920, p. 489	Haag, 1948, p. 141
Condylo-basal length (mm)								
Number	1	1		1	6		1	4
Mean	173	117.6		157	165.3		138	129
Range					155–178			117–144
Snout length (mm)								
Number	1	1		1	6		1	4
Mean	79.8	48.3		66.8	68.6		64	59.6
Range					63–75			52.5–67.7
Ratio	2.17	2.44		2.35	2.39		2.16	2.16
Length of lower molar row (mm)								
Number	2	1	1	1		1	1	4
Mean	35	27.6	32.5	33.2		30	32	29.1
Range								26.2–29.8
Height of mandible at M$_1$ (mm)								
Number	2	1	1	1		1		
Mean	22.8	15.4	24.1	24.6		18.3		
Range	22.6–23							
Estimated weight (g)	14 518	4971	14 378	11 084	12 791	9065	7750	6425
Humerus length (mm)								
Number	3	1	2		2	1		2
Mean	154.7	87.7	±134.8		138.5	111.5		135
Range	153–156		±131.6–138		130–147			114–156
Estimated height at shoulder (mm)	595	337	±518.5		533	429		519

Chapter 2 references, p. 38

TABLE 2.3

(Extension b)

Site name	Green-house	New Dun-geness	Madison-ville	Kodiak Is.	Kodiak Is.	Bonasila
Location	U.S., LA	U.S., WA	U.S., OH	U.S., AK	U.S., AK	U.S., AK
Time period	A.D. 1300–1500	A.D. 1000–1500		A.D. 500–1000	A.D. 1000–1750	20th century
Ref.	Haag, 1948, p. 205	Haag, 1948, p. 158	Allen, 1920, p. 459	Haag, 1948, p. 179	Haag, 1948, p. 179	Haag, 1948, p. 174
Condylo-basal length (mm)						
Number	1	3	3	6	13	6
Mean	145	154	168.7	149.3	169.8	190.3
Range		148–161	163–172	138–157	154–189	179–198
Snout length (mm)						
Number	3	3	3	6	13	6
Mean	69.3	70.9	75.2	67	76.6	86.0
Range	65–73	67–75.2	74–77.5	60.4–70.5	66–89	72–92
Ratio	2.09	2.17	2.24	2.23	2.22	2.21
Length of lower molar row (mm)						
Number	4	1	1		3	
Mean	32.1	31.5	33.5		34.5	
Range	30.2–33.3				32.3–35.9	
Estimated weight (g)	8892	10 505	13 530	9643	13 775	18 902
Humerus length (mm)						
Number	5		2			
Mean	126		162.5			
Range	115–137		162–163			
Estimated height at shoulder (mm)	485		625			

of the development and spread of diverse aboriginal dog types. Perhaps in the light of the new finds another generalization can be added to those proposed by Haag, namely, that the number of types of dogs increases with time.

Characteristics of the coat, the hair length and colour, are preserved archaeologically only in a few cases. Some mummified specimens with hair intact have been recovered from deserts and caves. A moderately long-haired, piebald dog from Arizona is illustrated by Allen (1920, plate 6). Remains of a young dog with yellow fur were recovered from the desert coast of Peru. This information on dog coat types is supplemented by a few artistic representations of dogs such as the spotted dog illustrated in Fig. 2.1. Accounts of coat types also come from the reports of the early chronicles. They describe both short- and long-haired varieties, spotted, piebald, dark red, and yellow (Allen, 1920). Hairless dogs are described from both Mexico and Peru. This trait, hairlessness, does not breed true. Potentially it could be traced archaeologically as the hairlessness is believed to be linked to congenital loss of the premolars (Wright, 1960). Although premolars are often missing, total absence of premolars and association of this with hairlessness has not been reported from archaeological remains. The hairless dog, or

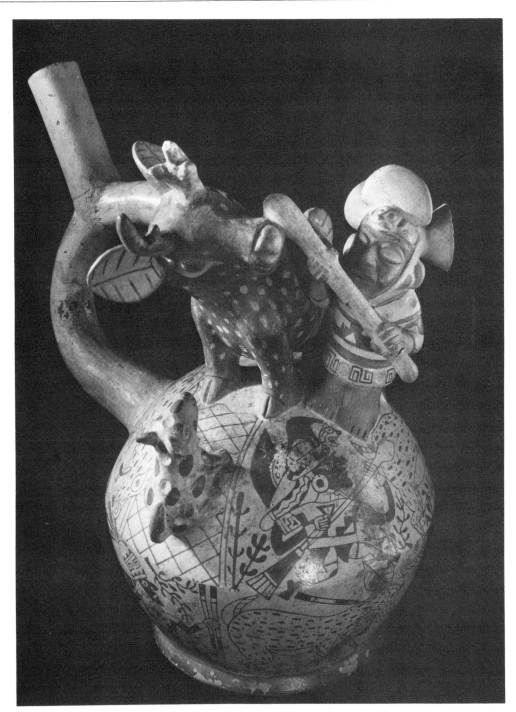

Fig. 2.1. Mochica period vessel depicting a dog assisting in the hunt of a deer (Peabody Museum, Harvard).

xoloitzcuintli as it is called in Mexico, is thought to have antiquity in Mexico, an interpretation based on what appears to be wrinkled naked skin on some pottery Colima dogs. This is open to question. At the other end of the spectrum, the dogs of the Northwest Coast were raised, at least in part, for their wool which was used in textiles.

Dogs initially may have associated themselves with people as camp scavengers. This symbiotic relationship developed into a whole array of goods and services provided by dogs in exchange for their care and nurture. Some of

Chapter 2 references, p. 38

these services can only be postulated, whereas we have good evidence for others. This care and nurture ranged from passively allowing the dogs to scavenge discarded food and waste to the suckling of dogs by women (Smith, 1968, p. 27). Companionship which is carried to the extreme in some sectors of the Americas today was also undoubtedly of great importance in the past. Companionship during life as well as in the afterlife was central to the wide-spread belief that dogs were required to ease the journey into the ultimate resting place (Allen, 1920). Archaeological evidences interpreted to give time depth to this belief are the association of dog burials with human and per-haps, in place of a dog, the realistic pottery dogs of Colima in human graves.

It is likely that dogs were also widely used for protection. This protection was typically in the form of advanced warning of danger. Protection may also have been extended to possessions and the role of the sheep dog may be thought of as protection of herd animals. No good evidence exists for the early use of dogs in herding the Andean Camelids, although dog remains have been identified with material from sites occupied by herders (Pires-Ferreira et al., 1976).

Dogs were sometimes used for hunting. The little dog represented in the Mochica pot may just be giving his master moral support but he may well have also taken active part in following the scent of the deer and given chase (Fig. 2.1).

Dogs of the North American plains and far North were, and among the Eskimos still are, used as draught animals. Travois were pulled by single dogs in the plains whereas sleds are pulled by teams of dogs over ice and snow.

As well as serving their masters in many ways, dogs themselves were eaten. In some cases dogs were only eaten as famine food. Other cultures had strong taboos prohibiting consumption of dogs under any circumstances. Yet in other cultures dogs were raised primarily for food, providing meat that

Fig. 2.2 Colima dog holding an ear of corn in its mouth (photograph from a reproduction of specimen in the Museo Nacional de Mexico).

was much esteemed. Dogs were raised for food in Mexico and some areas of Peru. Even as late as the 16th century there are accounts of the continued value of dogs for food. Duran (1967) describes more than 400 dogs for sale at the market at Acolman near Mexico City; this, he was told, was a relatively small number of dogs being offered for sale. Apparently the people willingly paid high prices to have the dogs butchered as a meat prepared for feasts celebrating baptisms and weddings. Consumption of a carnivore would appear to be a costly proposition; however, the dogs raised for food were given very little if any meat in their diet. Strontium analysis of the bones of these dogs indicates that their diet contained less animal protein than did their masters' (A.B. Brown, personal communication, 1979). One Colima dog is depicted with an ear of corn in its mouth (Fig. 2.2), a further intimation that corn was the dog's food.

A case has been made for the domestication of a native and now extinct canid, *Dusicyon australis* (Clutton-Brock, 1977). The physical characteristics of this animal and its isolation on the Falkland Islands a great distance from the South American mainland all point to the intervention of man in its breeding and distribution. A point brought out by this study concerns the propensity of man to tame animals. 'Domestication will only follow, however, if the social behavioral patterns of the tamed animals are sufficiently well developed to allow successive generations to breed in captivity, isolated from the wild species' (Clutton-Brock, 1977, p. 1342).

4. SOUTH AMERICAN DOMESTICATES

The highland of western South America is the only real centre of animal domestication in the New World. The animals that were domesticated in this region are two South American Camelids, alpaca and llama, and an hystricomorph rodent, the guinea pig. At the present time it appears that the process of domestication resulting in these domesticates occurred some time around 4000 B.C. Otherwise, the pattern of domestication of the herd animals and the rodent are quite different.

4.1 Camelids

The first evidence of intensive hunting of Camelids is associated with archaeological sites located in the Puna, or tundra life zone, found at elevations over 4000 m. The evidence is a relatively great abundance of Camelid remains replacing deer, the earlier dominant animal in faunas excavated from the Junin Valley (Pires-Ferreira et al., 1976). This shift occurred sometime after 5000 B.C. By 4000 B.C. over 80% of the animal remains were Camelid. A similar sequence of change is described for the animal remains from Lauricocha Cave located to the north of Junin. This almost exclusive use of Camelids was probably possible as a result of the increasing degree of control the people had over these animals.

Increased dependence on Camelids accompanied a change in the age structure of the animals used. A sample of bones from Pachamachay Cave, Junin, is composed almost entirely of Camelids of which a third to a half are juveniles. The majority are new born animals, reflecting a high mortality during the rainy season from December through to March. Normally this age group is not readily accessible and new-born animals are rarely found in such large numbers in archaeological contexts. This high mortality may be the result of the spread of disease through the confines of the corral (Pires-Ferriera et al., 1976).

At the present time, the greatest amount of information concerning this

Chapter 2 references, p. 38

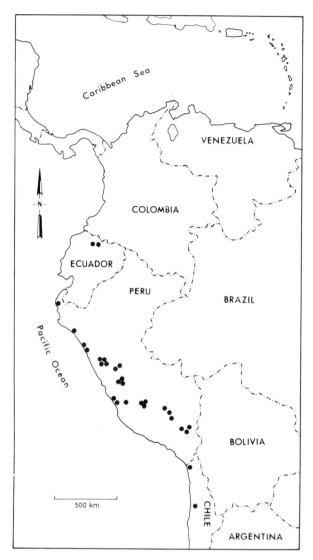

Map 2.1. Distribution of archaeological sites from which domestic Camelids have been identified. Dots represent site or sites.

early phase in the use of Camelids comes from the Junin Valley (Pires-Ferreira et al., 1976). This is proposed as the centre of Camelid domestication; comparable data must still be gathered to support this proposal. The sequence of spread of what by then were domesticated Camelids tends to support a Central Peruvian origin for domestic Camelids. Camelids become a dominant element of the faunas of the Ayacucho Valley, south of Junin, by about 3000 B.C. Camelid remains appeared for the first time in a coastal site, Chilca, by about 2000 B.C. but did not become important until about the time of Christ. Incorporation of domestic Camelids in the economies of people north in the Callejon de Huaylas and south in Chile is evident by 1000 B.C. (Wing, 1977; Pollard et al., 1975). Further spread north into Northern Ecuador did not occur until about A.D. 1000 (Oberem, 1975).

At the time of the Incan Empire, A.D. 1476–1534, the herds were greater than before or since. Two distinct types of animals composed these herds. The llama (*Lama glama*) is the larger primarily burden bearing animal and the alpaca (*Lama pacos*) is the smaller wool producing animal. Two breeds of alpaca, the huacaya and suri, were also developed for the special qualities of their wool. These domestic animals all interbreed with each other and with

Fig. 2.3. Llama pack train. Bags on the llamas backs and ropes are made of llama wool.

their wild presumed progenitor the guanaco (*Lama guanicoe*) and produce fertile offspring. In this way variability was maintained in the herds. This variability contributes to the difficulty that is encountered in trying to distinguish osteologically between the different types of Camelids. Future work will have to address itself to tracing the origins of the two domestic forms, the llama and the alpaca, as well as the identification and interpretation of additional faunal assemblages to provide detail to this outline of Camelid domestication.

These herd animals provide their masters with a variety of goods and services. Their most basic use has been to provide both meat and wool. The meat is used today primarily by the Indians of the Andes. The blood and entrails are used as well as the meat which is preserved by drying. The coarser wool of the llama is used for utilitarian purposes such as ropes and bags, whereas the fine warm wool of the alpaca is valued for clothing and is an important element in the modern economy of Peru.

Llama were used primarily as beasts of burden. Representations of llamas bearing burdens extend back to before the time of Christ. No evidence exists of their use as draught animals. They were only rarely ridden. Trains of llamas played an important role in trade between Andean communities and those in the Selva to the east and Pacific Coast to the west. This role is largely taken over today by mules and motor vehicles.

The herd animals are a measure of wealth and prestige. Ceremonies are held each year during which some of the animals of a herd may be redistributed among the younger members of the community. During the Incan Empire, herd animals were given in tribute to the state (Murra, 1965). Animals, preferably white ones, were also sacrificed to the gods. Prehistoric evidences of such sacrifices are the remains of entire animals associated with prominent architectural features such as entrances to temples.

4.2. Guinea pigs

The guinea pig (*Cavia porcellus*), the well-known laboratory animal, is a

Chapter 2 references, p. 38

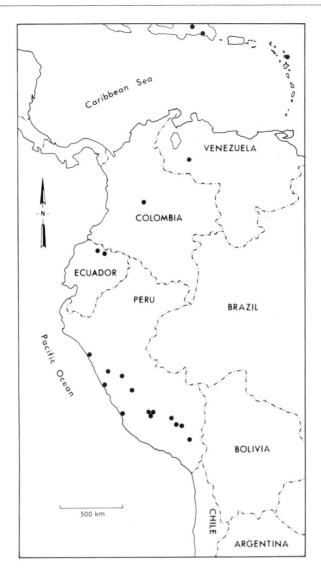

Map 2.2. Distribution of archaeological sites from which domestic guinea pigs have been identified. Dots represent site or sites.

gregarious rodent originating in South America. Wild members of this genus occur in a crescent extending from Eastern Brazil and Paraguay west across Argentina and north along the Andean chain to Colombia. There is no real agreement on which species gave rise to the domestic form although *Cavia tschudii* and *Cavia aperae* have been proposed as being close to the ancestor of the domestic form.

The pattern of domestication, as revealed by faunas studied thus far, indicates at least two centres of guinea pig domestication. One of these centres is in a highland valley such as Ayacucho in Central Peru. Guinea pig remains are very abundant in a number of early horizons of the Piki-machay and Puente sites in Ayacucho. The remains of guinea pigs indicate that by about 4000 B.C. they were probably domestic. Evidence that is interpreted as human control of these animals is the greater variability and increase in size of the animals represented (Wing, 1977). At this same time guinea pigs interpreted as domestic have been identified from the site of El Abra in the Valley of Bogota, Colombia (G.F. Ijzereef, 1978). By 3000 B.C.

domestic guinea pigs appeared in the Central Coast of Peru at the site of Culebras and were abundant north of Ayacucho at the site of Kotosh in Huanuco. Much later, about A.D. 750, guinea pigs first appeared in sites of Northern Ecuador. Remains of guinea pigs are associated with horizons that date just prior to contact with Europeans, indicating that they were introduced and kept in Venezuela and the Lesser Antilles. Guinea pig is also reported from Hispaniola which may also have been an Indian introduction (Wing et al., 1968). Domestic guinea pigs were used as far south as Central Chile at the time of the Incan Empire.

Guinea pigs are still kept by the Andean Indians of Peru and the neighbouring highland countries of Ecuador and Bolivia (Gade, 1967). These animals are kept in small numbers in peasant homes where they have free run of the shelter and retreat under the hearth for warmth and safety. They are fed food refuse as well as freshly cut legumes and barley raised for them. They are used primarily for food and as such may be the predominant source of animal protein in the diet. Their use for sacrifice and diagnosis and therapy in folk medicine still persists.

The initial association between man and guinea pigs may have been one of mutual benefit. Guinea pigs may have been attracted to human habitation by the warmth and sources of food found there. They are easily penned as they cannot climb well. Vestiges of what appear to be such enclosures were found in early strata of a cave site in the Ayacucho Valley. From the late Preceramic (2500–1500 B.C.) to Incan times elaborate enclosures with runways and tunnels, littered with guinea pig remains, have been excavated from sites on the coast as well as in the highlands. In exchange for care and feeding, guinea pigs provided food. They were also used for sacrifical offerings. One mummified guinea pig wrapped in a fabric was associated with remains dating from the time of Christ in an Ayacucho Valley cave site. Presumably, use of guinea pigs for medicine and divination has a great antiquity.

5. TROPICAL DOMESTIC AND TAME BIRDS

Although birds were widely used for food and other purposes few were domesticated. The two major domesticated birds are the turkey and muscovy duck. Of these, the turkey was the most intensively used in Southwestern United States and Mexico. Muscovy duck were used less frequently. In addition to these native domesticates, circumstantial evidence exists which suggests that the chicken may have been introduced in Northwestern South America before the arrival of the Spanish in the New World.

Taming small song birds, parrots, monkeys and other small mammals is a widespread custom in Latin America today. Presumably in Precolumbian times such animals were also tamed and kept around the home. This assumption is not supported by evidence such as skeletal remains from archaeological sites or graphic representations, with the exception of macaws.

5.1. Scarlet macaws

The work of Hargrave (1970) has thoroughly documented prehistoric keeping of scarlet macaws, *Ara macao*, in Southwestern United States. Remains of 145 macaws have been identified from a number of archaeological sites in Arizona and New Mexico. These sites are a great distance north of these birds' natural range, which is in the tropical lowland forest of Mexico from Southern Tamaulipas, and Oaxaxa south through Central

Chapter 2 references, p. 38

America. Osteological evidence indicates that these birds were buried without removal of their flesh and that most were immature, newly fledged, and adolescent birds. These factors support Hargrave's contention that these birds were traded into the southwest as young birds and breeding populations were never established. This trade took place from about A.D. 1100 to 1375 and then was revived in the Rio Grande Valley of New Mexico briefly after the conquest of Mexico. It is presumed these birds were used primarily for their plumage.

5.2. Turkeys

Turkeys, *Meleagris gallopavo*, are a highly esteemed feast food in the Americas. They are reserved for the main harvest feast, Thanksgiving, in North America and for Christmas, weddings and baptisms. Turkeys have been held in high regard for a long time. Prior to the time of the Spanish conquest turkeys were given in tribute payment to the Mexican lords. A measure of the degree to which they were valued is that in some towns in Mexico the right to eat turkey was reserved exclusively for the lords.

The domestic turkey was first domesticated in Mexico and is derived from the Mexican subspecies *Meleagris gallopavo* (Schorger, 1966). The earliest published record of domestic turkey is about A.D. 180. This find is associated with the Classic or Palo Blanco Period in the Tehuacan Valley (Flannery, 1967). From that time on domestic turkey became increasingly important in the economies of the cultures living in Mexico and the Southwestern United States.

Turkeys were introduced to Europe early in the 16th century. By the end of the 16th century they had spread to most Western European countries and they were introduced by the English to the Eastern United States, both Massachusetts and Virginia (Schorger, 1966, p. 479). From here domestic turkey gradually spread both west and south in North America.

5.3. Muscovy ducks

Muscovy ducks, *Cairina moschata*, are distributed in the coastal lowlands of Middle America, Northern South America, and the Caribbean. They are and undoubtedly were prehistorically an important local resource but little evidence points to widespread or great cultural importance of these ducks. They were first encountered as domestic ducks on the West Indian Island of Guadeloupe during the second voyage of Columbus in 1493 (Sauer, 1966, p. 71).

5.4. Chickens

The domestic chicken, *Gallus gallus*, originated in India and Southeast Asia. From there it spread north into China and west to the Near East and Europe. Carter (1971) has brought together quite a variety of data which support the interpretation that domestic chickens were transported across the Pacific and introduced into Northwestern South America from Asia prior to A.D. 1492. These data include: the very fast diffusion rate for chickens in Latin America if they were introduced only after European discovery of the New World; the names of chickens in America are derived from Asian names; the peacomb and pigmented egg shell suggest biological affinities with Asian chickens; and the uses of chickens in some parts of South America are more closely similar to Asian uses than European. Verification of this hypothesis would be the identification of chicken remains from Precolumbian archeological sites. Thus far, no such remains have been reported.

6. CONTEMPORARY DOMESTICATION AND MANAGEMENT

In the past several hundred years relatively few new species of animals have come under human control and then been modified for human economic purposes. One 20th century domestication is of the musk ox, *Ovibos moschatus*, in Alaska and Quebec, Canada (Wilkinson, 1972). The taming and control of these animals has not been difficult. The purpose of establishing herds of musk oxen has been to provide the fine wool of these animals for a new textile industry for the people of the extreme north. Use of the animals will be diversified in the future.

Other recent domestications include animals raised for a variety of purposes. Among these several fur bearing mammals were domesticated and are farmed for their pelts. Two of these raised for fur in North America are the chinchilla, *Chinchilla laniger*, originating in the Andes of South America, and the coypus, *Myocastor coypus*, from Argentina. Sporadic attempts have been made to domesticate the wapiti, *Cervus canadensis*, as a draught animal. A number of different aquatic organisms, including oysters, catfishes, bass, and trout, are now being farmed or controlled so that they can be intensively harvested.

A great portion of the native game animals, both mammals and birds, of the Americas are managed. This management ranges in the degree to which the animal populations are protected or their movements controlled. The vicuña is considered an endangered species and therefore is legally completely protected from human use. Reserves have been established for the vicuña, making enforcement of their protection possible. Many other game animals are confined to game reserves or parklands. Hunting of such animals as deer, bighorn sheep, mountain goats, bison, quail grouse, and waterfowl is restricted to certain parts of the year. The status of these animals is carefully monitored and the precise management practices modified to maintain vigorous populations.

7. EUROPEAN INTRODUCTIONS

Within a decade of the discovery of the New World, European domesticates were introduced into this new land. Cattle, horses and pigs were herded on the Island of Hispaniola where they increased rapidly. Within a very few years these animals, as well as sheep and goats, spread on this island and to other West Indian islands and then to the mainland. In each of the new environments these animals flourished. Some animals became an integrated part of Indian culture as, for example, the important position attained by horses among the Plains Indians or the sheep among the Navajos.

8. SUMMARY

As described, a number of animals became domesticated in the Americas. These animals provided food, bone, and hides and played a part in the secular and sacred functions of the people. Locally, domesticated animals were of great importance and were incorporated thoroughly into the cultures of the people who controlled them. Inspite of the Precolumbian importance many of these domesticates have been replaced by introduced European domestic forms. The European stock rapidly adapted to New World conditions and filled a gap in many of those Indian cultures which had no native domesticates. European domesticates have also gradually eroded the

Chapter 2 references, p. 38

importance of the native domesticates with which they came in contact. The
significance of the native domesticates as we see it today is greatly diminished
from the position they held prior to the discovery of the Americas.

9. REFERENCES

Allen, G.M., 1920. Dogs of the American Aborigines. Bull. Mus. Comp. Zool., Harv.
    Univ., 63: 431—517, plates 1—12.
Browman, D.L., 1974. Pastoral nomadism in the Andes. Current Anthropol., 15: 188—
    196.
Bullen, R.P. and Adelaide, K., 1976. The Palmer Site. Fla. Anthropol. Soc. Publ. No. 8,
    pp. 1—55, plates 1—22.
Carter, G.F., 1971. Pre-Columbian chickens in America. In: C.L. Riley, J.C. Kelley, C.W.
    Pennington and R.L. Rands (Editors), Man Across the Sea: Problems of Pre-Columbian
    Contacts. University of Texas Press, Austin, pp. 178—218.
Clutton-Brock, J., 1977. Man-made dogs. Science, 197: 1340—1342.
Duran, F.D. (Escritor del Siglo XVI), 1967. Historia de las indias de Nueva Espana y Islas
    de Tierra Firma. Edition introduced by J.F. Ramirez. Editora Nacional, Mexico.
Flannery, K.V., 1967. Vertebrate fauna and hunting patterns. In: D.S. Byers (Editor),
    The Prehistory of the Tehuacan Valley, Vol. 1, Environment and Subsistence. Univer-
    sity of Texas Press, Austin, pp. 132—175.
Gade, D.W., 1967. The guinea pig in Andean folk culture. Geogr. Rev., 57: 213—224.
Gross, R., ms. An analysis of dog remains from South Indian Field. 1971 paper on file
    at Florida State Museum.
Haag, W.G., 1948. An osteometric analysis of some aboriginal dogs. Univ. Ky. Rep.
    Anthropol., 7: 107—264.
Haag, W.G., 1954. A mummified dog from the Lighthouse site, Supe. In: G.R. Willey and
    J.M. Corbett, Early Ancón and Early Supe Culture. Columbia University Press, New
    York, NY, pp. 138—143.
Hargrave, L.L., 1970. Mexican macaws: comparative osteology and survey of remains
    from the southwest. Anthropol. Pap. Univ. Ariz., 20: 1—67.
Hickerson, H., 1965. The Virginia deer and intertribal buffer zones in the Upper Mississippi
    Valley. In: A. Leeds and A.P. Vayda (Editors), Man, Culture and Animals: the Role
    of Animals in Human Ecological Adjustments. American Association for the Advan-
    cement of Science, Washington, DC, Publ. 78, pp. 43—65.
Hill, F.C., 1972. A Middle Archaic Dog Burial in Illinois. Foundation for Illinois Archae-
    ology, Evanston, IL, 8 pp.
Ijzereef, G.F., 1978. Faunal remains from the El Abra rock shelters (Colombia). Palaeo-
    gr., Palaeoclimatol., Palaeoecol., 25: 163—177.
Jarman, M.R., 1972. European deer economics and the advent of the Neolithic. In: E.S.
    Higgs (Editor), Papers in Economic Prehistory. Cambridge University Press, Cam-
    bridge, pp. 125—147.
Larson, Jr., L.H., 1970. Aboriginal subsistence technology on the southeastern coastal
    plain during the late prehistoric period. University of Michigan Dissertation, 345 pp.
Lathrap, D.W. and Marcos, J.G., 1975. Informe preliminar sobre las excavacions del sitio
    Real Alto por la mision anthropoligica de la Universidad de Illinois. Arqueologia. Rev.
    Univ. Catolica, Quito, Ecuador, 10: 41—64.
Lawrence, B., 1967. Early domestic dogs. Z. Saugetierkd., 32: 44—59.
Lawrence, B., 1968. Antiquity of large dogs in North America. Tebiwa. J., Idaho State
    Univ. Mus., 11: 43—49.
Linares, O.F., 1976. 'Garden hunting' in the American tropics. Hum. Ecol., 4: 331—
    349.
Martin, R.A. and Webb, S.D., 1974. Late Pleistocene mammals from the Devil's Den
    Fauna, Levy County. In: S.D. Webb (Editor), Pleistocene Mammals of Florida. Univer-
    sity Presses of Florida, Gainsville, FL, pp. 114—145.
McMillan, R.B., 1970. Early canid burial from the Western Ozark Highland. Science, 167:
    1246—1247.
Murra, J.V., 1965. Herds and herders in the Inca State. In: A. Leeds and A.P. Vayda
    (Editors), Man, Culture, and Animals: the Role of Animals in Human Ecological
    Adjustments. American Association for the Advancement of Science, Washington, DC,
    Publ. 78, pp. 185—215.
Oberem, U., 1975. Informe de Trabajo sabre las Excavaciones de 1964/1965 en Cochasqui,
    Ecuador. Bonner Amerikanistische Studien, Bonn, pp. 71—79.
Pires-Ferreira, J.W., Pires-Ferreira, E. and Kaulicke, P., 1976. Preceramic animal utiliza-

tion in the Central Peruvian Andes. Science, 194: 483—490.

Pollard, G.C. and Drew, I.M., 1975. Llama herding and settlement in prehispanic Northern Chile: application of an analysis for determining domestication. Am. Antiq., 40: 296—305.

Sauer, C.O., 1966. The Early Spanish Main. University of California Press, Berkeley, CA, 306 pp.

Schorger, A.W., 1966. The Wild Turkey: Its History and Domestication. University of Oklahoma Press, Norman, 625 pp.

Smith, B., 1968. Mexico, a History in Art. Doubleday and Co. Inc., New York, NY, 296 pp.

Smith, B.D., 1975. Middle Mississippi exploitation of animal populations. Mus. Anthropol., Univ. Mich. Anthropol. Pap. No. 57, pp. 1—233.

Swanton, J.R., 1922. Early History of the Creek Indians and their Neighbors. Bur. Am. Ethnol., Washington, Bull. 73.

Turnbull, P.F. and Reed, C.A., 1974. The fauna from the Terminal Pleistocene of Palegawra Cave, a Zarzian occupation site in Northeastern Iraq. Fieldiana, Anthropol., 63: 81—146.

Von den Driesch, A., 1976. A guide to the measurement of animal bones from archaeological sites. Peabody Mus. Bull., Harv. Univ., No. 1, 136 pp.

Wilkinson, P.F., 1972. Current experimental domestication and its relevance to prehistory. In: E.S. Higgs (Editor), Papers in Economic Prehistory. Cambridge University Press, Cambridge, pp. 107—118.

Wing, E.S., 1977. Animal domestication in the Andes. In: C.A. Reed (Editor), Origins of Agriculture. Mouton Publishers, The Hague, pp. 837—859.

Wing, E.S., Hoffman, Jr., C.A. and Ray, C.E., 1968. Vertebrate remains from Indian sites on Atigua, West Indies. Carib. J. Sci., 8: 123—139.

Wright, N.P., 1960. El Enigma del Xoloitzcuinthi. Instituto Nacional de Antropologia e Historia, Mexico, pp. 11—102.

Chapter 3

Prehistoric Man and Animals in Australia and Oceania

RICHARD CASSELS

1. INTRODUCTION

When Europeans first reached the Australasian region (Australia, New Guinea and the Pacific Islands to the east), they found two quite different kinds of human economic systems. In Oceania (New Guinea and the Pacific Islands) the people practised horticulture and had three domestic animals — the pig, dog and chicken. In Australia, on the other hand, the Aborigines lived by gathering, fishing and hunting; they did not plant crops, and the dog (the dingo) was their only domestic animal.

This distinction between agricultural and non-agricultural economies must be qualified to some extent by noting the similarities between Australia and Oceania such as in shellfishing, fishing, plant gathering (Golson, 1971) and hunting. However, it is clear that the prehistoric horticultural systems of Oceania, based on root- and tree-crops, lacking any major cereals such as rice (except in West Micronesia), and ultimately derived from Southeast Asia, did not spread to Australia despite limited contact between Northern Australia and New Guinea (Beckett, 1972; Mulvaney, 1975). Nor did the Aborigines develop agriculture themselves, apart from minor local exceptions (Campbell, 1967; Allen, 1974).

Only three domestic animals — pig, dog and chicken — were introduced into the Australasian region in prehistory; they all came from the west, from Asia. In the case of the pig and the dog, they may have been introduced more than once before Europeans arrived. In addition, rats were spread by prehistoric man to virtually all the islands of the Pacific.

The domestic animals served mainly as a supply of meat; the dog was used in some areas for hunting. Nowhere in prehistoric Australasia were animals used for traction or riding or milking. Nevertheless, the domestic animals were important socially, symbolically and ritually, particularly in Oceania, despite their irregular contribution to the diet as a whole.

Cranstone's general remarks apply aptly to the domestic animals of Oceania:

> Among peoples who practise shifting cultivation and who lack the plough the number of animals kept and their economic importance tend to be relatively small, for the level of agricultural technique is not sufficiently high to produce a surplus of food to support them. It is true that the animals often forage for themselves ... but they usually return at night or at least at intervals to be fed. In these conditions domestic and wild or feral animals often interbreed, and in appearance may be indistinguishable ... Since few animals can be kept there is little or no deliberate use of manure in the fields (Cranstone, 1969, pp. 247--248).

The native land fauna of the Australasian region was dominated by marsupials in Australia and New Guinea (although rodents and bats, the only terrestrial placental mammals to cross the eastern boundary of Wallacea, were numerous (Darlington, 1957)). Only a few land mammals occurred naturally on the islands east of New Guinea. There were no flightless land mammals at all east of the Solomons (Darlington, 1957), and so birds, and to a lesser extent bats, were the prominent element of many island faunas, having radiated most spectacularly in New Zealand.

This fauna proved very vulnerable; it was badly depleted around the time of, and probably as a result of, the first prehistoric human colonization of the different land masses. Notable among the extinctions were many of the largest forms of the Australian marsupials, and the moas of New Zealand, giant flightless birds that had disappeared within a thousand years of the first human settlement.

Not one species of the native fauna was ever domesticated — in the sense of regularly breeding while under the control of man — although individual animals were sometimes tamed or kept alive in captivity.

The native fauna that survived the initial period of adaptation to man was hunted wherever it occurred; however, only New Guinea and Australia possessed a large enough fauna for this to be an activity of major economic significance. And even in New Guinea, as in the larger Pacific Islands where they occurred, feral pigs were often a greater source of meat than the native game.

The settlement of the remoter and smaller Pacific Islands was made possible by horticulture and fishing; animals were of little importance or, in the case of most of the atolls, of no importance at all (Oliver, 1951).

This chapter covers the prehistoric period, which in this case is the period before the arrival of Europeans. Information on the pre-European use of animals in Australasia comes from archaeology and from the records of early European visitors — explorers, missionaries and ethnographers. A major problem with the written European record is that European contact very rapidly altered the indigenous economies, particularly by immediately introducing alien animals, so that in many regions, but particularly Melanesia, pre-European practices are hard to reconstruct with confidence.

2. THE PLEISTOCENE PERIOD: MAN AND ANIMALS 40 000—10 000 YEARS AGO

At times of low sea-level, such as 20—10 000 years ago, Australia, New Guinea and Tasmania were joined into one great landmass; however, even at the lowest sea-level, 20–18 000 years ago, there were considerable spans of water separating Greater Australia from Southeast Asia, to cross which required some form of water transport. It was probably this requirement that prevented human populations, who may have been in Southeast Asia for as long as 2 million years, from reaching Greater Australia until perhaps 40 000 years ago (Mulvaney, 1975; Jones, 1976).

Orthodox archaeological opinion holds that man had not domesticated any animals anywhere in the world until the late Pleistocene. Whatever the truth of this, it seems at present that the first settlers of Greater Australia failed to introduce successfully any animals (with the possible exception of some rats)[1].

---

[1] While the majority of writers assume that the rats of Australia and New Guinea arrived unassisted by, and probably long before, prehistoric man, it is possible that some of the rats which are evidently more recently evolved made use of the watercraft of the early human settlers. Groups of rats likely to be involved in any Pleistocene movements are Tate's (1951) *Rattus assimilis* and *R. lutreolus* divisions, not the *R. exulans* division,

The earliest archaeological sites in Australia have not provided large faunal assemblages that are unequivocally the work of man. However, one site at least, the Mungo site in New South Wales, contained the remains of shellfish, fish, and smaller mammals, a subsistence pattern generally similar to that observed in the region over 20 000 years later by European visitors (Bowler et al., 1970).

At some stage in the late Pleistocene a number of the larger Australian marsupials, including some remarkable giant forms, became extinct. They included the family Thylacaleontidae (marsupial 'lions'), the Diprotodontidae (the largest of which, *Diprotodon*, was the size and shape of a rhinoceros), *Phascolonus* (a wombat 1.5 m long), giant kangaroos of the Macropodidae, and the kangaroo-like Sthenurids; all together at least 20 genera and numerous species became extinct; Merrillees estimates that one-third of the larger marsupial fauna became extinct. Also, *Genyornis*, a large flightless bird bigger than an emu, disappeared (Jones, 1968, 1975; Merrillees, 1968; Calaby, 1971; Mulvaney, 1975; Milham and Thompson, 1976).

It has been suggested that the giant forms all became extinct at about the same time, and perhaps as the result of a similar cause; neither of these two hypotheses have yet been convincingly supported. Further, Merrillees and Jones (following Tindale, 1959) suggested in the late nineteen-sixties that prehistoric man may have played a major role in causing these extinctions, both by hunting and more particularly, by the use of fire which drastically affected the environment by subjecting it to a fire regime of a much greater intensity than normal. These papers coincided with the publication of an influential American symposium in which some authors argued strongly that early man was a major factor in the extinction of 'megafauna' in several different parts of the world (Martin and Wright, 1967). Many archaeologists now believe that man may have played an important part in the Australian extinctions, although alternative explanations have been proposed (Merrillees, 1968). Recently Jones concluded as follows:

> [answers still elude us to the questions of] the precise dating of this episode (or episodes), the determination of whether or not it was a sudden event of mass extinction or a slower species by species depauperisation over an appreciable length of time, and lastly the possibility of the coexistence of man and the extinct forms ... the trend of recent results ... supports the view that the main or last episode of extinctions occurred 'a geologically short time' before 25 000—30 000 years ago, with the local survival of a few species (within the Sthenurids and Macropods) down to about 20 000 years ago in some places (Jones, 1975, pp. 28—29).

A recent find has demonstrated the contemporaneity of man with some extinct giant macropods on Kangaroo Island (Hope et al., 1977) supplementing the less conclusive evidence known earlier from Western New South Wales (Tindale, 1964; Tedford, 1967; Jones, 1968).

It is likely, although there is very little evidence yet, that New Guinea experienced a process of megafaunal extinction similar to that of Australia. However, the Thylacine (marsupial wolf) and a giant bat are the only extinct fauna that have yet been found in archaeological deposits (Van Deusen, 1963; White, 1972; Bulmer, 1975; Hope and Hope, 1976; Menzies, 1977).

Generally the Pleistocene relationships between man and animals in Australasia remain poorly documented.

---

whose man-aided expansion into the Pacific is much more recent. The classification of Australasian rats is still a matter of controversy (Ride, 1970, p. 134) and the history of their arrival and radiation is certainly a complex one.

Chapter 3 references, p. 57

3. THE RECENT PERIOD: 8000 B.C. UNTIL EUROPEAN CONTACT

With the end of the last glacial period and the consequent melting of the ice caps, world sea-levels rose; and from this time on, Australia, Tasmania and New Guinea have separate prehistories. Apart from the possibility of earlier settlement of the larger Melanesian islands near New Guinea, present archaeological evidence does not document the settlement of the majority of the Pacific Islands before the last 4000 years: their populating was a slow process, with the remotest islands such as New Zealand being reached perhaps less than 1000 years ago.

3.1. Tasmania

Tasmania was isolated from mainland Australia around 12 000 years ago, and there is no sign that it was ever again contacted by outsiders until the coming of the Europeans. The Tasmanians lacked all domestic animals, notably the dog, thus strengthening the argument that the dog did not reach Australia until after the end of the Pleistocene (Jones, 1970). Archaeological evidence shows that major components of the prehistoric Tasmanian diet were elephant seals, *Miroungia leonina* (which, however, became virtually locally extinct by A.D. 1000, perhaps due to aboriginal hunting), seals, wallabies and kangaroos, bandicoots, possums, native cats, Tasmanian devils, sea birds, shellfish and fish. It is notable that fish seem to have totally disappeared from the diet, for some unknown reason, by 3500 years ago (Jones, 1967, 1968). European historical records show that vegetable foods (not usually preserved in archaeological contexts) were also a major component of the diet, equivalent in importance to shellfish and macropods (Hiatt, 1967).

3.2. Australia

The later prehistory of the Australian mainland suggests a general continuity with the subsistence patterns established in the Pleistocene, perhaps by 30 000 years ago, by which time the majority of the megafauna seem to have disappeared, and the initial adaptations of the Australian environment to man (and vice versa) had been established. There are some exceptions to this suggested continuity, such as: the appearance of stone grinding tools, probably for processing seeds, by about 15 000 years ago; the appearance after 3000 B.C. of microlithic tools (miniature stone implements); and the appearance of the dingo (*Canis familiaris*) by 1000 B.C. (Mulvaney, 1975).

There were also various local changes in the composition of the fauna, notably the extinction of the marsupial 'wolf' (*Thylacinus cynocephalus*) on many parts of the mainland by the time of European contact. These changes have been variously attributed to the effects of climatic change, human hunting and the spread of the dingo, but none of these hypotheses has been clearly established. On present evidence it appears that changes in the composition of the fauna after 30–20 000 B.P. were not as major as the extinction of the giant fauna before this date, although future, more detailed work may alter this view (Mulvaney et al., 1964; Merrillees, 1968; Dortch and Merrillees, 1971; Flood, 1974; Mulvaney, 1975; McCarthy, 1976; Milham and Thompson, 1976).

In general, the limited number of faunal reports from archaeological sites indicate a pattern of hunting that is similar to that recorded by European explorers and ethnographers (Gould, 1968; Lampert, 1971a, b). On a more precise level, there is a difference between sites that reveal a pattern of hunting a wide range of the smaller marsupials, and those that show the hunting of a narrow range of one or two of the larger species such as kangaroos. The former pattern is much more common at archaeological sites than the latter, and it is possible that this reflects a trend through time where 'big-game' hunting developed later in prehistory (H. Allen, 1972). There is some evidence that European expansion into Australia favoured the larger kangaroos — it certainly had a drastic effect on the smaller marsupials of the grasslands and savannah woodlands (Tyndale-Biscoe, 1973) — and in this way the patterns of Aboriginal hunting seen in some areas by Europeans, especially some time after first contact, may differ from the prehistoric patterns.

European records describe Aboriginal subsistence as follows. With the exception of the association between man and dingo, there were no domestic animals, and food was obtained by gathering, fishing and hunting. Most groups collected a wide range of foods. Hunting, traditionally a male activity, often did not provide as much food as the gathering of plant foods, small animals and shellfish by the women (Hiatt, 1974); even in the cooler temperate regions where meat was more important, the diet was mainly vegetarian. Mulvaney (1975) has suggested that a major reason for the unimportance of meat was the Aborigines' inability to preserve it. Nevertheless, virtually all animals were taken, from emu to kangaroo to mice and lizards, with the macropods (kangaroos, wallabies, etc.) being the most productive of the larger game.

There was considerable regional variation in subsistence; on tropical coasts the basic sources of protein were snakes, crabs and shellfish, whereas land mammals were more important elsewhere. In inland areas the diet was up to 80% vegetarian. Most areas experience seasonal variations which are reflected in seasonal changes of aboriginal food-gathering strategies (Thompson, 1939; Gould, 1969; Lawrence, 1971; Mulvaney, 1975).

In terms of the two polarized stereotypes of the hunter-gatherer and the farmer, the Australian Aborigines are clearly hunter-gatherers. The hunter-gatherer is commonly imagined to have no major effect on, or control over, his prey beyond killing what he can lay his hands on. The farmer, however, is contrasted with the hunters: the farmer creates the habitat for his domestic animals, controls their breeding, alters their behaviour etc. It is now clear that many technologically-primitive people fall somewhere between these two extremes, and the Aborigines are among them. In particular, Aboriginal man manipulated his environment, and the animals in it, by the use of fire — probably since the first arrival of man in Australia c. 40 000 B.P. (Kershaw, 1975). Jones (1969) has even talked of 'fire-stick farming'. Mulvaney (1975) has gone furthest in describing aboriginal burning as essentially a 'pastoral control or incipient animal husbandry'. There is an extensive literature on the role of fire which is claimed to have many effects: to create a suitable habitat for man, to improve human mobility, to create environmental diversity, to favour plants and animals sought by man, to increase environmental productivity, to provide an extra hunting 'weapon' to attract animals to certain localities, and to make hunting easier (Anell, 1960; Jones, 1968, 1969, 1975; Gould, 1969, 1971; Mulvaney, 1975).

While the use of fire may be a way of manipulating animals, nevertheless one does not find in Australia anything that a Western European would confidently label 'domestication'; the dingo debate will be mentioned shortly; one example from Cape York, of cassowaries being tethered, is probably post-European. Often Aborigines were seen to keep animals, especially young ones, as pets but this never developed into a method of food production. There has been considerable debate on why Australia lacked animal and plant domestication when New Guinea has both. There probably were domestic pigs on some of the islands of the Torres Straits, but

Chapter 3 references, p. 57

not on the Australian mainland (Campbell, 1967; White, 1971; Beckett, 1972; Allen, 1974; Mulvaney, 1975).

3.3. New Guinea

Present evidence shows that the New Guinea Highlands were inhabited by 25 000 years ago, although there is every reason to expect a greater antiquity of settlement, equal to that of Australia. Rising sea-levels separated New Guinea from Australia between 6500 and 8000 years ago. At the time of European contact, the people of New Guinea practised horticulture, which was most intensively developed in the Highlands. They possessed pig, dog, and chicken, of which pig was by far the most important (Bulmer, 1975).

Archaeological evidence from sites in the Highlands establishes that the macropods (tree-kangaroos, wallabies, etc.) were important foods; also eaten were pigs, phalangers (a group including the cuscus), possums, flying fox, cassowaries and rodents; archaeological evidence for plant food is almost totally lacking. Based on recent work (Bulmer and Bulmer, 1964; Powell, 1970; Brookfield and Hart, 1971; Bulmer, 1975; Jones, 1976; Golson and Hughes, 1976) one can suggest the following hypothetical sequence of economic development in the Highlands.

(1) A hunting and gathering period, from the time of the first human settlement until the development of agriculture (the date of which is at present unknown). Pandanus nuts were eaten and may have been an important plant food. Pigs were present by 10 000 years ago, but it is not known if they were domesticated.

(2) An early agricultural phase, prior to the introduction of the sweet potato; important crops may have been taro, *Pueraria*, bananas, and yams. One would guess that pigs were domesticated, although this may be hard to prove.

(3) A late agricultural phase, with sweet potato, which may have been introduced in the 16th century A.D., perhaps indirectly from the Spaniards. The introduction of the sweet potato may have led to more intensive pig husbandry (Watson, 1977).

The well-known recent intensive sweet-potato horticulture of the Highlands may therefore not be very old; current archaeological investigations have not yet clarified this, although some form of horticulture and, probably, pig husbandry almost certainly go back over 5000 years ago (Powell, 1970; Golson and Hughes, 1976).

In the mid-altitude zone of New Guinea, especially in areas of low population density, hunting was more important than in the Highlands; wild pigs were a greater source of meat than domestic ones; domestic stock were kept but in a very loose association with man (Dornstreich, 1973; Morren, 1977).

The coastal lowlands of New Guinea had a pattern of subsistence which differed from that of the Highlands, and resembled rather that of the rest of island Melanesia. Shellfish, fish and sea mammals provided a major alternative source of protein to the pig; archaeological evidence confirms the antiquity of the exploitation of these — faunal remains from sites include turtle, dugong, fish and shellfish as well as pig and macropods such as wallabies. Horticulture, however, provided most of the food. Food and other material items were sometimes exchanged over long distances.

3.4. The Pacific Islands

Knowledge of the prehistory of the Pacific Islands has increased enormously over the last 20 years, particularly knowledge of the more distant regions of Eastern Melanesia and of Polynesia. Unfortunately little is known of the earliest settlement of the larger islands of West Melanesia, the islands closest to New Guinea. Melanesia may have been first settled by populations lacking agriculture; at present the archaeological evidence documents only the existence of agricultural populations on the Pacific Islands by 1000 B.C. By this time communities were established from New Guinea to Western Polynesia (Samoa and Tonga). Human populations expanded into Eastern Polynesia over the last 2000 years (Golson, 1972; Howells, 1973; Pawley and Green, 1973; Shutler and Shutler, 1975).

At the time of European contact, horticulture was the basis of the economy of most Melanesian islands, with some coastal peoples emphasizing marine exploitation. It was the Polynesians, and especially the East Polynesians, who particularly developed the art of fishing which they combined with horticulture. Hunting could not be very productive in the Pacific Islands, there being no flightless land mammals east of the Solomons. Birds, bats and reptiles were the only game beyond there. Pigeon hunting was particularly important on some islands. Most of the larger, high islands of the Pacific possessed all three domestic animals — the pig, dog and chicken — with such exceptions as New Caledonia and Easter Island which had only the chicken, New Zealand which had only the dog, and the New Hebrides and Marquesas which lacked the dog (McKern, 1929; Urban, 1961; Green and Davidson, 1969, 1974).

Archaeological evidence, however, now shows that the domestic animals were in fact introduced to some islands but failed to persist there (Cram, 1975b). A few islands, particularly atolls, had no domestic animals at all, and their inhabitants lived by fishing and horticulture. Most remarkably, the inhabitants of the Chatham Islands had no domestic animals and did not practice horticulture; sea mammals and fish were particularly important foods for them.

4. THE PIG

The pig is not native to the Australasian region, although its presence in New Guinea by 10 000 years ago (S. Bulmer, personal communication, 1977) might be the result of a natural immigration. With this possible exception the 'wild' pigs of Oceania are all feral descendants of domestic stock, and are classified by most authorities as *Sus scrofa*, that is the same as the widespread wild pig of the Eurasian mainland, from which stock most of them undoubtedly came (Laurie and Hill, 1954; Urban, 1961; Zeuner, 1963).

Unfortunately the classification of the pigs of island Southeast Asia is not agreed upon, and the picture is further confused by the likely role of prehistoric human populations in transferring animals from island to island with the possible consequence of native species being replaced by the introduced species (Medway, 1972). However, in the Southeast Asian region four groups of pigs are usually recognized: (a) *Sus scrofa* (*Sus cristatus* of the Indian Region and *Sus vittatus* of Sumatra now being regarded as subspecies of *S. scrofa*); (b) *Sus barbatus*, the Bearded Pig of Malaya, Sumatra and Borneo; (c) *Sus verrucosus* (including *Sus celebensis*) of Java, Celebes and Moluccas; (d) *Babyrousa babyrusa* of the Northern Celebes and Buru (Lauri and Hill, 1954; Sanderson, 1955; Ellerman and Morrison-Scott, 1966; Medway, 1969, 1972; Grzimek, 1972).

Within the Australasian region pigs were distributed continuously from New Guinea through island Melanesia to Fiji (but not New Caledonia) at the time of European exploration (Urban, 1961). They were found on the high islands of Polynesia, but not Easter Island nor the more southerly high islands

Chapter 3 references, p. 57

of Polynesia. Pigs were absent in Australia and New Zealand, on nearly all atolls and on most raised coral islands; they were probably absent in Micronesia (Birket-Smith, 1956) although there is the possible exception of the Chamorros (Urban, 1961). In contrast, the European-introduced *Sus scrofa* has been successful in many of these places.

Archaeologically, pig bones are known earliest from the sites of Yuku and Kiowa in Highland New Guinea, over 10 000 years ago (S. Bulmer, personal communication, 1977), although they are not common in archaeological sites there until after 6000 years ago (Bulmer, 1966, 1975; White, 1972). Elsewhere in the Pacific the earliest pig bones are found in Melanesia and Polynesia (Tonga), about 2500—3000 years old, associated with the Lapita cultural complex (Specht, 1968; Garanger, 1972; Egloff, 1975; Davidson, 1976; Green, 1976, 1979). These later dates for the Pacific Islands are consistent with the evidence for domestic pig in mainland Asia (Thailand, North China) by 6000 B.P. (Chang, 1968; Barnard, 1975; Green, 1979; Bayard, ms) and in Timor by 5000 B.P. (Glover, 1971).

4.1. Use of the pig

The intensive pig 'complex' of the New Guinea Highlands has been studied in detail; information is less complete for island Melanesia and Polynesia.

4.1.1. New Guinea

To summarize the salient features of recent New Guinea Highland pig husbandry: many societies keep large numbers of pigs; they are fed on the products of horticulture, and at times recieve a larger proportion of such crops as sweet potato than do the people themselves; pigs are allowed to forage for themselves which means that gardens must be protected by fences, ditches, banks, etc. Often the number of pigs kept by any group goes through a repeated cycle, at the end of which large numbers are killed at spectacular feasts; after a feast pig numbers are low for a while, and it will be several years before there are enough pigs, or before the pigs are large enough, for another feast (Vayda et al., 1961; Brookfield and Brown, 1963; Rappaport, 1967; Waddell, 1968; Bulmer, 1968; Vayda, 1972). These cycles form part of various systems of exchange throughout the Highlands which serve, among other things, to earn prestige or political power for their holders; thus the pig cycle plays a major role in social, ritual and political life (Strathern, 1972; Ryan, 1972a, b; Meggitt, 1974). Pig meat is not eaten regularly on a day-to-day basis but only at feasts or on occasions of misfortune or occasions requiring certain rituals. Piglets are sometimes suckled by women, and may become cherished pets, although this does not prevent them from being killed.

Feral pigs are found in most areas below 5500 ft. and they interbreed with the domestic stock, in fact forming one breeding population (Bulmer, 1968).

In many areas a much looser form of pig husbandry is found where all male domestic pigs are castrated and the domestic sows are served by wild boars. Some areas replace their stocks, or indeed derive most of their stock, by capturing wild piglets (Hughes, 1970; Chowning, 1973; Dornstreich, 1973; Morren, 1977). Often the hunting of wild pigs is much more important in such areas. Hughes (1970) describes one example of a less intensive system where pigs are confined to limestone valleys, and are fed by having sago palms cut for them. This use of sago palms is a common way of exerting a loose form of control over pig populations. In such areas an individual pig may be wild, domestic and semi-domestic at different stages of its life.

As with the dingo, there has been extensive debate whether Highlands pig husbandry is economic or not, this example forming one of the battle-grounds (along with the sacred cow of India) for a general debate on the adaptiveness or otherwise of human cultural institutions. Taking as their starting point Linton's (1955) statement that the pigs are uneconomic, and a drain on resources and labour, many authors have suggested ways in which the pigs may be of value, for example as:

a valuable source of high-grade protein and fat;

a source of high-quality nutrition eaten at times of stress;

a hedge against uncertainty, for example crop failure caused by irregular climate;

a transportable, negotiable and divisible store of surplus, or generally a good way to 'store' surplus vegetable produce;

a way of making use of residual sweet-potato plants not harvested by man;

a natural 'cultivating machine';

a source of fertilizer;

a disposer of garbage; and

an agent in establishing and maintaining a desirable tall grass fallow

(Vayda et al., 1961; Brookfield and Brown, 1963; Rappaport, 1967; Hughes, 1970; Vayda, 1972).

4.1.2. Island Melanesia and lowland New Guinea

The pig is kept as a domestic animal throughout the rest of Melanesia, but nowhere in sufficient numbers to be used regularly as food. In most areas it is eaten only at feasts (Blackwood, 1935; Cranstone, 1961; Laycock, 1975). In Melanesia pigs usually forage for themselves, as well as being fed on crops (Forde, 1963); they feed greedily on human faeces (Baker, 1929), they are said to eat snakes sometimes, as well as other small animals, and they often forage on reefs. Where an island is large enough to support feral pig populations, these normally interbreed with the domestic stock. Feral pigs are hunted everywhere but in most areas are not sufficiently numerous to provide much food. In New Guinea pigs are hunted occasionally by large drives, sometimes employing fire and/or dogs. Anell (1960) and Bulmer (1968) describe the hunting techniques involved.

Domestic pigs are very important everywhere in economic and social life. Some authors, writing of the role of pigs in Melanesian society, have described them as 'standard currency' (Baker, 1929), or as the equivalent of money and capital (Layard, 1942). Pigs have been compared to the shell 'money' of parts of Melanesia (Laycock, 1975). Regional peculiarities of custom which have attracted the attention of anthropologists are as follows:

(a) The relatively widespread practice of tusk elongation. This is effected by knocking out the upper canine so that the lower one grows round in circles, eventually re-entering and emerging from the lower jaw; examples are known of pigs surviving until three circles have been formed. The antiquity of this practice is not known, its purpose is nominally a ritual one, as circle-tuskers are prized for sacrifice, but these animals in fact play a major part in social and economic life. The practice is found in the Banks Islands, North and Central New Hebrides, and two other isolated examples; good descriptions are provided by Baker (1929) and Layard (1942).

(b) The breeding of intersex pigs in parts of the New Hebrides and Banks Islands (Baker, 1929). Very high proportions (10—20%) of intersex pigs to normal males are born; they are fundamentally different from intersex pigs of European breeds. These animals were highly valued. Seven forms of inter-sexuality are recognized, and the natives operate on piglets of one form to allow them to pass urine.

Chapter 3 references, p. 57

4.1.3. Polynesia

In contrast to Melanesia, pigs in Polynesia were never the focus of such intense ritual and social attention. Several features of man—pig relationships in Polynesia are probably due to the generally smaller size of Polynesian islands. These include the more general practice of tethering or confining pigs near the home, rather than letting them forage for themselves on the land, and the lack of feral pig populations on many islands (particularly before the introduction of European strains) and hence lack of pig hunting (Hiroa, 1930; Anell, 1960; Forde, 1963).

Pigs in Polynesia were commonly fed on the coconut, breadfruit and sweet potato (Yen, 1973). In many areas they foraged on the sea shore for food and refuse. Piglets were often pets and were suckled by lower-class women (Luomala, 1960). Pigs seem to have reached many parts of Polynesia with the earliest settlers, e.g. Hawaii (Kirch, 1971; Cordy, 1974) and possibly also Tonga (Cram, 1975b).

Everywhere in the Pacific there is the problem of the damage pigs can cause to crops, and every ethnography describes the way each society tries to resolve this problem. Solutions include the use of natural enclosures to control pigs (Hughes, 1970), construction of ditches, fences, banks, etc., around the crops to keep pigs out (especially Melanesia), construction of sties, walls, fences, etc. to keep pigs in (especially Polynesia) (Hiroa, 1930; Blackwood, 1935; Laycock, 1975); the creation of separate gardens to draw off the pigs from the main gardens (Brookfield and Hart, 1971); the blinding or nose-skinning of individual animals to discourage rooting (Cram, 1975b), and the hunting and trapping of wild pigs to keep their numbers down, especially around planting time (Anell, 1960; Harding, 1967).

5. THE DOG

At present there is no evidence that the dog reached Australasia in the Pleistocene period; this means that Tasmania, Australia and New Guinea are likely to have been separate when it arrived.

It is widely held that the Australian dingo (which is a feral dog) is an ancient form of dog that has probably been in Australia for a long time; it may be that the Pacific Island dog is a more recent introduction. However, at present the oldest uncontested archaeological evidence for dog in Australia (1000 B.C. at two sites in southeast Australia) is hardly older than that for the dog in Oceania. However, older ages, in the order of 8—5000 years ago, have been claimed for dingo remains in Australia (Mulvaney, 1975).

In Oceania dog remains have been recovered from archaeological contexts as follows: from sites in Southern Papua (Papua New Guinea) back to 2000 B.P. and on Buka (Northern Solomons) and Bellona (Solomons) at nearly the same date[2]; dog has been claimed for several Lapita sites (about 3000 years old) in Oceania but either the identification or the context of these finds can be questioned (Green, 1979).

An extremely conservative point of view then would be that only one dog population is involved, which reached Australia about 1000 B.C., New Guinea a little later, and the Pacific Islands yet later. A viable alternative would be that the dingo was established in Australia sometime after 8000

[2] Nebira 4, Papua, older than 1760 ± 90 B.P. (J. Allen, 1972; Bulmer, 1975, p. 54).
Hangan-Tsonoma Beach, Buka, 1860 ± ? B.P. (Bulmer, 1975, p. 65).
Sikumango, Bellona 2070 ± 80 B.P. (Cram, 1975a, pp. 248—250).
Oposisi, Yule Island, Papua New Guinea 1920 ± 80 B.P. (Vanderwal 1972, p. 155; Bulmer, 1975, p. 49).

B.C., and that a later group of domestic dogs, perhaps originating ultimately in Southeast Asia, were distributed as far as Polynesia by 1000 B.C. At present the archaeological evidence is too limited to make any definite inference, although the latter of the two possibilities would be preferred by most archaeologists.

The dingo undoubtedly came from Asia and was transported to Australia by man. Thus it is likely to have come from domesticated, or at least tamed, stock; given the lack of genetic isolation characteristic of domesticated animals, it will almost certainly be impossible to locate a definite 'ancestor' surviving among the modern domestic or feral dogs of Asia. Some writers have tried to identify the wild ancestor of the dingo-group dogs; here an important point is that Southeast Asia lacks any suitable wild ancestor, there being only the jackals of Thailand and the dhole (*Cuon*) further south, and the closest candidate would be the wolf of India (or Northern Asia). The extinct Tengger dog of Java was claimed by some to be a wild species related to the dingo, but it is very inadequately known and was almost certainly (given the distribution of the canids) a feral dog (Jentink, 1896; Zeuner, 1963; Ellerman and Morrison-Scott, 1966; Medway, 1969).

Archaeologically, domestic dogs have been identified in late Pleistocene Japan (Jones, 1969), in Northern China by 4000 B.C. (Chang, 1968), in Thailand perhaps by 3500 B.C. (Bayard, ms), and in Timor around 1000 B.C. (Glover, 1971). It is not known whether these animals resemble any Australasian dogs.

The affinities of the dingo have been hotly debated. Dingoes have been described as: closely related to the New Guinea 'singing' dogs and to Indian pariah dogs, probably all directly descended from the Indian wolf; undoubtedly of northern wolf ancestry; not closely related to the Indian wolf or pariah or New Guinea dogs, but a distinct form; and descended from the New Guinea dog (Longman, 1928; Wood-Jones, 1929; Clutton-Brock, 1963; Fiennes and Fiennes, 1968; Troughton, 1971; MacIntosh, 1972, 1975; Clutton-Brock et al., 1976).

The introduction of the dingo may be related to the extinction of the marsupial carnivores, the Tasmanian wolf and devil, but this has been debated (Mulvaney et al., 1964; Gill, 1971; MacIntosh, 1975; Mulvaney, 1975; Milham and Thompson, 1976; McCarthy, 1976).

In New Guinea at least three groups of dogs have been identified; one of these, the 'wild' dog, or 'singing' dog of the Highlands, has been claimed to be a relative of the dingo, and hence possibly a relict of a more ancient dog population. As well as this feral dog, there are domestic dog populations in both the Highlands and the lowlands. It has not been ascertained how separate these groups are in reality nor have they been systematically compared with each other; the picture is further confused at present by the introduction of dogs by Europeans. Thus the claims that the wild dog of the Highlands is an ancient form and that the domestic dog of the lowlands a later prehistoric introduction to New Guinea remain unsubstantiated at present (Bulmer and Bulmer, 1964; Bulmer, 1968).

The 'wild' dog of the Highlands (often referred to as 'the New Guinea dog') is, of course, a feral dog (Titcomb, 1969), and must originally have been introduced by man. Troughton (1971) names this dog *Canis hallstromi* but there are no good grounds for assigning it to anything but *Canis familiaris* (Van Deusen, 1972).

With the exception of MacIntosh (1972), most authors group the Highlands 'wild' dog with the dingo rather than with the Pacific Island dog, and it is often described as resembling a small dingo. Others have stated that it resembles neither the dingo nor the Pacific Island dog (Troughton, 1955,

Chapter 3 references, p. 57

1971; Titcomb, 1969). MacIntosh is the only worker to have made serious reference to osteological data.

At present therefore one can state that there is a feral dog in the New Guinea Highlands, which is not completely isolated from the domestic dog population. Superficially it has some resemblances to the dingo; as yet archaeological evidence sheds no light on its antiquity.

Dogs were found on many but not all of the Pacific Islands east of New Guinea. Archaeological evidence shows that they were introduced to some islands, but did not survive there up to the time of European contact (e.g. the New Hebrides, the Marquesas, Tonga, Nukuoro and Bellona). At this time there were probably dogs on some of the Micronesian islands (Ponape and Truk); like some of the atolls (e.g. some of the Tuamotus) further east, Micronesia had dogs but no pigs. Dogs were not seen in the Chatham Islands, New Caledonia, the Loyalties, Easter Island and various of the smaller Polynesian islands. They were the only domestic animal in New Zealand (Urban, 1961; Titcomb, 1969; Davidson, 1971; Cram, 1975b).

Almost nothing is known about the pre-European dog of island Melanesia. In contrast, the dog of Polynesia was well described by the earliest European explorers; it was small, long-bodied, and short-legged with a bushy tail and large head. Its colour was variable and there may have been more than one breed, both long- and short-haired dogs being reported from New Zealand and the Tuamotus (Luomala, 1960; Titcomb, 1969).

The Polynesian dog is now 'extinct', that is, it has probably been swamped by the gene pool of European dogs. In the absence of any comparative studies, and of suitable archaeological material from Melanesia or Southeast Asia, the affinities and possible origins of the Polynesian dog are simply unknown at present.

5.1. Use of the dog

In Australia the dingo survived successfully as a wild animal; however, some dingoes were also associated with some aboriginal groups, in a relationship that has been described as 'minimal symbiosis' (MacIntosh, 1975). There has been considerable debate over the role the dingo played in aboriginal society; the picture has been confused by the rapid adoption of European breeds of dog by Aborigines.

MacIntosh (1975, p. 97) states that dingoes 'cannot be domesticated and never were'. Yet it appears that: semi-tame dingoes were sometimes used in hunting, especially in forested regions; however, they were a nuisance in other kinds of hunting, such as stalking, and were not used for this; dingo puppies were often kept as pets and were treated affectionately until they grew up, when they had a tendency to leave human company; they were of course, more than capable of fending for themselves; sometimes they used to hang around camps, and were occasionally given food; people would sleep next to them, using them as 'blankets'; wild dingoes were sometimes followed and robbed of their prey; wild dingoes were also hunted and eaten (Meggitt, 1965; Hamilton, 1972; Hayden, 1975). Hamilton (1972) has argued that pet dingo puppies were important outlets for nurturant behaviour; Meggitt (1965) argued that dingoes were generally of no economic value, except in forested regions where they aided hunting.

In New Guinea most dogs were clearly domestic, with the exception of the 'wild' dog of the Highlands. New Guinea seems to have been the only part of Oceania where pre-European dogs were feral.

The use of domestic dogs in New Guinea varies from area to area but there can be no doubt that it was a key element in hunting animals in forest (Dornstreich, 1973). Bulmer (1968, 1972) reports that dogs are sometimes used in the chase or drive, sometimes in combination with the use of fire; they are used to discover or flush game, to run down or 'tree' it. Pigs and phalangers were the main game hunted with dogs. In other regions, however, dogs were clearly not used for hunting (De Miklouho-MacLay, 1881). At present dogs are widely eaten in the Highlands, although some particular ethnic groups emphatically will not eat them; there is as yet no archaeological evidence for the eating of dogs in the Highlands, as there is now in the lowlands (Bulmer, 1975). Dogs used to be kept for their teeth which were widely used as ornaments in Melanesia (Rappaport, 1967). As for all animals, extensive use was made of bones, teeth, hides, etc.

The dog was kept in many parts of island Melanesia; it was sometimes used for hunting, in some areas it was eaten, while in others it was not. Generally it was not as important as the pig, which played such a dominant role in Melanesian life (Urban, 1961).

At least some of the islands of Micronesia seem to have possessed dogs before European contact; dogs on Ponape are said to have been castrated and fattened for eating (Urban, 1961; Titcomb, 1969).

In Polynesia, New Zealand seems to have been the only area where the dog was (sometimes) used for hunting. Elsewhere in Polynesia it was clearly primarily a source of food, albeit often spending its life as a pet (Colenso, 1877; Luomala, 1960). With the exception of some dogs in New Zealand the Polynesian dog was overwhelmingly described by the early European explorers as timid and stupid. It seems that its diet was mainly vegetarian (being fed on coconuts, breadfruit, yams and other vegetables) with the following exceptions: in New Zealand it was also fed on fish in the prehistoric period, as well as eating rats and birds; in the Tuamotu atolls the dogs used to catch fish for themselves; in coastal areas generally it seems likely that they were fed on fish (Colenso, 1877; Allo, 1971; Yen, 1974).

Dogs were used in sacrifices in parts of Polynesia: pet dogs were sometimes buried with their owners; puppies were sometimes suckled by women; in some islands women could not eat dog meat, but generally the dog was a food animal (Luomala, 1960). Dog teeth were frequently used as ornaments; use was made of bone, skin, and hair. In New Zealand dog-skin cloaks were highly prized, and the dogs themselves were highly valued possessions of chiefs.

It seems likely that all the pre-European dogs of Australasia did not bark, but rather howled, whined or 'sang' (Colenso, 1877; Urban, 1961; Titcomb, 1969).

To summarize: the dingo of Australia lived mainly as a feral animal but was associated with man in some areas; New Guinea was the only other area where feral dog populations existed. The dog was used for hunting in New Guinea, the larger islands of Melanesia, and in New Zealand, but not in Polynesia where there was little to hunt anyway. In Polynesia (and Micronesia?) it was kept mainly for eating; in island Melanesia dogs were eaten in some areas but not in others.

6. THE GOAT

Goats have been identified in archaeological sites in Timor by 1000 B.C. and possibly earlier (Glover, 1971). However, there is no evidence that they reached Australasia before European contact. Bulmer (1975) notes that goat spread to parts of New Guinea after the first visits of Europeans, but before most areas were directly contacted.

Chapter 3 references, p. 57

7. THE DEER

Deer (*Cervus timorensis*) were introduced into Timor about 3000—2500 B.C. (Glover, 1971), but do not seem to have reached Australasia before their introduction into New Guinea in the first decade of this century (Bulmer, 1975).

8. THE CHICKEN

The chicken is today found feral on many Pacific Islands where it was originally introduced by prehistoric man. There is no evidence that it was in Australia at the time of European contact.

Archaeologically the earliest records of chicken are with sites of the Lapita complex off New Britain, in the Solomons, New Hebrides, Tonga and Samoa, about 1000 B.C. (Gifford, 1951; Specht, 1968; Davidson, 1976; Green, 1976, 1979; Cave, ms). It was therefore probably introduced together with the pig. At present it has not yet been identified in New Guinea archaeological sites before 800 years ago, and may not have been introduced to parts of the New Guinea Highlands until even more recently (Bulmer and Bulmer, 1964; Bulmer, 1975).

The Oceanic chicken is *Gallus gallus*, undoubtedly originating from Southeast Asia. In Polynesia as late as the eighteen-forties, chickens maintained a form and plumage almost identical to the jungle fowl of Southeast Asia (Ball, 1933); the present-day heterogeneity of fowl is the result of European introductions. The birds that originally came from Southeast Asia could even have been wild ones, although it is probable they were tame, if not fully domesticated.

Chickens were mostly feral or free-ranging and were widely hunted. Domestic birds were also kept. In Polynesia chickens were mainly valued for cock-fighting (also in Micronesia) or various ritual uses such as sacrifices. They were never a major food source except on Easter Island where they were the only domestic animal, and where they were kept in thatched scoops in the ground. In the Marquesas and elsewhere they were particularly prized for their feathers, one of the main values placed on birds throughout Oceania (Ball, 1933; Blackwood, 1935; Urban, 1961; Forde, 1963; Cave, ms). Generally chicken eggs were rarely eaten, and often ignored. On the whole then, they were, surprisingly, not an important source of meat or eggs.

9. RODENTS

Australia and New Guinea have many rat species. In Oceania east of New Guinea most of the *Rattus* species are assigned to the *Rattus exulans* group (including the old *R. concolor* group), with the following exceptions:*Rattus rennelli*, on Rennell Island, two subspecies of *Rattus ruber* in the Solomons; and *R. rattus mansorius* in the Philippines, Marianas and Carolines; a large rat in archaeological deposits on Nukuoro may represent the eastern-most extension of this latter rat into the Pacific, but the species of this find has not been ascertained (Tate, 1951; Laurie and Hill, 1954; Davidson, 1971).

There is a remote possibility that the Pleistocene arrival of some rat species in Australia and New Guinea was in some way associated with the arrival of man (Troughton, 1965). However, there is no doubt at all that the spread of the Polynesian Rat or Little (Burmese) Rat, *Rattus exulans*, from its native Southeast Asia to virtually every island of the Pacific, was the result of intentional or unintentional transport by prehistoric man. It probably spread with the boats of the first agricultural settlers (Tate, 1951; Laurie and Hill, 1954;

Darlington, 1957; Ellerman and Morrison-Scott, 1966; Medway, 1969; Cram, 1975b).

The rat was an important source of food on Mangaia (Cooks) and was definitely eaten on Niue and in New Zealand — all islands that lacked the pig. There is little doubt that it was probably also eaten in many other parts of Oceania although the inhabitants of some islands specifically deny eating rats (Loeb, 1926; Hiroa, 1932, 1938, 1944; Beaglehole and Beaglehole, 1938). However, its significance in man-animal relationships undoubtedly lies primarily in its probably-devastating effect on the indigenous faunas (particularly birds) of small and/or remote islands. There is little hard evidence but a wealth of speculation about this, inspired to some extent by the better documented and certainly disastrous effect of European rats (G.R. Williams, 1973; J.M. Williams, 1973).

Other rats of the genera *Uromys* and *Melomys* are found in Australia, New Guinea and Melanesia and possibly one can detect the hand of man in some of their Melanesian distributions (Carter et al., 1945; Tate, 1951).

10. THE INDIGENOUS FAUNA OF OCEANIA

The three species of cassowary (*Casuarius* spp.), native to New Guinea and Northern Australia, were captured and kept alive in New Guinea, being reared to adulthood for trade, slaughter and plumes (Bulmer, 1968). They do not breed in captivity (Rappaport, 1967). The presence of the cassowary in New Britain may well have resulted from intentional introduction by man. Cassowaries were the largest land animal hunted in New Guinea (Darlington, 1957; R. Bulmer, personal communication, 1977).

Occasionally other animals were kept as pets, but this did not develop into an established economic practice (e.g. cuscus, ring-tail possum and various birds, particularly parrots, in New Guinea; parrots and frigate birds in island Melanesia; and parrots and pigeons throughout Polynesia) (Urban, 1961; Bulmer, 1968).

In New Guinea a form of larvae was 'farmed' by felling sago palms, splitting open the bark and punching some holes in it. The trunks were then colonized by the larvae of Longicorn beetles; later these colonies were harvested, thus providing a nutritious food to complement the sago (Dornstreich, 1973).

In parts of Melanesia, notably New Britain, the eggs of megapodes (or brush turkeys, Megopodidae) are regularly harvested from the mounds or tunnels which these birds build. The mounds may be owned by particular individuals or villages, and restraints on gathering ensure the continuation of the supply (Codrington, 1891; Chowning, 1973; Kwapena, 1975).

The transfer of animals other than the true domesticates may have occurred more commonly in prehistory than is generally realized. In addition to the possibility of the cassowary being introduced to New Britain by man, other possible examples of human introductions are: the wallaby, e.g. *Thylogale* sp. and *Macropus agilis*, in the Bismarck Archipelago and Trobriands (Egloff, 1975), and some of the phalangers of island Melanesia and westwards to Timor and the Celebes (Carter et al., 1945; Glover, 1971; R. Bulmer, personal communication, 1977); frogs on Fiji and on Palau, a small snake, *Candoia*, that is distributed as far east as Tonga and Samoa, and the lizards of Polynesia (Darlington, 1957); fruit bats in Polynesia (held in high regard on many islands) (Collocot, 1921); and parrots and parakeets in various islands (Greenway, 1958).

In addition ecological changes caused by man undoubtedly enabled new

Chapter 3 references, p. 57

species, particularly birds adapted to open or disturbed habitats, to colonize new areas, and sometimes replace existing species.

Virtually all the native fauna was hunted by man, with the occasional exception of species reserved for chiefs to hunt. Generally speaking, land mammal hunting was only a major activity in New Guinea; east of New Guinea the number of land animals tails off to nothing very rapidly (Darlington, 1957) and animal hunting is increasingly replaced by the hunting of birds, particularly pigeons, and bats where available. However, the generally small size and remoteness of the Eastern Pacific Islands meant that even the bird faunas were small, until man reached the Hawaiian Islands and New Zealand. Once east of New Guinea the sea was the main source of protein, a tendency that increases with decreasing island size. It is possible that the tameness of birds on islands not previously settled by man provided a temporary abundance of meat for the first visitors, but it is certain that this virgin abundance was very short-lived.

Archaeology in New Zealand has produced evidence of extensive bird hunting which, according to some authors, was once thought to provide almost the only case, apart from the Chatham Islands, where Oceanic populations lived by hunting without agriculture (Duff, 1956). The evidence for this is the indisputable presence of large numbers of bones of moas (Dinornithidae and Anomalopterygidae) and other extinct, mostly flightless, birds in early archaeological sites. At one stage the early Polynesian settlers of New Zealand were labelled 'the moa-hunters' and it was thought that they lived mainly by hunting these giant birds, some larger than ostriches. However, recent archaeological evidence has made it clear that the earliest settlers of New Zealand were primarily horticulturalists cultivating the sweet potato (Leach, 1976). They certainly did hunt moas, but even then they obtained greater food value from seals and dolphins than from moas (Shawcross, 1968). Recent C14 dates from New Zealand make it clear that moas survived for several hundred years in some areas.

There can be no doubt that Polynesian man had a major role in the extinction of the large and flightless birds of New Zealand, although other factors may also have been important such as climatic deterioration, the introduction of the dog and the rat (*Rattus exulans*) and, possibly, avian disease. The effect of man was certainly felt both through direct hunting and also the extensive use of fire (Fleming, 1962; Martin and Wright, 1967; Ambrose, 1968; G.R. Williams, 1973; R. Scarlett, personal communication, 1977).

The effect of prehistoric man on the bird faunas of other Pacific islands is not yet well documented (but see Kirch (1973) on the Marquesas). At present, in the absence of evidence to the contrary, one may suppose that the New Zealand scenario is an amplified version of a drama played out on many Pacific islands, and recent finds in Hawaii suggest this picture may be repeated. In the case of New Zealand at least, one is tempted to see a much later and better documented parallel to the megafaunal extinctions of Australia.

Prehistoric man also hunted the sea-animals he encountered, particularly the porpoise, turtle and dugong in tropical waters and the seals of the New Zealand islands (Bulmer, 1975; Cram, 1975b; Green, 1976). In New Zealand it seems likely that seals ceased to breed on many mainland, particularly North Island, rookeries as a result of hunting. Turtle was a highly prized food among Polynesians: there are indications that turtle meat may have been easier to obtain in the early prehistoric period than later on, but as yet there is no proof that local extinctions were caused. In New Zealand the tuatara, *Sphenodon punctatus*, a large and primitive reptile, seems to have become extinct on the mainland during the prehistoric period, and now survives only on offshore islands (Robb, 1973).

11. CONCLUSION

Man—animal relationships can be crudely divided into four categories: (a) domesticated; (b) regularly hunted; (c) occasionally exploited; and (d) over-exploited — where the relationship fails when the animal becomes extinct. All four types of relationship can be found in prehistoric Australasia.

Three domesticated animals, the pig, chicken and dog, were supported by the horticultural system of Oceania. The diversified gathering economies of Australia also accommodated the dog, particularly in forested regions. These three animals were introduced to the region by man.

Hunted animals were important only on the large landmasses of Australia and New Guinea; to a limited extent they were replaced by birds in some of the Pacific Islands.

The giant marsupials of Greater Australia, and the large flightless birds of New Zealand (and perhaps other islands of the Pacific) are notable among the animals that were not able to adapt themselves to the presence of man. It is clear that major changes occurred among the faunas of Australasia, particularly in the early stages of prehistoric human colonization. Later the coming of the Europeans entailed a second, equally drastic readaptation.

12. ACKNOWLEDGEMENTS

I would particularly like to thank the following for the advice (not always taken!) which they gave me: R. Bulmer, S. Bulmer, R. Green, R. Hide, H. Allen, G. Rogers and J. Stretton at Auckland; R. Jones, J. Mulvaney, P. Bellwood and C. Groves from Canberra.

13. REFERENCES

Allen, H., 1972. Where the Crow Flies Backwards: Man and Land in the Darling Basin. Unpublished Ph.D. Thesis, Australian National University, Canberra.

Allen, H., 1974. The Bagundji of the Darling Basin: cereal gatherers in an uncertain environment. World Archaeol., 5: 309—322.

Allen, J., 1972. Nebira 4. An early Austronesian site in Central Papua. Archaeol. Phys. Anthropol. Oceania, 7: 92—124.

Allo, J., 1971. The dentition of the Maori dog of New Zealand. Rec. Auckland Inst. Mus., 8: 29—45.

Ambrose, W., 1968. The unimportance of the inland plains in South Island prehistory. Mankind, 6: 585—593.

Anell, B., 1960. Hunting and trapping methods in Australia and Oceania. Studia Ethnographica Upsaliensia, XVIII.

Baker, J.R., 1929. Man and Animals in the New Hebrides. George Routledge, London.

Ball, S.C., 1933. Jungle fowls from the Pacific Islands. Bernice P. Bishop Mus. Bull. No. 108.

Barnard, N., 1975. The First Radiocarbon Dates from China, 2nd edn. Monographs on Far Eastern History, No. 8. Research School of Pacific Studies, Australian National University, Canberra.

Bayard, D.T., ms. Recent developments in the prehistory of Mainland Southeast Asia and South China. Anthropology Dept., University of Otago.

Beaglehole, E. and Beaglehole, P., 1938. Ethnology of Pukapuka. Bernice P. Bishop Mus. Bull. No. 150.

Beckett, J.R., 1972. The Torres Strait Islanders. In: D. Walker (Editor), Bridge and Barrier. The Natural and Cultural History of Torres Straits. Dept. of Biogeography and Geomorphology, Australian National University, Canberra.

Birket-Smith, K., 1956. An ethnological sketch of Rennel Island. Danske Historisk-Filologiske Meddelser, 35 (3).

Blackwood, B., 1935. Both Sides of Buka Passage. Clarendon Press, Oxford.

Bowler, J.M., Jones, R., Allen, H. and Thorne, A.G., 1970. Pleistocene human remains from Australia: a living site and human cremation from Lake Mungo, western New South Wales. World Archaeol., 2: 39—60.

Brookfield, H.C. and Brown, P., 1963. Struggle for Land. Oxford University Press, Melbourne.

Brookfield, H.C. and Hart, D., 1971. Melanesia: A Geographical Interpretation of an Island World. Methuen, London.

Bulmer, R., 1968. The strategies of hunting in New Guinea. Oceania, 38: 302—318.

Bulmer, R., 1972. Hunting. In: P. Ryan (Editor), Encyclopedia of Papua New Guinea. Melbourne University Press, Carlton.

Bulmer, R. and Bulmer, S., 1964. The prehistory of the Australian New Guinea Highlands. Am. Anthropol., Spec. Publ. 66: 39—76.

Bulmer, S., 1966. Pig bone from two archaeological sites in the New Guinea Highlands. J. Polynesian Soc. 75: 504—505.

Bulmer, S., 1975. Settlement and economy in prehistoric Papua New Guinea. J. Soc. Océanistes, 31: 7—75.

Calaby, J.H., 1971. Man, fauna and climate in Aboriginal Australia. In: D.J. Mulvaney, and J. Golson (Editors), Aboriginal Man and Environment in Australia. Australian National University Press, Canberra.

Campbell, A.H., 1967. Elementary food production by the Australian Aboriginies. Mankind, 6: 206—222.

Carter, T.D., Hill, J.E. and Tate, G.H.H., 1945. Mammals of the Pacific World. Macmillan, New York, NY.

Cave, J.B.J., ms. The cultural significance of the distribution of *Gallus gallus* in the Pacific Islands. Anthropology Dept., University of Otago.

Chang, K.-C., 1968. The Archaeology of Ancient China, revised edn. Yale University Press, New Haven.

Chowning, A., 1973. An Introduction to the Peoples and Cultures of Melanesia. Addison-Wesley Module in Anthropology 38.

Clutton-Brock, J., 1963. The origins of the dog. In: D. Brothwell and E. Higgs (Editors), Science in Archaeology. Thames and Hudson, London.

Clutton-Brock, J., Corbet, G.B. and Hills, M., 1976. A review of the family Canidae, with a classification by numerical methods. Bull. Br. Mus. (Nat. His.), Zool., 29 (3).

Codrington, R.H., 1891. The Melanesians. Clarendon Press, Oxford.

Colenso, W., 1877. Notes, chiefly historical, on the ancient dog of the New Zealanders. Trans. Proc. N. Z. Inst., 10: 135—155.

Collocot, E.E.V., 1921. Notes on Tongan religion. J. Polynesian Soc., 30: 152—163.

Cordy, R.H., 1974. Cultural adaptation and evolution in Hawaii. J. Polynesian Soc., 83: 180—191.

Cram, C.L., 1975a. Prehistoric fauna and economy in the Solomon Islands. In: R.W. Casteel and G.I. Quimby (Editors), Maritime Adaptations of the Pacific. Mouton, The Hague and Paris.

Cram, C.L., 1975b. Osteoarchaeology in Oceania. In: A.T. Clason (Editor), Archaeozoological Studies. North Holland Publishing, Amsterdam and Oxford.

Cranstone, B.A.L., 1961. Melanesia. A Short Ethnography. British Museum, London.

Cranstone, B.A.L., 1969. Animal husbandry: the evidence from ethnography. In: P.J. Ucko and G.W. Dimbleby (Editors), The Domestication and Exploitation of Plants and Animals. Duckworth, London.

Darlington, P.J., Jr., 1957. Zoogeography, The Geographical Distribution of Animals. John Wiley, New York, NY.

Davidson, J., 1971. Archaeology on Nukuoro Atoll. Bull. Auckland Inst. Mus., No. 9.

Davidson, J., 1976. The prehistory of West Polynesia. In: La Préhistoire Océanienne Colloque XXII, IX Congress. Union Internationale de Sciences Préhistoriques et Protohistoriques, Centre National de la Recherche Scientifique, Paris.

De Miklouho-MacLay, N., 1881. Remarks about the circumvolutions of the cerebrum of *Canis dingo*. Proc. Linn. Soc. N.S.W., 6: 624—626.

Dornstreich, M.D., 1973. An ecological study of Gadio Enga (New Guinea) subsistence. Microfilmed Ph.D. Thesis, Columbia University.

Dortch, C.E. and Merrillees, D., 1971. A salvage excavation in Devil's Lair, Western Australia. J. R. Soc. West. Aust., 54: 103—113.

Duff, R., 1956. The Moa-Hunter Period of Maori Culture. Government Printer, Wellington.

Egloff, B.J., 1975. Archaeological investigations in the coastal Madang area and on Eloaue Island of the St. Mathias Group. Rec. Papua New Guinea Public Mus. Art Gallery, 5: 13—31.

Ellerman, J.R. and Morrison-Scott, T.C.S., 1966. Checklist of Palearctic and Indian Mammals, 1758—1948. British Museum (Natural History) London.

Fiennes, R. and Fiennes, A., 1968. The Natural History of the Dog. Wiedenfeld and Nicolson, London.

Fleming, C.A., 1962. The extinction of Moas and other animals during the Holocene period. Notornis, 10: 113—117.

Flood, J., 1974. Pleistocene man at Cloggs Cave: his tool kit and environment. Mankind, 9: 175—188.

Forde, C.D., 1963. Habitat, Economy and Society. Methuen, London.

Garanger, J., 1972. Archéologie des Nouvelles-Hebrides. Publications de la Société des Océanistes 30, Musée de l'Homme, Paris.

Gifford, E.W., 1951. Archaeological excavations in Fiji. University of California Anthropological Records, 13, No. 3, Berkeley and Los Angeles.

Gill, E.D., 1971. The Australian Aborigines and the Tasmanian Devil. Mankind, 8: 59—61.

Glover, I.C., 1971. Prehistoric research in Timor. In: D.J. Mulvaney and J. Golson (Editors), Aboriginal Man and Environment in Australia. Australian National University Press, Canberra.

Golson, J., 1971. Australian Aboriginal food plants: some ecological and culture-historical implications. In: D.J. Mulvaney and J. Golson (Editors), Aboriginal Man and Environment in Australia. Australian National University Press, Canberra.

Golson, J., 1972. The Pacific Islands and their prehistoric inhabitants. In: R.G. Ward (Editor), Man in the Pacific Islands. Clarendon Press, Oxford.

Golson, J. and Hughes, P.J., 1976. The appearance of plant and animal domestication in New Guinea. In: Colloque XXII, La Préhistoire Océanienne, IX Congrès. Union Internationale des Sciences Préhistoriques et Protohistoriques, Centre National de la Recherche Scientifique, Paris.

Gould, R.A., 1968. Preliminary report on excavations at Puntutjarpa Rockshelter, near the Warburton Ranges, Western Australia. Archaeol. Phys. Anthropol. Oceania, 3: 161—185.

Gould, R.A., 1969. Subsistence behaviour among the Western Desert Aborigines of Australia. Oceania, 39: 253—274.

Gould, R.A., 1971. Uses and effects of fire among Western Desert Aborigines of Australia. Mankind, 8: 14—24.

Green, R.C., 1976. Lapita Sites in the Santa Cruz Group. In: R.C. Green and M.M. Cresswell (Editors), Southeast Solomons Islands Cultural History. Bull. R. Soc. N.Z., No. 11.

Green, R.C., 1979. Lapita. In: J.D. Jennings (Editor), The Prehistory of Polynesia. Harvard University Press, Cambridge and London.

Green, R.C. and Davidson, J.M. (Editors), 1969, 1974. Archaeology in Western Samoa. Bulletins of the Auckland Institute and Museum, Nos. 6, 7.

Greenway, J.C., 1958. Extinct and Vanishing Birds of the World. Special Publication 13, American Committee for International Wild Life Protection, New York, NY.

Grzimek, B. (Editor), 1972. Grzimek's Animal Life Encyclopedia. Van Nostrand Reinhold, New York, NY.

Hamilton, A., 1972. Aboriginal man's best friend? Mankind, 8: 287—295.

Harding, T.G., 1967. Ecological and technical factors in a Melanesian gardening cycle. Mankind, 6: 403—408.

Hayden, B., 1975. Dingos: pets or producers? Mankind, 10: 11—15.

Hiatt, B., 1967. The food quest and the economy of the Tasmanian Aborigines. Archaeol. Phys. Anthropol. Oceania, 38: 99—133.

Hiatt, B. 1974. Woman the gatherer. In: F. Gale (Editor), Woman's Role in Aboriginal Society. Australian Institute of Aboriginal Studies, Canberra.

Hiroa, Te Rangi, 1930. Samoan material culture. Bernice P. Bishop Mus. Bull. No. 75.

Hiroa, Te Rangi, 1932. Ethnology of Manahiki and Rakahanga. Bernice P. Bishop Mus. Bull. No. 99.

Hiroa, Te Rangi, 1938. Ethnology of Mangareva. Bernice P. Bishop Mus. Bull. No. 157.

Hiroa, Te Rangi, 1944. Arts and crafts of the Cook Islands. Bernice P. Bishop Mus. Bull. No. 179.

Hope, J.H. and Hope, G.S., 1976. Palaeoenvironments for man in New Guinea. In: R.L. Kirk and A.G. Thorne (Editors), The Origin of the Australians. Australian Institute of Aboriginal Studies, Canberra.

Hope, J.H., Lampert, R.J., Edmondson, E., Smith, M.J. and Van Tets, G.F., 1977. Late Pleistocene faunal remains from the Seton Rock Shelter, Kangaroo Island, South Australia. J. Biogeography, 4: 363—385.

Howells, W., 1973. The Pacific Islanders. A.H. and A.W. Reed, Wellington.

Hughes, I., 1970. Pigs, sago and limestone: the adaptive use of natural enclosures and planted sago in pig management. Mankind, 7: 272—278.

Jentink, F.A., 1896. The dog of Tengger. Notes Leyden Mus., 18: 217—220.

Jones, R., 1967. Middens and man in Tasmania. Aust. Nat. Hist., 1967: 359—364.

Jones, R., 1968. The geographical background to the arrival of man in Australia and Tasmania. Archaeol. Phys. Anthropol. Oceania, 3: 186—215.

Jones, R., 1969. Fire-stick farming. Aust. Nat. Hist., 1969: 224—228.

Jones, R., 1970. Tasmanian Aborigines and dogs. Mankind, 7: 256—271.

Jones, R., 1975. The Neolithic, Palaeolithic and the hunting-gardeners — man and land in the Antipodes. Bull. R. Soc. N.Z., 13: 21—34.

Jones, R., 1976. Greater Australia: research into the Palaeo-ecology of early man 1974/1975. Early Man News, International Association on Quaternary Research (INQUA), Tubingen.

Kershaw, A.P., 1975. Late Quaternary vegetation and climate in north-eastern Australia. Bull. R. Soc. N.Z., 13: 181—187.

Kirch, P.V., 1971. Halawa Dune site (Hawaiian Islands). A preliminary report. J. Polynesian Soc., 80: 228—236.

Kirch, P.V., 1973. Prehistoric subsistence patterns in the northern Marquesas Islands. Archaeol. Phys. Anthropol. Oceania, 8: 24—40.

Kwapena, N., 1975. A Wildlife Programme in Papua New Guinea. Wildlife Branch, Dept. of Agriculture, Stock and Fisheries, Papua New Guinea.

Lampert, R.J., 1971a. Coastal Aborigines of Southeastern Australia. In: D.J. Mulvaney and J. Golson (Editors), Aboriginal Man and Environment in Australia. Australian National University Press, Canberra.

Lampert, R.J., 1971b. Burrill Lake and Currarong. Terra Australis, No. 1, Prehistory Dept., Research School of Pacific Studies, Australian National University, Canberra.

Laurie, E.M.O. and Hill, J.E., 1954. List of the Land Mammals of New Guinea, Celebes and Adjacent Islands, 1758—1932. Br. Mus. (Nat. Hist.), London.

Lawrence, R., 1971. Habitat and economy: a historical perspective. In: D.J. Mulvaney and J. Golson (Editors), Aboriginal Man and Environment in Australia. Australian National University Press, Canberra.

Layard, J., 1942. Stone Men of Malekula. Chatto and Windus, London.

Laycock, D., 1975. Butchering pigs in Buin. J. Polynesian Soc., 84: 203—212.

Leach, H.M., 1976. Horticulture in prehistoric New Zealand. Unpublished Ph.D. Thesis. Anthropology Dept., University of Otago, Dunedin.

Linton, R., 1955. The Tree of Culture. Knopf, New York, NY.

Loeb, E.M., 1926. History and traditions of Niue. Bernice. P. Bishop Mus. Bull. No. 32.

Longman, H.A., 1928. Notes on the dingo, the Indian wild dog and a Papuan dog. Mem. Queensl. Mus., 9: 151—157.

Luomala, K., 1960. The native dog in the Polynesian system of values. In: S. Diamond (Editor), Culture in History. Columbia University Press, New York, NY.

Martin, P.S. and Wright, Jr., H.E. (Editors), 1967. Pleistocene Extinctions: the Search for a Cause. Yale University Press, New Haven and London.

MacIntosh, N.W.G., 1972. Radiocarbon dating as a pointer in time to the arrival and history of man in Australia and Islands to the North-West. Proc. 8th Int. Conf. Radiocarbon Dating, Wellington, 1: 44—56.

MacIntosh, N.W.G., 1975. The origin of the dingo: an enigma. In: M.W. Fox (Editor), The Wild Canids, their Systematics, Behavioural Ecology and Evolution. Van Nostrand Reinhold, New York, NY.

McCarthy, F.D., 1976. The Tasmanian Devil in rock art. Mankind, 10: 181—182.

McKern, W.C., 1929. Archaeology of Tonga. Bernice P. Bishop Mus. Bull. No. 60.

Medway, Lord, 1969. The Wild Mammals of Malaya. Oxford University Press, London, Kuala Lumpur, Singapore.

Medway, Lord, 1972. The Quaternary mammals of Malesia: a review. Trans. 2nd Aberdeen—Hull Symp. Malesian Ecol. University of Hull, Dept. of Geography, Misc. Series, No. 13.

Meggitt, M.J., 1965. The association between Australian Aborigines and dingos. In: A. Leeds and A.P. Vayda (Editors), Man, Culture and Animals. Publication 78, American Association for the Advancement of Science, Washington D.C.

Meggit, M.J., 1974. Pigs are our hearts! The Te exchange cycle among the Mae Enga of New Guinea. Oceania, 44: 165—203.

Menzies, J.I., 1977. Fossil and subfossil fruit bats from the mountains of New Guinea. Aust. J. Zool., 25: 329—336.

Merrillees, D., 1968. Man the destroyer: late Quaternary changes in the Australian marsupial fauna. J. R. Soc. West. Aust., 51: 1.

Milham, P. and Thompson, P., 1976. Relative antiquity of human occupation and extinct fauna at Madura Cave, southeastern Western Australia. Mankind, 10: 175—180.

Morren, G.E.B., 1977. From hunting to herding: pigs and the control of energy in Montane New Guinea. In: T.P. Bayliss-Smith and R.G.A. Feachem (Editors), Subsistence and Survival: Rural Ecology in the Pacific. Academic Press, London.

Mulvaney, D.J., 1975. The Prehistory of Australia, revised edn. Penguin Books, Harmondsworth.

Mulvaney, D.J., Lawton, G.H. and Twidale, C.R., 1964. Archaeological excavation of Rock Shelter No. 6, Fromms Landing, S.A. Proc. R. Soc. Victoria, 77: 479—516.

Oliver, D.L., 1951. The Pacific Islands. Harvard University Press, Cambridge, MA.

Pawley, A. and Green, R., 1973. Dating the dispersal of the Oceanic Languages. Oceanic Linguistics, 12: 1—67.

Powell, J.M., 1970. The history of agriculture in the New Guinea Highlands. Search, 1: 199—200.

Rappaport, R.A., 1967. Pigs for the Ancestors. Yale University Press, New Haven and London.

Ride, W.D.L., 1970. A Guide to the Native Animals of Australia. Oxford University Press, Melbourne.

Robb, J., 1973. Reptiles and amphibia. In: G.R. Williams (Editor), The Natural History of New Zealand. A.H. and A.W. Reed, Wellington.

Ryan, D.J., 1972a. Mok-Ink. In: P. Ryan (Editor), Encyclopedia of Papua New Guinea. Melbourne University Press, Carlton.

Ryan, D.J., 1972b. Te. In: P. Ryan (Editor), Encyclopedia of Papua New Guinea. Melbourne University Press, Carlton.

Sanderson, I.T., 1955. Living Mammals of the World. Hamish Hamilton, London.

Shawcross, W., 1968. Energy and ecology: thermodynamic models in archaeology. In: D.L. Clarke (Editor), Models in Archaeology. Methuen, London.

Shutler, R. and Shutler, M.E., 1975. Oceanic Prehistory. Cummings Modular Program in Anthropology, Menlo Park.

Specht, J., 1968. Preliminary report on excavations on Watom Island. J. Polynesian Soc., 77: 117—134.

Strathern, A. J., 1972. Moka. In: P. Ryan (Editor), Encyclopedia of Papua New Guinea. Melbourne University Press, Carlton.

Tate, G.H.H., 1951. The rodents of Australia and New Guinea. Bull. Am. Mus. Nat. Hist., NY 97 (4).

Tedford, R.J., 1967. The fossil Macropodidae from Lake Menindee, New South Wales. Univ. of Calif. Publ. Geol. Sci., No. 64.

Thompson, D., 1939. The seasonal factor in human culture. Proc. Prehistoric Soc., 5: 209—221.

Tindale, N.B., 1959. Ecology of primitive aboriginal man in Australia. In: A. Keast, R.L. Crocker and C.S. Christian (Editors), Biogeography and Ecology in Australia. Junk, The Hague.

Tindale, N.B., 1964. Radiocarbon dates of interest to Australian archaeologists. Aust. J. Sci., 27: 24.

Titcomb, M., 1969. Dog and man in the Ancient Pacific. Bernice P. Bishop Mus. Spec. Publ., No. 59.

Troughton, E., 1955. A new native dog from the Papuan Highlands. Proc. R. Zool. Soc. N.S.W., 1955—1956: 93—94.

Troughton, E., 1965. Furred Animals of Australia, 8th edn. Angus and Robertson, Sydney, London and Melbourne.

Troughton, E., 1971. The early history and relationships of the New Guinea Highland Dog (*Canis hallstromi*). Proc. Linn. Soc. N.S.W., 96: 94—97.

Tyndale-Biscoe, H., 1973. The Life of Marsupials. Edward Arnold, London.

Urban, M., 1961. Die Haustiere der Polynesier. Hantzschel, Gottingen.

Vanderwal, R.L., 1972. Prehistoric studies in central Papua. Ph.D. Thesis, Australian National University, Canberra.

Van Deusen, H., 1963. First New Guinea record of *Thylacinus*. J. Mammalogy, 44: 279—280.

Van Deusen, H., 1972. Mammals. In: P. Ryan (Editor), Encyclopedia of Papua New Guinea. Melbourne University Press, Carlton.

Vayda, A.P., 1972. Pigs. In: P. Ryan (Editor), Encyclopedia of Papua New Guinea. Melbourne University Press, Carlton.

Vayda, A.P., Leeds, A. and Smith, D.B., 1961. The place of pigs in Melanesian subsistence. Proc. 1961 Annu. Spring Meet. Am. Ethnol. Soc.

Waddell, E.W., 1968. The dynamics of a New Guinea Highlands agricultural system. Ph.D. Thesis, Australian National University, Canberra.

Watson, J.B., 1977. Pigs, fodder and the Jones effect in Postipomoean New Guinea. Ethnology, 16: 57—70.

White, J.P., 1971. New Guinea and Australian prehistory: the Neolithic Problem. In: D.J. Mulvaney and J. Golson (Editors), Aboriginal Man and Environment in Australia. Australian National University Press, Canberra.

White, J.P., 1972. Ol Tumbuna. Terra Australis, No. 2. Prehistory Dept., Research School of Pacific Studies, Australian National University, Canberra.

Williams, G.R., 1973. Birds. In: G.R. Williams (Editor), The Natural History of New Zealand. A.H. and A.W. Reed, Wellington.

Williams, J.M., 1973. The ecology of *Rattus exulans* (Peale) reviewed. Pac. Sci., 27: 120—127.

Wood-Jones, F., 1929. The cranial characters of the Papuan dog. Am. J. Mammalogy, 10: 329—333.

Yen, D.E., 1973. The origins of Oceanic agriculture. Archaeol. Phys. Anthropol. Oceania, 8: 68—85.

Yen, D.E., 1974. The sweet potato in Oceania. Bernice P. Bishop Mus. Bull. No. 236.

Zeuner, F.E., 1963. A History of Domesticated Animals. Hutchinson, London.

14. POSTSCRIPT

Since this chapter was written, several new developments have occurred. Horton (1980) has made a strong argument that the extinction of the Australian giant marsupials was caused by climatic change rather than the effect of man, particularly in view of the new finds from Lancefield which have demonstrated the survival of some extinct kangaroos until at least 26 000 years ago (Gillespie et al., 1978). A large number of extinct bird species has been identified in Holocene deposits in Hawaii (P.V. Kirch, personal communication, 1981). Pig bones have been found in prehistoric deposits on Kapingamarangi (B.F. Leach, personal communication, 1980) and Tikopia (P.V. Kirch, personal communication, 1981). Downie and White (1978) have reported *Macropus agilis* or *Dendrolagus* from New Ireland.

Downie, J.E. and White, J.P., 1978. Balof shelter, New Ireland — report on a small excavation. Rec. Aust. Mus., 31: 762—802.

Gillespie, R., Horton P.R., Ladd, P., Macumber P.G., Rich T.H., Thorne R. and Wright R.V.S., 1978. Lancefield Swamp and the extinction of the Australian megafauna. Science, 200: 1044—1048.

Horton, D.R., 1980. A review of the extinction question. Man, climate and Megafauna. Archaeol. Phys. Anthropol. Oceania, 15: 86- 97.

Chapter 4

Indigenous Domesticated Animals of Asia and Africa and their Uses

H. EPSTEIN

1. CATTLE

Cattle have a wide ecological range in Asia and Africa. They are bred nearly everywhere, except in deserts, the Arctic, and parts of the tropical rain forest, and even occur in regions in which other bovines, such as buffalo, yak, gayal and Balinese, are at home. There is a fundamental division between hot environments which are stocked mainly with humped cattle, and temperate or cool environments where humpless cattle are bred.

In Asia, humped cattle extend from Southern Arabia and Southern Iraq through Iran, Afghanistan, the Indian subcontinent and Southeast Asia to Southern and Central China, with minor extensions into the southern republics of the U.S.S.R. In Africa they are ubiquitous south of the Sahara.

Humpless cattle prevail north of the humped cattle territory, from Israel, Lebanon, Jordan, Syria and Anatolia on the west of Asia through Northern Iraq and Iran, the whole of Tibet and Mongolia to Manchuria, Korea and Japan on the east, and throughout the major part of Soviet Asia as far north as cattle can be economically maintained.

In Africa, humpless native cattle are found in Egypt, the Atlas countries and parts of West Africa. Most of these are shorthorned, a very few long-horned. In Southern Africa many European breeds have been introduced by settlers of European derivation.

Humped cattle are classed, in accordance with the position of the hump, into thoracic-humped and cervico-thoracic-humped types. The thoracic-humped cattle are commonly referred to as 'zebu', although not all of them are zebu proper; some, such as the longhorned Fulani cattle of West Africa, include a humpless long horn strain. Cervico-thoracic-humped cattle occur in, or are derived from, contact areas of thoracic-humped zebu and humpless cattle. In crossbreds of humpless and thoracic-humped zebu cattle the hump is usually cervico-thoracic. In East and South Africa, cervico-thoracic-humped cattle are referred to as 'sanga'. In Bhutan, the Siri cattle, derived from Indian zebu and Tibetan humpless cattle, are occasionally also cervico-thoracic- rather than thoracic-humped. In Central China all cattle breeds between the area of the humpless Mongolian and Manchurian types in the north and the South China zebu in the south are cervico-thoracic-humped. In the more northern breeds, where the influence of humpless cattle is greatest, the hump is present only on the neck of the bull, but towards the south, where the influence of thoracic-humped zebus predominates, the hump becomes more pronounced in both sexes and more caudally situated (Epstein, 1972a).

In Asia and Africa, cattle are kept for milk, work and meat, more rarely for social prestige, sacrifice or sport.

Chapter 4 references, p. 91

Fig. 4.1. Kikuyu zebu cow.

Fig. 4.2. Ankole sanga bull and cow.

1.1. Milk

Few of the native breeds, whether humped or humpless, have been specially bred for milk, although among most of them individual cows can be found which produce fairly high yields under superior conditions of feeding and management. Among the zebu cattle of Asia several breeds, such as the Red Sindhi, Ongole, Hariana and Sahiwal, are distinguished by relatively good yields. The Red Sindhi gives about 1500 kg in a lactation of 9 months; selected cows produce 2000 kg. The Ongole yields 1200—1500 kg with 5% fat on average in a lactation of 10—11 months, the Hariana yields 1500 kg in 10 months, and the Sahiwal, which is probably the best dairy breed of zebu, produces 2300 kg with 4.3—6.0% fat. Selected Sahiwal cows have given 4000 kg milk in 300 days (Joshi and Phillips, 1953). The milk yields of most of the other zebu breeds of the Indian subcontinent and of those of Southeast and Southwest Asia are considerably lower.

The zebu cattle of West and East Africa also give only small quantities of milk under the usual conditions of feeding and maintenance. At experiment stations the following average lactation yields have been recorded (Joshi et al., 1957): Adamawa Ngaundere 680 kg; Adamawa Banyo 730 kg; Adamawa Yola 960 kg, with 1400—1700 kg for the best yields; Azaouak 250—300 kg; Maure 600—700 kg; Shuwa 1000—1200 kg; Sokoto 1000 kg with 5.7% fat; Kenana 1800 kg with 4.7% fat; and Butana 1900 kg with 5.5% fat. East African zebu breeds have given the following average yields at experiment stations: Boran in Tanzania 850 kg and in Kenya 1650 kg; Bukedi at experiment stations 800—1100 kg with 5.8% fat and in farmers' herds 230 kg in addition to the milk consumed by the calves; Lugware 250—300 kg with 6.0—6.5% fat in addition to the milk taken by the calves; Nandi cows 1000 kg milk with 5.1% fat; Southern Sudan Hill zebu 530 kg in 300 days with calves suckling; Tanganyika zebu 600 kg with 4.9% fat in a lactation of 8 months, selected Tanganyika cows 940 kg; Toposa-Murle 900 kg in addition to the milk consumed by the calves; Madagascar zebu 150—200 kg with 4.5% fat in a lactation of 6 months.

The thoracic-humped Fulani breeds of West Africa, descended from zebu and humpless longhorn, give even less milk than most of the zebu breeds, namely, 400—500 kg with 4.8—5.5% fat in 5—7 months, exclusive of the quantity taken by the calves. The sanga cows of East and South Africa are similarly poor milkers: Ankole 300 kg with 4.5% fat in 7 months; Nguni 160—250 kg with 3.6—4.2% fat in 5 months; Nilotic sanga with calf suckling 400—500 kg; and the Nganda, of zebu—sanga derivation, 1000 kg with 6.0—6.5% fat.

The cervico-thoracic-humped cattle of China are rarely milked. Phillips et al. (1945) reported an average yield of 119 kg with 5.9% fat for Szechwan cows, and Van Fu-ĉžao (1958) reported 115 kg with 3.1—8.2% fat in a 180-day lactation for Chinchwan cows. Nanyang cows give 4.5—7.0 kg milk per day with 5.0—5.7% fat in a lactation period of 180—300 days, and Hwangpei cows produce 3—4 kg daily during a short lactation (Hofmann, 1959—1960).

The common humpless shorthorn cattle of Syria, Lebanon, Jordan and Anatolia yield between 400 and 800 kg milk per lactation, testing 4—5% fat, the Jaulan breed 700—800 kg, Lebanon or Beirut cattle 1000—1500 kg with 3—4% fat, and the Damascus or Shami, which is a single-purpose dairy breed, 2000—4000 kg with approximately 4% butterfat (Epstein, 1971). North African shorthorn cattle, kept at experiment stations, produce about 1300 kg milk with 3.2% fat per lactation, while ordinary farm cows yield much less. West African humpless shorthorn cows give only 125—225 kg milk with 3.3% fat in a lactation of 4—6 months. In several regions of West Africa the cows are not milked but suckle their offspring until they dry off in the course of nature.

The milk yields of the humpless longhorn breeds of Africa, i.e. N'Dama

Chapter 4 references, p. 91

Fig. 4.3. Libyan humpless shorthorn cow.

and Kuri, are also low. The N'Dama gives 350—400 kg in a lactation of 5—6 months, and the Kuri produces 600—700 kg in 6—10 months (Joshi et al., 1957).

Humpless Mongolian cattle on ordinary pasture yield 500—700 kg milk, and with improved feeding up to 1100 kg. The fat content varies between 4.6 and 6.0%. Manchurian cattle were not milked in the past. Now the milk left by the calf is drawn twice a day, yielding 600—800 kg per lactation with approximately 6% fat. The improved Pinchow of Manchuria and the Sanho of Mongolia give 1500—3500 kg milk, depending on feed and climate, with 3.6—4.8% fat (Epstein, 1969).

The cattle breeds of Soviet Asia produce in accordance with the improved breeds to which they have been upgraded and the quality of the available feed. The European dairy breeds of South Africa yield according to the degree of their adaptation and the level of maintenance and feeding.

1.2. Draught and transport

Nearly all zebu breeds are used for transportation and field work such as ploughing, harrowing, threshing, drawing water from wells and working oil mills. Some are employed for pack work and a few also for riding. Among Indian zebus several breeds, e.g. Kankrej and Hariana, are fast yet very powerful; a pair of bullocks can haul 600—900 kg in an iron-tired cart on a rough road, and up to 1800 kg in carts with pneumatic-tired wheels on good roads, covering a distance of 40—60 km in 8—10 h. Other breeds, e.g. Kherigarh and Amrit Mahal, are suitable for fast light draught, while others again, such as the Bhagnari and Mewati, are slow but useful for heavy ploughing or cart work. In China, Hainan zebus are excellent draught animals; a Hainan ox can plough 0.1 ha land in 4—6 h or carry a load of 150 kg a distance of 30 km in a day.

In Africa, too, zebu cattle are widely used for work. Adamawa and Sokoto oxen, for example, make good draught animals for farming operations on easily worked soil. Azaouak and Maure zebus are excellent pack oxen, carrying loads of 80—100 kg 4—5 km/h or 40 km in a 10—11-h day. Northern Sudan bulls are employed as riding animals, covering 40 km in a day; a pair can till 0.28 ha of heavy clay in 5—6 h. The Madagascar zebu is used for the preparation of land for the rice crop by trampling the soil. A team of six Madagascar oxen can plough 0.5 ha in a 5-h working day. Two oxen harnessed to a cart are able to draw a load of 350—500 kg over a distance of 4 km/h or about 30 km in the course of a working day.

Fulani cattle are more rarely used for work. The M'Bororo are wild and intractable; only a few young males are castrated and trained for pack work. White Fulani bullocks are even-tempered but slow; a pair are capable of hauling a cartload of 450—550 kg at 3.25 km/h for 6—8 h a day.

Sanga oxen are not normally worked for more than 4 h/day. Ankole bullocks make active draught animals; a pair can haul a load of 800 kg in a rubber-tired cart, travelling 3.5 km/h or approximately 16 km in a working day. Barotse oxen, although docile, have only a limited working capacity because of the poor durability of their hoofs. Nilotic, Nganda and Nguni cattle are all employed for work, although Nilotic bulls and oxen are rather slow. The Hottentots use oxen for riding and pack work.

In China, the cervico-thoracic-humped Chinchwan ranks among the best draught breeds. The Nanyang and Chowpei are not inferior; a Nanyang cow can draw a load of 450 kg, and a Chowpei ox 500—700 kg, a distance of 30—35 km/day. Shantung cattle are also employed for work; a pair of fully grown oxen are capable of hauling 1500 kg over 40—50 km in a working day.

A humpless Mongolian trek ox can draw a two-wheeled cart, weighing 300 kg, with a load of 1000 kg for 30 km per day. Manchurian working bulls are not castrated. Two bulls or two cows can plough 1 acre of land/day. One bull can pull a load of 600 kg in hilly areas, or 1000 kg on a straight road, a distance of 35 km a day (Epstein, 1969).

In Egypt, humpless shorthorn cattle are almost exclusively employed on tillage operations, threshing, and lifting water by means of the water wheel. In the Atlas countries the cattle are also trained for draught; their walking pace covers between 2.4 and 2.8 km/h. In West Africa, N'Dama oxen are used for haulage and tillage, usually in large teams worked only in the cool hours of the morning (Doutressoulle, 1947). Kuri cattle are unsuited for draught, but some are used as pack animals (Joshi et al., 1957).

1.3. Meat and blood

In India and Nepal beef is not eaten, and cattle are allowed to die a natural death. Among the Masai and several other peoples of East Africa cows are also never slaughtered; they may, however, be eaten on their death. But blood is an important element of the Masai diet, being obtained by piercing the jugular vein with a special arrow, a practice similar to that employed by the Sherpa of Nepal with yak. This suggests a previous hunting culture, since all people who live by hunting regard blood as an important part of their diet (Linton, 1956).

Apart from these exceptional instances, beef is generally consumed throughout Asia and Africa, although in many pastoral areas mutton or goat is preferred. There are only a few native breeds, such as the Boran of Kenya and the Africander of South Africa, which have been developed for meat production by European settlers. In the grassland areas of Soviet Asia and the savannas of Southern Africa European beef breeds or local stock crossed or upgraded with these are widespread. In general, native cattle, whether humped or humpless, yield a carcase of rather low quality, for in areas where cattle are used for work and the cows are milked, only aged animals are commonly slaughtered. But some Indian breeds, e.g. Gir, Kankrej and Ongole, which are not used for meat in their home country, show good potentialities for beef production and have been used for this purpose elsewhere (Joshi and Phillips, 1953).

In China, Hainan zebu oxen have an average dressed weight of 51.3%, and 37.4% boneless meat; in Kwangsi zebus the dressing percentage ranges from 44 to 54% (Balezin, 1959). The cervico-thoracic-humped draught breeds of Central China make rather poor beef cattle. Although adult Nanyang cows in good condition have an average liveweight of 400 kg, and particularly large cows may weigh up to 600 kg, their dressing percentage is only about 43% (Van Fu- čžao, 1958), that of Chowpei cattle 45—48% (Hofmann, 1959—60). Humpless Mongolian cattle have considerably higher carcase yields, 50% on average, and in good seasons with plentiful rainfall up to 58% in autumn (Epstein, 1969).

In Africa, Adamawa zebu oxen at 4 years of age and a liveweight of 580 kg have a dressed weight of 51%, and 52% at 5 years with a weight of 680 kg. Four-five-year-old Maure zebus, weighing c. 350 kg, dress out at 45—50% with 30% boneless meat. Shuwa and Sokoto zebus, 4—6 years old, yield a carcase weight of 50% of the liveweight. Aged Northern Sudan zebu bulls of poor quality have a dressing percentage varying between 40 and 50%.

In humped longhorned Fulani cattle the following dressed weights have been recorded: Nigerian Fulani 50%; Senegal Fulani bullocks — prime 48—51%, poorer quality 42—45% with 26% boneless meat; Sudanese Fulani in fair condition 46—47%; 5-year-old White Fulani, weighing 500—550 kg, 50—55%; M'Bororo 40—42% of coarse poor quality meat with a high percentage of bone.

Sanga cattle are of variable quality as meat producers. The dressing percentage of Nilotic steers is reported to be about 50%. The Ankole do not fatten easily on the grasslands of their breeding area. Bulls and castrates are

Fig. 4.4. Kuri humpless giant-horned bull.

usually sold for slaughter when they have two permanent teeth. The dressing percentage averages 45%; at 3.5 years they dress out at about 50%, though 55% is not unusual. The meat is well marbled. Nioka cattle thrive well on grassland and produce excellent beef; they are generally slaughtered at 3 years of age with a liveweight of about 330 kg, killing out at 50%. Barotse cattle, slaughtered at 5 years with a liveweight of 550 kg, have a dressing percentage of almost 53 with 18% bones in the carcase. Fattened Nguni steers, 3.5—4.5 years old, dress out at 57—62%, with 87% beef and 13% bone.

Libyan humpless shorthorn cattle kill out at 45—55%; their meat is of reasonably good quality, but the fat is mainly distributed subcutaneously and around the kidneys. Brown Atlas cattle dress out at 45—49%, and West African shorthorns at as much as 55%.

The humpless N'Dama of West Africa yields a high quality carcase of close-grained meat, though without marbling fat, with a dressing percentage of 45—50% (Doutressoulle, 1947). The Kuri is also a good beef breed; at 5 years of age and a liveweight of 500—600 kg the dressing percentage averages 50% (Joshi et al., 1957).

1.4. Cultural values

In many regions of Asia, and particularly of Africa, cattle form the emotional and cultural centre of their owners' lives. Cattle have a prestige value and herds are judged by their mere size. In East and South Africa, 'lobola' (bride price) is always paid in cattle. Several sanga breeds are especially prized because of their gigantic horns. Among the Bakosi and neighbouring tribes in Cameroun milking is unknown, and cattle are kept only for lobola and sacrifice in the cult of the dead, when the corpse is rolled up in the hides of the slaughtered animals (Curson and Thornton, 1936; Staffe, 1938). Among the Naga tribes of Assam the sacrifice of cattle is an essential part of their feasts of merit. Cattle serve also as currency for ceremonial payments, and among the Apa Tanis cattle are the only accepted medium of exchange in the purchase of land. The restriction of cattle to sacrifice extends over many parts of Southeast Asia. Until a few years ago the sacrifice of oxen was widely and publicly practised by several tribal populations of Peninsular India (Von Fürer-Haimendorf, 1963). In Thailand and Java bulls are used for racing (Fischer, 1967; Rouse, 1970). In Nepal, where slaughter of cattle is prohibited, their almost sole contribution to agriculture is dung (Cockrill, 1974).

2. BUFFALO

The water buffalo's wild progenitor, i.e. the arni, still survives in isolated marsh and jungle regions of India and the Kosi river area of Nepal. The domestic buffalo is better suited than cattle to hot humid regions; its skin has a very thick epidermis. The sebaceous gland volume is four to eight times as great as that of cattle, but there is a paucity of sweat glands. Cooling ability by sweating is therefore low, and the major role in cooling must be played by conduction to the water in wallows, evaporation of the wetted skin, and pulmonary evaporation. The hair density is only about 1/10 that of cattle (Mason, 1974).

In Southeast Asia the area of distribution of the buffalo extends north to the Yangtze valley in China and west as far as Assam, including Southeast China, Burma, Assam, Laos, Kampuchea, Vietnam, Thailand, Malaysia, Indonesia and the Philippines, with extensions into Nepal and Sri Lanka. In

Chapter 4 references, p. 91

Fig. 4.5. Swamp buffalo cow, Nepal.

Nepal buffaloes are pastured in summer for a few weeks near the snow line as high as 4500 m (Epstein, 1977).

Most of the buffaloes of East Asia are of the Swamp type, which is morphologically closer to the arni than are the buffaloes of India and Pakistan. The latter and those farther west belong to the River type, which has been selected to form improved breeds (Mason, 1974). The Swamp buffalo is usually dark grey with white or pale grey stockings and one or two light bands below the neck. In the River buffalo black is the most common colour, but unimproved stock may be grey with white stockings like the Swamp type. The two types are believed also to differ in chromosome number (Ulbrich and Fischer, 1967, 1968; Fischer and Ulbrich, 1968). Swamp buffaloes generally prefer static water to streams or rivers, which are favoured by the River type.

West of Pakistan, buffaloes are discontinuously distributed in Afghanistan, Iran, Iraq, Syria and Anatolia. In Soviet Asia they are most common in Azerbaijdzhan and Gruziya, contiguous with their breeding area in Northwest Iran.

In Egypt two varieties are recognized which are distinguished mainly by size. There have been several attempts at introducing Indian buffaloes into other parts of Africa. In the past, most of these have failed, e.g., in Tunisia, Zaire, Congo, and Madagascar. But recent importations into Uganda, Tanzania and Mozambique promise greater success (Cockrill, 1974).

Buffaloes are used for work, milk, meat and hide, religious ceremonies, and sport.

2.1. Draught and transport

In the large majority of countries where buffaloes are kept they are employed for work on small farms. In particular, the buffalo is the ideal work animal for paddy fields owing to its aquatic predeliction and its large hoofs which enable it to move easily in deep mud. The buffalo is a very docile animal, easily trained. Much of the responsibility of handling lies with young children who exert a remarkable degree of control over their charges and act as competent herdsmen.

The working life of the buffalo commonly lasts 25—30 years, in rare instances up to 40 years. The average buffalo works about 5 h/day and can plough 0.1 ha during this time. In Kiangsu province, China, a fully grown castrate can plough 0.25—0.35 ha of irrigated land in a working day of 8—10 h, broken by five intervals of rest. In Hupeh province a castrate can do all the work on 2.3 ha of cultivated land; but the ploughing capacity of smaller animals is considerably less (Epstein, 1969). In many countries buffaloes are rested for long periods between planting and harvesting, or are used intermittently for other tasks. They may spend most of their time in the forest and are used merely for a few weeks in the year to plough and harrow the paddy fields. Buffaloes can survive forest life, while cattle fall victim to predators. In Java, Sumatra and the northeastern region of Thailand there are many accounts of buffaloes having killed tigers.

In several countries the buffalo is used as a riding animal. It is also quite widely employed for pack work. In some islands of Indonesia the rider mounts from the rear, pulling himself up by the tail. In China the Tanyang type of Swamp buffalo can carry a load of 100—150 kg a distance of 25 km in a day. Heavier breeds carry up to 250 kg.

2.2. Milk

The greatest number of the world's dairy buffaloes are located in India and Pakistan, and the largest part of the world's buffalo milk is produced and consumed in these two countries. Although buffaloes constituted only 30% of all domestic bovines in India in the nineteen-sixties, they contributed 55% of the total milk supply of the country (Amble et al., 1965). Buffalo cows can be milked profitably until 15 years of age. The major reason why the buffalo rather than the zebu cow is preferred in India as a dairy animal is the aura of sanctity and veneration that surrounds all bovines with the single exception of the buffalo, and the taboo on the killing and eating of cattle.

Buffalo milk is always white, even with high carotene feeding, thick and somewhat ropy in consistency. It has a high fat content, ranging from 7.5 to 15%, and c.18% total solids. Butter made from it is white and does not turn rancid easily. Most of the butterfat is made into ghee, which in several countries of Asia forms an important export article.

The best dairy breeds of buffalo, such as the Murrah, Nili-Ravi and Kundi, have been developed in Northwest India and Pakistan, and the Surti, Mehsana and Jafarabad in Gujarat. These are commonly massive, stocky animals with deep and wide frames, short limbs and heavy bones. The horns of the buffaloes of the Murrah group are closely curled, those of the Gujarat breeds curled or sickle-shaped. The udders are large with conspicuous milk veins and long squarely placed teats. The liveweight of adult males varies between 450 and 800 kg and of females between 350 and 700 kg. The lactation yield of good cows ranges from 1600 to 2500 kg with approximately 8% butterfat; outstanding animals give above 3000 kg. Their placid, docile nature facilitates their adaptation in many parts of Asia where they have been introduced. Cows with high milk records can also be found in other Indian breeds, but generally their yields are lower. Buffalo cows of the Swamp type in Southeast Asia do not commonly give more than 250—350 kg milk per lactation in addition to that consumed by the calves, but the butterfat content of the milk exceeds 10%. In several countries, e.g. China, Burma and Thailand, they are not milked, and all milk is left to the calves.

Chapter 4 references, p. 91

2.3. Meat and hide

Buffalo meat is derived from male calves not required for work or breeding, but mostly from aged animals past their profitable working or milking life. In Sri Lanka, for example, buffaloes may not be slaughtered below the age of 12 years, but here and in other countries they are usually much older at slaughter. The meat of such animals is dark red, coarse and stringy and devoid of marbling fat. Carcase yields of 4-month-old calves range from 55 to 60%, while aged animals yield 45—50%. The fat is white, for the major part intermuscular, less subcutaneous. In the Chin hills of Burma feral buffaloes are hunted for meat. In Southeast Asia buffaloes no longer fit for work are often allowed to live their lives out instead of being sent for slaughter.

In Timor and many parts of Thailand buffalo hide specially prepared and cooked is a popular delicacy. In Nepal and several other countries of Asia buffalo meat is sold with the hide attached, or the hide may be sold separately for human consumption. Some of the meat and hide is dried, smoked and stored for later consumption. In Keningau and other areas of Malaysia, soup is made from the hide and bones by prolonged boiling, but in some villages the hides are fed to pigs. Bones are carefully scraped and dried for use as fuel.

2.4. Cultural values

In Pulilan in the Philippines a buffalo festival is held at the feast of the patron saint of farmers. In parts of Indonesia buffaloes are maintained solely for sacrifice, fighting or funeral ceremonies. In Bali buffalo cart racing, and in Thailand buffalo fighting and racing are popular sports. In Malaysia buffaloes are paid by the bridegroom to the father of the bride. Buffaloes may be sacrificed at ceremonies connected with the building of a new house, at funerals or weddings. Buffalo skulls and horns are often planted in fields and on graves. In Nepal there is much ritual slaughter of buffaloes by decapitation with one downward cut of the sacrificial kukri. The Muruts of Sarawak sacrifice buffaloes in the same manner; if the head is not completely severed by one stroke, the meat will not be eaten (Cockrill, 1974).

3. YAK

The domesticated yak is bred on the high plateaus of Central Asia, including Tibet, Northern Nepal, Kashmir and Northeast Afghanistan, the mountain ranges between Tadzhikistan and Buryatskaya in the U.S.S.R., the Mongolian People's Republic, and the western regions of Sinkiang and Inner Mongolia in China. Conditions most suitable for its existence are an elevation of 4000—5000 m, where the animals can be pastured throughout the year. Below 3000 m yak are apt to lose vigour. Yak are used for many products and purposes: milk, work, meat, wool, hair and leather.

Yak cows are poor milkers. In the Eastern Pamirs the average milk yield during a lactation period of 150—170 days was estimated at 300 kg in addition to 200 kg taken by the calf. In an experimental herd in Kirgiziya the mean yield during a lactation of 8 months was 608 kg. The milk has a rich golden colour with 6.5% fat (Schley, 1967).

Yak attain full body weight at 6—7 years of age. In the Altai mature bulls weigh 380—400 kg and cows 260—270 kg. In Eastern Tibet an average weight of 225 kg has been recorded for cows. The dressing percentage of bulls averages 40%, and that of oxen 46—47%. Owing to its high haemoglobin and low fat content, yak meat is of an intense red colour. The meat is coarse-fibred and devoid of marbling.

The yak is an excellent pack and riding animal for mountain travel, being capable of carrying a load of up to 150 kg. Even with relatively poor feed it can carry 50—75 kg for 13—16 km a day for months at a time and stay in good condition. In difficult country yak are superior to mules in finding their way, and are skilful in moving through snow (Phillips et al., 1946). Hornless yak are much in demand as riding animals (Ekvall, 1968).

The hide is thin, uneven and porous. Adult yak yield 750—1400 g hair and 500 g wool per year; yearlings yield over 800 g wool with a fibre diameter ranging from 15 to 19 μm. The hair of the outer coat is used for tent covers, grain sacks or ropes, and the wool of the under-coat is made into felt.

Dried yak dung provides the most common fuel on the high plateaus of Central Asia.

3.1. Yak—cattle hybrids

Hybrids of yak and cattle are found in the entire yak-breeding area. The females are fertile, but the males are not. The hybrids excel in hardiness, size, strength, working ability and milk production. Liveweights of up to 50% above the average weights of the parent stocks, and a 5% higher killing-out weight, have been recorded in the U.S.S.R. In Tibet, hybrid oxen, being gentler than the rather stubborn yak, are highly valued as beasts of burden and draught, and are generally preferred for ploughing.

Hybrid cows reach sexual maturity earlier and yield more milk than yak cows. Hybrids from Kazakh bulls and yak cows yielded 85% more milk than yak cows and 52% more than cows of the Kazakh breed.

There are differences in size, conformation, coat, tail development and temperament between reciprocal crosses, analogous to the differences between mules and hinnies. In Nepal, hybrids by a Tibetan humpless bull out of a yak cow are more easily maintained on cold summer pastures above an elevation of 4000 m than hybrids sired by a yak bull, which adapt better to relatively high temperatures and subtropical winter pastures below 2500 m. Hybrids from humpless bulls and yak cows yield up to 7 kg milk/day at the beginning of lactation, while those from yak bulls and zebu cows yield only 5 kg; the pure yak cow's yield is 3 kg at this period. The milk of hybrid cows sired by humpless bulls has a 0.5—1.0% higher fat content than the milk of hybrids from a yak bull and a zebu cow (Epstein, 1974, 1977).

4. GAYAL (MITHAN)

The gayal or mithan is the domesticated form of the gaur. It is bred in Northwest Burma and in Assam. The principal breeders are the Naga and other hill tribes among whom the mithan is the most highly valued sacrificial animal. Most mithan owners leave their animals in the jungle throughout the year and bring them to the village only for sale or slaughter. In order to improve the progeny of their cows in size and strength they encourage inter-breeding with wild gaur bulls by attracting the latter to the gayal cows by the provision of salt licks. Gayal are also hybridized with zebu cattle, and in some districts the hybrids are much in favour. Among the Mönbas of the Dirang Dzong area gayal and their hybrids are used for ploughing, and hybrid cows are milked as well as eaten. In Bhutan, gayal bulls from Assam are in great demand for hybridization and may be exchanged for five to six times the number of cattle (Von Fürer-Haimendorf, 1963).

Chapter 4 references, p. 91

Fig. 4.6. Balinese bull.

5. BALINESE

The Balinese of the Indonesian island of Bali is a domesticated banteng. A mature bull weighs 360 kg, a cow 275 kg. The Balinese is an excellent draught animal and is widely used for this purpose. It also yields superior beef.

Bali bulls have been exported to Malaysia for hybridization with Kelantan zebu cows. A similar hybrid type is found in Java (Rouse, 1970). Like yak— cattle hybrids, the female Balinese hybrids are fertile, while the males are sterile for two or three generations. On the island of Madura Balinese —cattle hybrids are smaller than the pure Balinese, an adult hybrid bull weighing about 250 kg and the cow 200 kg. Trained Madura bulls in teams of two are popular in races over grassy turf.

6. SHEEP

The main ecological regions of Asia and Africa are inhabited by different basic types of native sheep. In the hot and relatively dry country of Southern India, in equatorial Africa and the savanna belts of the Sudan and West Africa hairy thin-tailed sheep are found (Phillips, 1948). North of their range, a wide belt extending from Tunisia on the west through Egypt, the whole of Southwest Asia, Afghanistan and Baluchistan to Mongolia, China and Indonesia is occupied by coarse-wooled fat-tailed sheep. In Africa, fat-tailed sheep, partly coarse-wooled and partly hairy, have spread from the Horn along the east coast down to South West Africa and the Cape, whither they accompanied the early Hottentots. In Asia the fat-tailed Karakul extends from Afghanistan into Turkmeniya, Kazakhstan, Tadzhikistan and Uzbekistan, U.S.S.R., and several other fat-tailed breeds are found in the Pamir and Tadzhikistan.

In Asia the range of the fat-tailed type is bordered on the north by a belt

Fig. 4.7. Thin-tailed Baruwal ram, Nepal.

of coarse-wooled fat-rumped sheep, which extends, often discontinuously, from the Black Sea and the confines of Europe (Ukraine and North Caucasus) on the west through parts of Central Asia (Kazakhstan, Kirgiziya, Turkmeniya, Uzbekistan) to Sinkiang and Southeast Siberia on the east.

In Africa there is an enclave of short-haired fat-rumped sheep, unrelated to the Asian type, in Somalia and its western and southern border areas. From importations of the Somali breed into South Africa the so-called Blackhead Persian has been developed (Epstein, 1970).

In Soviet Asia the zone of fat-rumped sheep passes north into the fourth and final belt, which is occupied by wooled thin-tailed sheep. Sheep of this general type are also found in Central and Northern India, Nepal and Tibet.

In North Africa several breeds of wooled thin-tailed native sheep occur in Algeria, Tunisia and Morocco, from where offshoots have spread into the northernmost riverain area of the Sudan. In some parts of the Atlas countries Merino sheep are employed for upgrading.

In South Africa Merino sheep are widespread in the least arid grazing regions, while the semi-arid areas are stocked with Blackhead Persian and Karakul sheep.

Sheep are used for wool, meat and fat, milk, and fur, more rarely for work.

6.1. Wool

Several breeds of Merino type are kept for fine-wool production in Soviet Asia; some of these are dual-purpose wool-and-mutton sheep. In Sinkiang and Northern China local fat-tailed sheep have been graded up to Caucasian Rambouillet, Soviet Merino, Stavropol Merino and other fine-wooled breeds. The future sheep population of China is planned to be 40% fine-wooled. Large quantities of Merino wool are also produced in South Africa from introduced Merino stock.

The fat-tailed sheep of Asia and Africa and the fat-rumped sheep of Asia

Fig. 4.8. Zambian fat-tailed ram.

Fig. 4.9. Kirgiz fat-rumped ram.

produce carpet wool of variable quality. Carpet wool is also yielded by most
of the thin-tailed breeds of India, Nepal and Tibet. Among the thin-tailed
sheep of Africa some produce carpet wool and others merely hair.

6.2. Meat and fat

There are no pronounced mutton breeds in Asia and Africa, such as there
are in Britain, for instance. But surplus male lambs of all the different types
of sheep as well as aged animals are eaten when their productive life is over.
Among several breeds in the savanna belts of Africa and the desert of
Somalia, meat is the only purpose for which sheep are kept, and the same
holds for some of the sheep of Southern India, although their conformation
is the very opposite of a mutton type. Among the fat-tailed, and especially
among the fat-rumped sheep of Asia, the main breeding purpose may be fat,

followed by meat and carpet wool. In South Africa the Blackhead Persian with its short hairy coat is a single-purpose fat-and-mutton sheep.

6.3. Milk

In the whole of Southwest Asia, in parts of Mongolia and adjacent regions, in North Africa and a few areas in East Africa the ewes are milked. In the U.S.S.R. Karakul ewes are also frequently milked; the average lactation yield is slightly more than 60 kg. In Israel the local Awassi fat-tailed breed has been so highly selected for milk that the ewes of several large flocks now produce an average of 400 and even 500 kg/lactation in addition to the milk consumed by the lambs. The Hottentots taught their infants to suck the milk direct from the ewes' udders.

6.4. Fur

There are several fur breeds of sheep in Asia and Africa, the most important of which is the Karakul. Karakul pelts are produced on a large scale in Afghanistan, the U.S.S.R. (Uzbekistan and Turkmeniya), also in Iran, and in Namibia and South Africa. In China two similar types, the Ningsia Black and Kuche, are being improved by an infusion of Karakul blood. White pelts, known in the international fur trade as 'Chekiang' and prized for their light weight and great durability, are obtained from the Hu-yang, an extremely fertile breed in Southeast China including the northern part of Chekiang province. Another Chinese fur breed, the Tan-yang, occurs in several districts of Kansu, Shensi and Inner Mongolia. The fur of the lamb, slaughtered at 20—30 days of age, forms 8—9 cm long curls of a silky texture and beautiful lustre. In Northern Siberia and the Far Eastern Region the Romanov breed of sheep, introduced from its breeding centre in the Volga region, is now bred on a large scale; the fur, of greyish blue colour and high quality, is used for coats.

Fig. 4.10. Jumli wethers with pannier bags.

Chapter 4 references, p. 91

6.5. Transport

In Tibet and Nepal, sheep, along with goats, are employed for pack work. In summer they carry 12—14 kg salt from Tibet over the high passes in the snowy region to the northern parts of Nepal. In winter they make up to five trips between the hilly midlands and the tropical Tarai (Epstein, 1977). The poorer Tibetans on their travels attach their possessions to the back of a solitary sheep which they drive before them (Stamp, 1962).

7. GOAT

Save for the tundra, where only the reindeer can exist, and the heart of the desert, where the camel alone finds enough sustenance, goats are distributed throughout the entire Asian and African continents. In general, they are herded in flocks together with sheep; only where the latter cannot economically be maintained, as in the tropical rain forest, the semi-desert or rugged mountains, are goats pastured alone.

Goats are used for various purposes: milk, meat, mohair and cashmere. In exceptional cases their purpose is to destroy dense shrub for subsequent agricultural land use, to serve as a sacrifice, to do pack work or merely to give aesthetic satisfaction. In several parts of the world, e.g., Sudan, Ghana (Ashanti), Pakistan, Bangladesh, Nepal, Malaysia and Fiji, goat meat is preferred to mutton (Devendra and Burns, 1970).

7.1. Milk

In most countries goats are milked. Several single-purpose milch breeds are found in India; among these the Jamnapari is outstanding. From India goats of Jamnapari type extend through Iran and Iraq to Syria and Lebanon, where they are known as Damascus or Shami. From here they have been introduced into Egypt, where they are called Zaraibi, and hence into the

Fig. 4.11. Nubian goat, Sudan.

Fig. 4.12. Maltese she-goat, Libya.

Sudan and adjacent regions, where they are known as Nubian. Indian milch goats are kept in the towns of the east coast of Africa and also in Southeast Asia. In the Atlas countries the Maltese is the most widespread milch goat. European breeds are found in South Africa and sporadically in the larger coastal towns of China, a very few also in Northern India, Nepal and the Philippines.

7.2. Multi-purpose

In the steppes of Asia and the savannas of Africa many different breeds of goat are bred for milk and meat. Some of these are short-haired and others long-haired. A few short-haired breeds, among these the Sokoto of Nigeria, are famous for the high quality of their skins. The hair of the long-haired breeds is mainly used for tent covers and grain sacks. Goat skins are used as water containers.

7.3. Mohair and cashmere

In two areas of Asia and Africa goats are bred for mohair, namely, in Anatolia, the original home of the Angora goat, and South and South West Africa whither they were imported from Turkey during the 19th century.

In a large area of central Asia, extending from Kashmir through Pakistan, the Punjab, Bengal, Tibet, Sinkiang, Tsinghai, Mongolia and the Kirgiz steppe to Kurdistan, pashmina or cashmere goats are bred. These are covered with a thick outer-coat of moderately long, straight, rather coarse and medullated hair and an under-coat of short, soft, non-medullated down or pashm of a fineness equal to fine Merino wool. The under-coat is removed by combing in early summer, when the down has come loose from the skin. The hairy outer-coat may be white or pigmented and the under-coat white, grey or tan. White goats produce about 125 g cashmere, while black goats from the same

Chapter 4 references, p. 91

Fig. 4.13. Female cashmere goats, Inner Mongolia.

Fig. 4.14. Tibetan goat caravan, Dhauli Valley.

flock yield 175—200 g. The pashmina goat of the U.S.S.R. produces 400 g cashmere. Cashmere fibres are 10% weaker than the finest wool and 40% weaker than fine mohair. In some parts of China and the Mongolian People's Republic, but not in Inner Mongolia, cashmere goats are milked for a few weeks after the kids are 5 months old; the total yield obtained during the milking period is 15—20 kg (Epstein, 1969).

7.4. Dwarf

In adverse environmental conditions goats display a marked tendency to dwarfing. In Asia, dwarf goats are found in the Himalayan part of Tibet and Bengal, and in Africa in the Southern Sudan, the rain forest and on the west coast (Devendra and Burns, 1970). In the tropics of Africa they are used only for meat. Some of them are achondroplastic dwarfs, which are prized for their strange appearance.

7.5. Transport

In Tibet and Nepal, goats, along with sheep, are used for pack work, carrying salt and borax from Tibet to Nepal to be exchanged there for wheat, rice or manufactured goods. Each goat carries about 11 kg in balanced pannier bags of cloth and leather (Weir, 1952).

8. PIG

The principal breeding area of pigs in Asia comprises China, Japan and the southeastern part of the continent including the islands of Indonesia. In Mongolia and Tibet the animal is rare, but in the agricultural regions of the U.S.S.R. in Asia pig breeding is common. In India and Nepal pigs are kept only by some of the former low castes. The animal is absent throughout the sphere of Mohammedan influence in Central and Western Asia, but is bred in Lebanon.

Fig. 4.15. Mongolian unimproved sow.

Chapter 4 references, p. 91

Fig. 4.16. South China sow — lard type.

Prior to the introduction of Islam the pig played a prominent role in Egypt, the Sudan and in the economy of the Berbers of the Atlas countries. It is still bred by Copts in Egypt and also in the vicinity of the larger towns of Tunisia, Algeria and Morocco. A primitive type is found among several peoples of the Sudan, Zaire and West Africa. Some of the Bantu peoples of Southern Africa received the pig during the 19th century, but it occurs only in small numbers. In most parts of South Africa improved breeds are raised by European farmers.

In China, Southeast Asia and in the cold regions of Eastern Siberia the pig is important as a producer of lard. Before 1949, there were over 100 different breeds and varieties of pig in China; only about 40 of these are now considered to be of economic value. Large Whites and Berkshires have been used to improve several local breeds. Both castration and spaying are common. In Northern China many native breeds are distinguished by a coat of long coarse bristles which are an important export article. In Inner Mongolia, Tibet, India and Nepal the pigs live mainly on the garbage of human households. In the Asian part of the U.S.S.R. the majority of breeding zones in the regionalization plan for pigs are assigned to the Large White and Ukrainian White Steppe breeds.

In African villages the pigs find their sustenance by foraging in the bush. Only a few African peoples castrate the boars to facilitate fattening. The Bateke of Gabon are exceedingly fond of pigs, not necessarily to eat, but as pets; a favourite boar may be the friend of a whole village (Johnston, 1908). A similar custom obtained in Taiwan on the arrival of the Dutch in the early part of the 17th century; 'every aboriginal woman had a great Pig running after her, as we used to have a Dog' (Swinhoe, 1870).

Fig. 4.17. Mongolian riding ponies.

9. HORSE

In Asia and Africa horses are not employed for draught to the same extent as they were in Europe before the mechanization of transport and tillage. They serve mainly as riding animals, more rarely for pack work.

In Southern Asia the majority of horses are of the Oriental type, resembling the Arab in cranial and body conformation. In parts of Inner and Outer Mongolia, in Tibet and the eastern grassland steppe of the U.S.S.R. horses of Mongolian type predominate. These are distinguished from the Oriental mainly by their coarser heads and stockier build.

The Arabian horse has been developed in the harsh conditions of the desert. It excels in hardiness, staying power and speed. In Arabia the possession of a purebred mare gives prestige to its owner. Previously mares were important in the tribal raids of the bedouin. The mares would be led behind the camels in reserve for the actual moment of attack. Stallions could not be used for this purpose as they might neigh in the proximity of a mare and give the raiders away.

Since the Mohammedan expansion Arab horses have influenced the majority of breeds of Western and Southern Asia. The horses of Syria, Turkey and Northern Iraq are a mixture of local stocks with a large share of Arabian blood. In the Turkmen steppes of U.S.S.R. and the adjacent part of Northeast Iran the Turkoman horse is bred in two varieties, viz. Akhal Teke and Iomud. The Akhal Teke is also bred in Kazakhstan, Uzbekistan and Kirgiziya. It stands about 15 hands at the withers and is supreme among Soviet sporting horses. The Iomud is slightly smaller and coarser. Previously it was used by the Kirgiz Turkomans for their long-distance raids that used to be a prominent feature of their life. In Iran several local breeds are recognized in addition to the Turkoman and Arab. From Northwest Iran the Kurdi extends into the Kurdistan area of Iraq and Turkey. Most of the horses of India and Pakistan are a mixture of local stocks, introduced from the north in the course of nearly 4 millennia, with Turkoman and Waziri

Chapter 4 references, p. 91

horses from Iran and Kabuli and Herati ponies from Afghanistan, but especially with Arabs imported since the Mohammedan conquest (Epstein, 1972b). In the mountainous areas of Chitral, Hunza, Gilgit, the Spiti River Valley and Manipur, hill ponies, 12—14 hands high, are used for riding and pack work. In the towns of the Indian and Nepalese Tarai the weedy Tarai pony is employed to draw the tonga, a light two-wheeled vehicle.

In Southeast Asia small ponies of Oriental type are found in nearly every part. They range from the 13-hand Burma or Pegu pony of the Shan Hills to the very small primitive breeds in the islands of Indonesia, such as the Sandalwood and Batak, and the Timor pony which is just over 10 hands high.

In the Asian part of the U.S.S.R. horses are now little used for heavy draught. In the regionalization plan for horse-breeding the northern zone is allocated to ponies, since previous attempts to introduce larger horses there were unsuccessful. The steppes of Siberia and the Far Eastern Region are assigned to various improved breeds of trotters and riding horses. Russian horses have also been used to develop or improve some of the breeds of Manchuria and Inner Mongolia. After 1949 over 100 horses were imported into China from the U.S.S.R. for upgrading of local ponies.

The Buryat-Mongolians of the U.S.S.R., the Mongols of Outer and Inner Mongolia, the Kirgiz and related peoples keep large herds of ponies for riding, milk and meat. Unlike the Arabs, the Mongols do not ride mares; these are milked for 3—4 months after foaling. A mare yields about 180—200 kg milk/lactation in addition to the considerable amount consumed by the foal. The milk is fermented into koumiss, which is a very tasty and nourishing drink that will keep for a long time. A small part of it is distilled into a fairly strong alcoholic spirit, resembling vodka (Epstein, 1965). Horse flesh, especially that of young fat mares, is a highly prized meat. The paunch fat is salted and packed into pieces of gut and smoked for storage. Horse hide is made into tough leather straps and lines, while the hair is braided into stout cords for lassos, halters and yurt ropes (Forde, 1934).

Ponies of Mongolian type are also bred for riding and pack work through-

Fig. 4.18. Tibetan pony mare.

out Tibet, in the mountainous parts of Nepal and Western and Southern China, and in a few islands of Japan.

The horses of Africa north of the Equator are classed into three groups: (1) Oriental, (2) Barb and Dongola, (3) Ponies. All of these are riding horses.

Horses of Oriental type, derived from Syria and Arabia, are found mainly in the coastal plains of Libya, Tunisia, Algeria and Morocco, and are occasionally ridden by the Tuareg and other desert tribes. They average 14 hands in height.

The Barb and the closely allied Dongola are of Occidental type, partly derived from ancient Spanish stock. They are generally distinguished by rather coarse heads with a convex profile, light body, long legs and sloping croup. The Barb is bred in the hills and mountains of the Atlas countries and is found also in West Africa. With a withers height of 14—16 hands it is commonly larger than North African horses of Oriental type. For this reason many horse-breeding tribes of North Africa prefer the Barb to the Arab. The French formerly used it in large numbers in their colonial armies (Epstein, 1971). In the early years of this century several thousand Barb foals were purchased annually by Marseilles butchers for meat (Wentworth, 1945).

The Dongola is now nearly extinct in its original home, but is still bred west of it. Like the Barb it has been much crossed with Oriental stock. In mediaeval times the Dongola horse was very important in the Sudan. The great kingdoms which developed in the Eastern and Central Sudan were dominated by a mounted aristocracy, and mailed cavalry formed the nucleus of their armies (Linton, 1956).

The pony breeds of East and West Africa are descended from larger horses of Oriental, Barb and Dongola type, their small size being due to the unfavourable conditions near the Equator. Large parts of Central Africa are closed to horses by the tsetse fly. Previously African horse-sickness also prevented the spread of horses from the north to beyond the tropical forest belt.

The Basuto pony of Lesotho, now nearly extinct, is descended mainly from Arab and Thoroughbred horses imported into South Africa during the 19th century. Apart from the Arab and Thoroughbred, there are more than 20 different horse breeds in South Africa, among these the American Saddle Horse, Australian horses and Welsh ponies. Most of these are sporting horses.

10. ASS

The ass, being derived from the hot arid regions of Somalia and the Sudan, is bred mostly in the subtropical parts of Asia and Africa, and is especially frequent in the Mohammedan sphere of influence. It is absent in the cold north and the humid tropics. The ass is an extremely hardy animal adapted to inferior conditions of feeding and maintenance, and is generally the companion of the poorer classes of the population.

In the whole sphere of ass-breeding regions two main types of ass can be recognized, viz. the small grey or brown beast of burden, and the large riding or draught ass. The smaller type is a general utility animal. It shows little variation in type throughout its breeding area, except near the northern border of its distribution, as in Tibet, Mongolia and Manchuria, where it grows a long rough coat. Its height at withers varies between 80 and 120 cm; in accordance with its size it can carry a load of 40—60 kg. In several parts of its range it is used as a riding animal in towns and villages, by boys guarding cattle, sheep, goats or camels on pasture or by persons accompanying camel caravans. The wealthier Arabs in Syria and Jordan, the bedouin of Arabia, and the Indians are usually loth to ride an ass (Burton, 1855—1856).

Chapter 4 references, p. 91

Fig. 4.19. Kwanchung jack ass, China.

Fig. 4.20 Pack mule, Nepal.

The large type of ass, called Muscat after a town in Oman, is employed in Egypt and Western Asia for riding. Its coat is usually white with a black skin. This type of ass is also found in India and Pakistan, from where it has reached Nepal. In China the large ass is represented by the Kwanchung, which is, however, nearly black. It stands 130—145 cm at the shoulder and weighs 300—350 kg. The Kwanchung is an excellent draught animal, widely used in agriculture.

In at least two parts of the world she-asses are milked: in the country of the western Masai between Mt. Kilimanjaro and Lake Victoria, and in Mongolia. The meat of the ass is eaten by a few Eastern Bantu peoples; for this purpose the Wakamba specially fatten the few asses they possess (Kroll, 1929). Asses are also eaten in Mongolia.

In the grazing areas of Northeast China and Mongolia asses are frequently pastured in large droves. At night these must be accompanied by a stallion because asses will allow wolves to devour them, but a stallion will boldly attack a pack of wolves approaching the drove and drive them away (Epstein, 1969).

10.1. Mule

In countries with a hot, dry climate, where the horse suffers from the heat, and in conditions where the work demanded is too heavy for the ass, the mule replaces both. Mules are also less sensitive to cold than are asses. They vary in size in accordance with that of the mares and jack-asses used to breed them. Mules are employed for pack work and draught, occasionally as saddle animals in rough country.

10.2. Hinny

Hinnies are produced mainly in Ethiopia and in several ass-breeding areas of China. As in the case of mules, their size depends on that of the parent breeds. They are used for the same purposes as small mules. Stallions intended for breeding hinnies are not allowed to serve mares and are trained only to serve jennies. Fertility, both in natural service and artificial insemination, is lower than in mule breeding.

11. CAMEL

The camels are divided into two species: the one-humped dromedary or Arabian camel, and the two-humped Bactrian. In the Bactrian one hump is placed on the withers and one on the loins. It is stouter, with shorter legs, harder and smaller feet, and a heavier and thicker coat. Camels are fitted to live in arid regions by their ability to go without water for considerable periods, an ability which they owe to their remarkable tolerance of dehydration while keeping the blood volume normal. They can survive rainless seasons on the scantiest feed and exist in territory which cannot support even the modest goat. The Bactrian camel has a more northerly distribution and is better adapted to stand a bitterly cold winter, while the Arabian is able to thrive in the heat of the summer in subtropical desert countries. Again, the Arabian camel is more suited to flat country and the Bactrian to rocky hills. In the contact areas of the two species in Asia, the Bactrian is usually found in the hills and the Arabian in the plains.

The one-humped camel is bred throughout North Africa from the Red Sea to the Atlantic, the southern limit of its distribution being decided by the

Chapter 4 references, p. 91

Fig. 4.21. One-humped camel ploughing.

degree of humidity and the occurrence of trypanosomiasis. In Asia the territory of the dromedary extends from Arabia and the east coast of the Mediterranean to Pakistan and India, and north into Afghanistan, Turkmeniya, Uzbekistan, Tadzhikistan and West Sinkiang.

The Bactrian occurs in the Mongolian People's Republic and adjacent northern border areas in the U.S.S.R. up to about 55°N, further in Inner Mongolia and neighbouring provinces of Northern China, Tibet and the major part of Sinkiang. In Kirgiziya, Southern Turkmenia, Iran and Anatolia the Bactrian is found along with the one-humped camel. Here hybrids between the two species are common; the usual cross is of the one-humped male with the two-humped female, the progeny of the reciprocal cross being of lesser value.

Among the one-humped camels two main classes are distinguished: the riding and the baggage camels. These two classes are quite as distinct as the racer is from the carriage horse, or the hunter from the cart-horse (Leonard, 1894). The perseverance of desert-bred riding camels is immense; a horse must exert itself to follow in trot. A large heavy pack camel is capable of carrying a load of 300—500 kg, but lighter animals carry only 150—200 kg. The average load of a Turkestan one-humped camel is 200—250 kg with which they will travel 35 km/day. In the agricultural regions of Western Asia and Egypt the camel has occasionally to draw the plough, either alone or in a team with an ass or an ox.

In Egypt camel meat is in great demand by the poorer classes; the Cairo market is for slaughter camels what the Paris market is for slaughter horses. Camels destined for slaughter are usually gelded at 3 or 4 years of age.

Among the Arab and Berber tribes of the deserts of North Africa and the bedouin of Arabia she-camels are milked; indeed, in Somalia and Arabia milk

Fig. 4.22. One-humped camel, North Africa.

Fig. 4.23. Two-humped camel cow in the Gobi.

is the most important product of the camel. A she-camel in milk gives 3—4 l/day according to the quality of the pasture, supporting not only the children, women and men but the horses as well. A good one-humped she-camel of the Turkestan breed may produce up to 3000 kg milk with 4.6—4.7% fat in a lactation of 16—17 months (Kondrashov, 1958a).

Chapter 4 references, p. 91

The Bactrian camels are bred for wool, milk, work and meat. The outer-coat is coarse and may reach a length of 37 cm. The downy under-coat varies between 2.5 and 12.5 cm in length. The fleece of 12 months' growth weighs 2.5—3.5 kg. Generally, the milk yield of the Bactrian is lower than that of the one-humped camel. It amounts to 150—300 kg/lactation in addition to the milk consumed by the calf during a suckling period of 16—17 months. In the U.S.S.R. the daily yield averages 6.5 kg with 5.8% fat (Kondrashov, 1958b). The load of a Bactrian pack camel is 120—150 kg with which it will travel 35—40 km/day on long journeys, and up to 85 km on a short trip. In the Gobi, camels may carry loads of 240 kg for 30—60 km a day on journeys lasting several months. Bactrian camels are fit for work from the fourth year to 25 years of age, and may live up to 40 years. The cows calve until they are about 30 years old, normally every second year (Epstein, 1969).

12. REINDEER

Within and beyond the northern forests of Asia cattle and ponies become increasingly difficult to raise, and domestic reindeer take their place. There are wide variations in their importance and use in this large area from the western to the eastern limits of Asia. Some of the Tungus and Evenks live almost entirely off their reindeer herds (Forde, 1934). Milk is the most highly valued product. The female yields just over 0.5 l milk/day for a short period in addition to suckling her fawn. The total lactation yield is not more than 24 l (Teichert, 1941). The milk is deep yellow in colour, sweet and thick like cream, and contains 14—22% fat (Herre, 1955; Hançar, 1956).

Only a few bucks are kept for breeding, the remainder are gelded at 4 years of age for meat or work. Reindeer meat is consumed cooked, smoked or in a raw state, either fresh or dried. Animals in good condition dress out at nearly 40%; the average dressed weight of an adult gelding is about 40 kg.

The reindeer is an indispensable means of transport in the tundra and northern forests. Some of the reindeer breeders use sledges similar in construction to, but larger than, dog sledges. A reindeer can drag a sledge with a load of 45—135 kg, depending on trail conditions, and cover 30—40 km/day (Phillips, 1948). The average load of a baggage reindeer is 40 kg which the animal can carry for 4—5 h before resting (Herre, 1955). Strong adult bucks and geldings may carry 70 kg or more over a distance of 80 km in a day. In some parts of the breeding area the reindeer is ridden, the rider sitting on the shoulder, for the average reindeer's back is not strong enough to support the weight of an adult man. The forest reindeer of the Evenks is larger and stronger than the breed of the northern tundras, so that the rider may sit on the back.

Reindeer hide is impervious to water. The coat consists of coarse, harsh, brittle hair of white, black or brown colour. In winter a thin under-coat develops underneath the thick outer-coat. The fur has remarkable insulating properties, retaining the heat generated by the body under extremely cold conditions. The skins of does and fawns are used for boots, clothing and sleeping bags. Thick and heavy skins of adult bucks and geldings are made into bedspreads and tent, saddle or sledge covers (Epstein, 1969).

13. DOG

Dogs are found in every part of Asia and Africa. In the high north they are the sole domesticated animal; in the heart of the desert they occur along with the camel, and in the tropical rain forest together with goat, hen and pig.

In the Arctic, dogs are the essential draught animals in the sledge. In

Manchuria they are bred for fur, and in Tibet their hair is used, in mixture with yak hair, for felt. In parts of Southeast Asia their meat is cherished, and in the Western and Central Sudan several peoples still adhere to the ancient custom of sacrificing and eating dogs (Frank, 1965). In the steppes and deserts of North Africa and Southwest Asia, and eastward through Afghanistan, Pakistan and Northern India to China, greyhounds are employed in the hunt. In the grassland and mountain areas of Asia mastiffs serve as guardians of flocks and herds from predators. Throughout Africa and Southern and Central Asia the ubiquitous pariah dogs eke out a miserable existence as scavengers in villages and towns.

14. REFERENCES

Amble, V.N., Murthy, V.V.R. and Sathe, K.V., 1965. Milk production of bovines in India and their feed availability. Indian Vet. J., 35: 221—223.

Balezin, P.S., 1959. Životnovodstvo Kitaja (Animal Breeding in China). Seljhozgiz, Moscow, 159 pp.

Burton, R.F., 1855—1856. Personal Narrative of a Pilgrimage to El-Madinah and Meccah. Longman, Brown, Green and Longmans, London, 3 vols., xxviii + 1262 pp.

Cockrill, W. R. (Editor), 1974. The Husbandry and Health of the Domestic Buffalo. FAO, Rome, 993 pp.

Curson, H.H. and Thornton, R.W., 1936. A contribution to the study of African native cattle. Onderstepoort J. Vet. Sci. Anim. Ind., 7: 613—739.

Devendra, C. and Burns, M., 1970. Goat Production in the Tropics. Commonwealth Agricultural Bureaux, Farnham Royal, 184 pp., 48 illustrations.

Doutressoulle, G., 1947. L'Élevage en Afrique Occidentale Française. Éditions Larose, Paris, 298 pp., 97 illustrations.

Ekvall, R.B., 1968. Fields on the Hoof: Nexus of Tibetan Nomadic Pastoralism. Holt, Rinehart and Winston, New York, NY, 100 pp.

Epstein, H., 1965. Regionalisation and stratification in livestock breeding, with special reference to the Mongolian People's Republic (Outer Mongolia). Anim. Breed. Abstr., 33: 169—181.

Epstein, H., 1969. Domestic Animals of China. Commonwealth Agricultural Bureaux, Farnham Royal, 166 pp., 205 plates.

Epstein, H., 1970. Fettschwanz- und Fettsteissschafe. A. Ziemsen Verlag, Wittenberg-Lutherstadt, 168 pp., 111 illustrations.

Epstein, H., 1971. The Origin of the Domestic Animals of Africa. Edition Leipzig, Leipzig, Africana Publishing Corporation, New York, Vol. I, 573 pp., 670 illustrations; Vol. II, 719 pp., 627 illustrations.

Epstein, H., 1972a. Studies on the relationship between cattle breeds in Africa, Asia and Europe — Historical evidence. World Rev. Anim. Prod., 8: 25—32.

Epstein, H., 1972b. The Chandella horse of Khajuraho with comments on the origin of early Indian horses. Z. Tierz. Zuechtungsbiol., 89: 170—177.

Epstein, H., 1974. Yak and chauri. World Anim. Rev., 9: 8—12.

Epstein, H., 1977. Domestic Animals of Nepal. Holmes and Meier, New York, NY, 131 pp.

Fischer, H., 1967. Stierkampf in Thailand. Tieraerztl. Umsch., 22: 368—372.

Fischer, H. and Ulbrich, F., 1968. Chromosomes of the Murrah buffalo and its crossbreds with the Asiatic Swamp buffalo (*Bubalus bubalis*). Z. Tierz. Zuechtungsbiol., 84: 110—114.

Forde, C.D., 1934. Habitat, Economy and Society. Harcourt, Brace and Company, New York, NY, 500 pp., 108 illustrations.

Frank, B., 1965. Die Rolle des Hundes in afrikanischen Kulturen. Franz Steiner Verlag, Wiesbaden, 256 pp.

Hançar, F., 1956. Das Pferd in historischer und früher historischer Zeit. Verlag Herold, Wien-München, 651 pp., 18 plates.

Herre, W., 1955. Das Ren als Haustier. Akademische Verlagsgesellschaft Geest and Portig K.-G., Leipzig, 324 pp., 79 illustrations.

Hofmann, F., 1959—1960. Haustierveredelung in Mittelchina. Wiss. Z. Friedrich-Schiller-Univ. Jena, 9: 575—583.

Johnston, H.H., 1908. George Grenfell and the Congo. Hutchinson and Co., London, 2 vols., 990 pp.

Joshi, N.R. and Phillips, R.W., 1953. Zebu Cattle of India and Pakistan. FAO Agricultural Studies No. 19, 256 pp., 62 illustrations.

Joshi, N.R., McLaughlin, E.M. and Phillips, R.W., 1957. Types and Breeds of African Cattle. FAO Agricultural Studies No. 37, 297 pp., 102 illustrations.

Kondrashov, K., 1958a. Astrakhanskaja poroda verbljudov. (The Astrakhan breed of camels.) In: G.M. Shall (Editor), Porody Seljskohozjaĭstvennyh Životnyh. Ministry of Agriculture, Moscow, U.S.S.R., 38 pp.

Kondrashov. K., 1958b. Turkmenskaja poroda verbljudov. (The Turkmen breed of camels.) In: G.M. Shall (Editor), Porody Seljskohozjaĭstvennyh Životnyh. Ministry of Agriculture, Moscow, U.S.S.R., 39 pp.

Kroll, K., 1929. Die Haustiere der Bantu. Z. Ethnol., 60: 177—290.

Leonard, A.G., 1894. The Camel. Longmans, Green and Co., London, 335 pp.

Linton, R., 1956. The Tree of Culture. Alfred A. Knopf, New York, NY, 692 pp.

Mason, I.L., 1974. Species, types and breeds. In: W. Ross Crockrill (Editor), The Husbandry and Health of the Domestic Buffalo. FAO, Rome, pp. 1—47.

Phillips, R.W., 1948. Breeding Livestock Adapted to Unfavourable Environments. FAO Agricultural Studies No. 1, 181 pp., 71 illustrations.

Phillips, R.W., Johnson, R.G. and Moyer, R.T., 1945. The Livestock of China. Department of State, Publ. 2249 (Far Eastern Ser. 9), U.S. Government Printing Office, Washington, 174 pp.

Phillips, R.W., Tolstoy, I.A. and Johnson, R.G., 1946. Yaks and yak—cattle hybrids in Asia. J. Hered., 37: 162—170, 206—215.

Rouse, J.E., 1970. World Cattle. University of Oklahoma Press, Norman, 2 vols., 1046 pp.

Schley, P., 1967. Der Yak und seine Kreuzung mit dem Rind in der Sowjetunion. Giessener Abhandlungen zur Agrar- und Wirtschaftsforschung des europäischen Ostens, 44, 131 pp.

Staffe, A., 1938. Die Haustiere der Kosi. Z. Zuechtung, B. 40: 252—285, 301—342.

Stamp, L.D., 1962. Asia. Methuen and Co., London, 11th edn., 730 pp.

Swinhoe, R., 1870. Catalogue of the mammals of China (south of the River Yangtsze) and of the Island of Formosa. 72 *Sus taivanus* (Formosan Wild Boar). Proc. Zool. Soc. London, 38: 641—644.

Teichert, K., 1941. Weidewirtschaft und Milchtiere der Arktis. Milchwirtschaftlich-geographische Studien, 17. Molk.-Ztg. (Hildesheim, Ger.), 55: 1172—1174.

Ulbrich, F. and Fischer, H., 1967. The chromosomes of the Asiatic buffalo (*Bubalus bubalis*) and the African buffalo (*Syncerus caffer*). Z. Tierz. Zuechtungsbiol., 83: 219—223.

Ulbrich, F. and Fischer, H., 1968. Die Chromosomensätze des türkischen und südeuropäischen Wasserbüffels (*Bubalus bubalis*). Z. Tierz. Zuechtungsbiol., 85: 119—122.

Van, Fu-čžao, 1958. Želtyi skot Kitaja (The yellow cattle of China). Zhivotnovodstvo, Moscow, 20: 70—72.

Von Füre-Haimendorf, C., 1963. The social background of cattle domestication in India. In: A.E. Mourant and F.E. Zeuner (Editors), Man and Cattle. Royal Anthropological Institute, Occas. Paper No. 18, pp. 144—149.

Weir, T., 1952. High adventure in the Himalayas. The National Geographic Magazine, CII, 2: 193—234.

Wentworth, Lady, 1945. The Authentic Arabian Horse and his Descendants. George Allen and Unwin, London, 388 pp., 291 plates.

Chapter 5

Conservation of Animal Genetic Resources

R.G. BEILHARZ

1. INTRODUCTION

Conservation of genetic resources should be considered against the background of the evolution of all forms of life. The result of evolution is that different forms of life have become adapted to every possible environmental niche. The activities of man over the last 15 000 years have markedly altered the environment on the surface of the earth, and plants and animals have adapted to utilize the resulting new niches. Domestic plants have arisen through the propagation by man of those plants whose parts (seeds, fruits, roots) were of greater value to man than those of the original wild species. Domestic animals have utilized the protection from predation in the vicinity of man, and have at the same time been selected by man for traits useful to him, such as increased production of milk, meat, fibre, etc. The forces of evolution continue to act as strongly as they have always done. In the case of domestic plants and animals, however, the 'arbitrary' selection decisions of man, and the modified environments he has created, often impose greater selection pressures than the 'natural' pressures resulting from predation and the restrictions imposed by climate, topography and other environmental features.

Two distinct kinds of processes bring about evolution. They involve the production and the elimination of variation (Mayr, 1970, p. 161). Mutations and other rearrangements of the genetical material in the chromosomes are the ultimate sources of variation. Genetic recombination, such as occurs at meiosis and fertilization in sexually reproducing species, contributes further to variation by continually reshuffling the genetic material every generation. On the other hand, selection among variants, and the chance elimination of variants through genetic drift or during inbreeding lead to greater uniformity. Thus, the variety of living forms seen at any time is the result of a dynamic balance between processes that create and others that eliminate variation. Such a balance manifests itself at various levels, e.g., between individuals within a population, between populations, and between species. On a geological time scale extinction of species has been common (Mayr, 1970, p. 372). This has been balanced by radiations of new species from common stocks when new opportunities arose, e.g., after major climatic changes. Within species there are clines of variation related to, for example, altitude and longitude. These clines reveal that the forces of evolution are actively holding each form of life at an optimum point for its survival in every locality.

The problem of conservation of genetic resources should be seen as part of the continuing balance between variation-producing and variation-decreasing

Chapter 5 references, p. 104

forces. When planning breeding programmes and conserving variation, we are interested in a relatively short span of time, say 50—100 years. Over such a period the production of variation through the process of mutation is slow and very weak. Sexual reproduction will continue to reshuffle the genetic material that is present. On the other hand, man has the technology to rapidly reduce genetic variation if he applies his technology unwisely. There is thus a real possibility that, throughout the large part of the world over which man has exerted control of the environment and of the direction of evolution, a reduction of genetic variability will occur, particularly in domestic plants and animals.

Man makes use of his domesticated plants and animals to supply his many and various needs. As man's conditions change (standard of living, technology, fashion) his needs may alter greatly. It is thus wise to retain as large a variation as possible in the genetic material available to us to keep as many future options open as possible over the sort of time span (50—100 years) over which we might be interested in planning. It is little consolation to know that evolution will eventually repair our mistakes.

Many plants can be propagated without continual recombination of genetic material, either vegetatively or as completely inbred self-fertilizing strains. Such plants are likely to remain genetically unchanged over the periods of time that interest us, and specific, identified genetic material can be conserved. Seeds of plants often survive many years. This means that for many plants conservation of specific identifiable genetic material is accurate and relatively cheap.

Conservation of identified genetic material in animals is both less certain and more costly than in most plants. Except in some highly-inbred strains of laboratory animals, the continual recombinations of genes in all our domestic animal species means that we can only hope to conserve a continually mixing population of genes which, we hope, contains the specifically valuable genes we want. In order not to lose particular genes by chance through the effects of inbreeding, we must conserve large enough populations of each type of animal we wish to conserve. Conservation by keeping large groups of animals is very expensive and never completely reliable. Other methods of storage, such as of frozen semen, or frozen ova, therefore become important. It is against this background that the work done in conservation of animal genetic resources must be judged.

2. HISTORY

Concern with the conservation of animal genetic resources is a recent phenomenon. The Food and Agriculture Organization of the United Nations (FAO) has taken an interest in the subject from early in its existence. The first session of FAO's then Standing Advisory Committee on Agriculture, which met in Copenhagen in 1946 recommended that FAO should undertake work on the evaluation and conservation of both plant and animal genetic stocks. Between 1947 and 1966, FAO prepared publications and held international and regional meetings covering animal genetic resources. The 13th session of the FAO Conference (1965) recognized the need for increased activities in the field of animal and plant genetic resources. As one result an ad hoc study group was arranged in 1966. This study group reported generally on the evaluation, utilization and conservation of animal genetic resources (FAO, 1966). There followed a series of further ad hoc study groups which concentrated on cattle (FAO, 1968), pigs (FAO, 1971) and poultry (FAO, 1973). As part of the United Nation's Environmental Programme (UNEP) a pilot study on conservation of animal genetic resources, utilizing investigations in sheep and cattle breeds around the Mediterranean area (FAO—UNEP, 1975) gave specific recommendations for action immediately, in the short term and in the long term. More recently, the first FAO expert consultation on breed evaluation and crossbreeding (FAO, 1977) reported on Mediterranean cattle and sheep in crossbreeding and discussed the place of locally adapted breeds of animals in the total system of animal production.

Another major event contributing to the discussions of conservation of animal genetic resources was a Round Table on the subject at the first World Congress on Genetics applied to Livestock Production, held in Madrid in 1974 (Bowman, 1974; Epstein, 1974; Iglesia-Hernandez, 1974; Laurans, 1974; Maijala, 1974; Mason, 1974; Sanchez-Belda, 1974; Somes, 1974; Turton, 1974). Members of FAO had a major input to this Round Table but the experiences of a number of individual countries were also presented and discussed. Reviews of the subject by individuals (e.g., Maijala, 1970; Turner, 1971, 1972; Rendel, 1975) have complemented the two main sources of literature.

3. THE PROBLEM

Man has depended on domestic animals for his material, cultural and scientific needs. Material needs include food, clothing, buildings or other shelter, draught power, and household medical and industrial goods. Cultural needs include recreation and education. Scientific research contributes to both material and cultural needs. Scientific study of animals will also contribute to a greater understanding of man himself.

Can man live in a totally artificial environment of his own creation, or does he need plants and animals, i.e. 'nature'? We do not know. Clearly man is very adaptable. However, most people alive today will probably share my view that it is good for man, necessary for his physical and mental health, to be able to get away periodically from artificial city surroundings into more natural environments. 'Nature' will include unmodified forests, natural heath, grasslands, etc., but also the rural environments in which man has existed for thousands of years. There is at present a movement towards the creation of national parks, and even regional farm parks. This is a valuable step towards conservation of natural and rural areas to which city man can escape.

Recreational needs automatically ensure the survival of the pets, horses, etc., involved. The same applies to laboratory animals used in science and for production of medical needs. The position is different with the animals supplying man's material needs. Table 5.1, taken from Turner (1971), shows the contributions domesticated mammals have made to man's material needs. Will each of these contributions remain important in the future? Will all of the numerous breeds and species now used continue to be needed? Again we cannot answer these questions with certainty. It is likely that many animal products will be needed by man for a long time yet. Therefore, it is wise to ensure the continued availability of the genetic resources contained in our domestic animals.

Natural selection, particularly through climatic stresses, has interacted with man's selection to adapt populations of domestic animals to each environment. Certain species and breeds are well adapted to high altitudes, others to deserts, others to disease-prone areas, and so on. Adaptation to stressful environments has usually been achieved at the cost of lowered potential for production of milk, meat, wool or young. Within the last 100 years certain breeds of cattle, sheep, pigs and poultry have been selected in

Chapter 5 references, p. 104

TABLE 5.1

Types of domestic mammals and their contribution to man's material needs

Animal type	Contribution
Buffalo	Traction, transport, meat, milk, leather, horn
Camels:	
Alpaca	Textile fibre, meat
Camel	Traction, transport, meat, milk, fats, textile and other fibre, leather, fuel, fertilizer (bone and dung)
Llama	Transport, textile fibre, meat
Vicuna	Textile fibre
Cat	Hunting (of rodents)
Cattle	Traction, transport, meat, milk, fats, leather, textile and other fibre, fertilizer (bone and dung), fuel, horn, glue, blood (for food and for serum in biological products)
Dog	Transport, meat, hunting (for food), guarding (man or animals), working (cattle or sheep), guiding (the blind), tracking
Elephant	Traction, transport, ivory
Equines:	
Donkey	Traction, transport, meat, milk, leather, fuel
Horse	Traction, transport, meat, milk, leather, fibre, fuel, fertilizer, blood (for serum)
Mule	Traction, transport, fertilizer
Fur-bearers:	
Chinchilla, ermine, fox, marten, mink and sable	Fur
Goat	Meat, milk, textile and other fibre, leather, fertilizer, horn, blood (for serum)
Monkey	Harvesting (coconuts)
Pig	Meat, leather, bristle, fats
Reindeer	Traction, transport, milk, meat, leather, textile and other fibre
Rabbit	Meat, textile, fur, laboratory animal (biological products)
Rodents:	
Ferret	Hunting (for food)
Guinea pig	Meat, textile fibre, laboratory animal (biological products)
Hamster	Laboratory animal (biological products)
Mouse	Laboratory animal (biological products)
Rat	Laboratory animal (biological prodcuts)
Sheep	Textile and other fibres, meat, fats, leather, fur, fuel, fertilizer, horn, transport
Yak	Transport, textile fibre, meat, milk, leather

Source: Turner (1971), with kind permission of the Editor of *Outlook on Agriculture*.

environmentally favourable areas to give very high production. These animals have been transported widely around the agricultural areas of the world and have displaced the lower-producing, but more adapted local breeds. This process of displacement, either by physical substitution or by grading the local animals up to sires of the 'improved' breed, has accelerated markedly in the last 25 years with the introduction of new techniques of reproduction such as artificial insemination with frozen semen, with the emphasis placed on breeding for high production based on recorded performance, and with standardization of animal products in marketing. An example of the magnitude of this problem is given by the FAO studies (FAO—UNEP, 1975; Rendel, 1975). In Europe and the Mediterranean Basin only 33 recognizable local breeds of cattle are holding their own, while about 115 others are in danger of extinction. With some exceptions, such as the Shorthorn and the Ayrshire, the threatened breeds have not gone beyond their local environment. The extinction of local breeds is being brought about by a change towards Friesian cattle in practically all of the lowland areas of the European continent and the British Isles, and towards Simmental cattle in the moderately elevated areas of Central and Southeastern Europe. In Scandinavia, the red cattle, mainly founded on Ayrshire and Shorthorn crosses, still hold their own although there has been a decline of the Red Danish breed. In Central Europe the brown cattle, which had expanded greatly before and shortly after the last World War, now seem to be retreating slightly.

Clearly, present animal production policies will result in the loss of many breeds of domestic animals. Do these breeds contain genes that should be conserved? It is a major part of our problem that we just do not know enough about many of the existing breeds of animals. How well would they produce if they were treated in the same way as the 'improved' breeds? Do they show hybrid vigour if crossed? Is their particular adaptation to the local area of commercial importance? All the FAO studies and several contributors to the Round Table have stressed the importance of obtaining information about all aspects of performance of local breeds as the first step in a conservation programme.

Thus there is a real problem relating to animal genetic resources. The problem can be summarized as:

(1) our inability to predict future needs;

(2) the inevitable loss through substitution or upgrading under current policies, of many locally adapted breeds; and

(3) our ignorance about the real value of most local breeds, in their own right and as potential components of integrated systems of animal production.

There is also a sociological aspect of the problem of conservation of animal genetic resources. Many of the local breeds are part of an ethnically primitive form of animal husbandry. As development and 'progress' replace primitive forms of production the local animals often disappear with the ethnic culture. Either progressive development, or retention of ancient forms of husbandry might be a desirable social policy in any particular region. The point is that the decisions made for social and political reasons will automatically have severe side-effects on animal genetic resources. Arguments about conserving animals are unlikely to receive much support when opposed by sociological policies. Maijala (1974) has pointed out another 'human' problem. It is that potential gains made through conservation are only harvested years, perhaps generations, later. There is thus no commercial and very little scientific motivation towards conservation among the people who should be interested in it.

4. POSSIBLE SOLUTIONS

Once we decide to conserve animal genetic resources what are the methods available for doing this? Before we can answer this question it is necessary to decide just what is to be conserved. This decision must be made by responsible authorities such as governments or large breeding organizations. In some countries special organizations have recently been set up to

Chapter 5 references, p. 104

supervise conservation of animal genetic resources. These organizations, the Rare Breeds Survival Trust in the United Kingdom and the Société de Ethnozootechnie in France, are suitable bodies to help decide what resources should be conserved. FAO policy with regard to conservation has always been that it is not justified merely to preserve breeds for their own sake, but that preservation must be justified by the use to which breeds can be put in future (FAO, 1971, 1977; Mason, 1974). Animals useful at present will be preserved automatically through their commercial exploitation.

In any case, breeds must be evaluated before one can decide rationally which of them to conserve. FAO (1971) recommended that governments should study representative samples of clearly-defined genetic stocks exposed to equally well-defined environments. There should be common stocks present in the tests at the several locations. On the basis of such tests the breeds and breed combinations that offer the most efficient production of desirable products can be identified and arrangements made for their conservation, if necessary.

Bowman (1974) disagreed with FAO policy as expressed in 1971. He stressed that it is important, when possible, to determine the genetic relationships between breeds and to maintain those with distinctive characteristics indicative of unique genetic material, irrespective of present economic importance. The aim should be to conserve as big a range of the existing irreplaceable genetic variation as possible. Turner (1972) had also made the same suggestion.

Once we know which populations should be conserved we have a number of possible strategies.

(1) We can maintain populations of each breed or strain to be conserved. This is a very expensive process, so it is important that the correct breeding technology be applied. As the purpose is conservation of the existing genetic material we must breed so as to avoid genetic change. This is done by prolonging generation intervals, i.e., by breeding from older individuals, by eliminating directional selection, and by avoiding the chance effects of inbreeding. To reduce inbreeding, effective population sizes of at least 50 animals should be aimed at, although lower numbers have been mentioned as possibilities (e.g. Maijala, 1970). Effective population size is highest if the ratio of breeding males to breeding females is 1:1 and if each male is replaced by his son and each female by her daughter. Epstein (1974) suggested that each breed to be conserved should be represented in two sizable herds within the country as well as in small groups in zoological parks in other countries. Old breeds of cattle have been maintained successfully in parks on estates in the United Kingdom for many years. There has also been a recent trend in several countries towards creating farm parks for recreation and education. Such parks seem ideal places for maintaining populations of animals to be conserved.

(2) We can create gene pools which combine the genes of a number of breeds or strains, and then breed the populations containing the gene pools deliberately to avoid genetic change as discussed in (1) above. This strategy has been suggested particularly for poultry where a mixture of genes can be maintained in such a pool without taking up large facilities. It is less likely that a mixed genetic pool of animals will be attractive to visitors of farm parks. Maijala (1970) summarized the steps required in creating a gene pool as follows. The basic step is to make an inventory of all available stocks, utilizing the accumulated information on the characteristics of each stock. Then the most useful or interesting stocks are tried in a common environment, after which the final choice of strains to be conserved is made. According to Jaap (1964) it is inadvisable to combine more than two or three

populations into a pool, in order to keep gene frequencies at a usefully high level. It is clear that for reasons related to generation length, there may not be sufficient time for such a systematic evaluation prior to setting up a gene pool in the case of large animals.

(3) We can set up stores of frozen semen and, in future, frozen fertilized ova, or gonadal tissue. Semen stores already exist in beef and dairy cattle. It is relatively cheap to systematically store semen of all interesting breeds and strains. Again, a responsible authority will have to pay for the space required for this purpose because existing semen stores are usually commercial ventures, which must clear out stocks of semen no longer commercially valuable. Another difficulty is that, while semen is routinely collected from currently fashionable breeds and sires, there are many interesting local breeds, which should be considered for conservation, that are remote from semen collecting facilities. Even if one wanted to collect semen from them there is often no information on which to decide which are the best bulls for providing semen. However, these are minor problems. We have here a cheap and reliable technique for an organization that wants to conserve genetic resources to do so immediately and effectively in cattle. Research work in other species is making rapid progress which will soon lead to the same possibility in all other important domestic animals.

(4) We should replicate our conserved populations and semen stores at several locations, to guard against loss or technical failure at any one site. This is particularly relevant with stores of frozen semen or other tissues. High cost will work against replication of populations of live animals. Yet, if our conserved material consists of only a single population of animals, the risk of loss in the long term is very high. It will probably be found optimum to conserve animal populations or gene pools, as well as semen stores.

(5) If we heed Turner (1972) and Bowman (1974), we must also conserve some genetic stocks that do not contain genes for high production or for specific adaptations in the short-term future, in order to be able to come to an understanding of the physiological mechanisms through which genes act. Conservation of such 'inferior' stock is unlikely to receive much support.

(6) We may, in our national breeding policies based on recorded performance, deliberately reduce our selection differentials to allow us to keep the genetic base broad. Such a policy deserves consideration particularly in national breeding plans based on artificial insemination, as currently used in dairy-cattle populations. There is a real possibility that commercial emphasis on fashionable sires will reduce the genetic base in national herds to the point where chance effects of inbreeding will detrimentally reduce variation in all traits at the same time as the selection reduces variation in milk production (Beilharz, 1975; Barker, 1977).

There is relatively much more known about poultry stocks than is the case in the mammal species. This is because of the shorter generation length and because poultry breeding has become a commercial operation with individual breeders controlling many and large populations. Thus the FAO (1973) study group on poultry was able to consider genotype-environment interactions and their implications for conservation. The biological principles behind such interactions are exactly the same in larger animals. Their relative absence from discussions relating to large animals merely reflects the more advanced state of poultry breeding.

Chapter 5 references, p. 104

5. THE STEPS THAT HAVE BEEN TAKEN

The first FAO study group on evaluation, utilization and conservation of animal genetic resources (FAO, 1966) established priorities for action. Five different categories of animals were distinguished:

(1) farm animals in common use, which are the main sources of animal protein, and traction power (cattle, buffaloes, sheep, goats, pigs, poultry, horses and donkeys);

(2) animals adapted to particular environments (camels, llamas, alpacas, guanacos, vicunas, yaks);

(3) animals in (1) or (2) with special characteristics such as resistance to particular diseases (e.g. trypanosome-tolerant breeds of dwarf, short-horned cattle in West Africa, currently threatened by extinction);

(4) animals in (1), (2) and (3) under threat of extinction by replacement or gene dilution through upgrading programmes to other strains, with sufficient apparent promise to justify their conservation until they can be evaluated;

(5) small domestic animals useful in a variety of circumstances (e.g., domestic rabbits) and wild species, with potential future value in the wild or in domestication.

The species in the first category are of most importance to man. Many breeds will maintain themselves because of their commercial exploitation; yet much more information is needed even for the highly productive breeds, e.g., when used in less favourable environments. The second category contains animals important in particular regions of unfavourable environments. For the third category the importance of each case must be determined separately. The number of breeds in the fourth category could rapidly increase because of the considerable amount of movement of animals and frozen semen around the world at present. It was noted that some breeds that had been close to extinction have again become popular because of their special qualities, e.g., Finnish sheep for prolificacy.

Many studies have called for the collection of information as the most immediate and urgent task so that informed decisions on conservation can be made (FAO, 1966, 1968, 1971; Turner, 1971; Mason 1974). Turton (1974) discussed the problem of collecting, storing and disseminating information on breeds of livestock. He pointed out that information was available in the conventional scientific literature (books, journals, bibliographies, etc.), in non-conventional literature (unpublished research reports, data held at research stations, etc.) and on computer storage media, e.g. in herd-recording organizations. Access to conventional literature is simple with appropriate indexing of the material. With non-conventional literature the problem is to find that the particular information exists and then to gain access to it. The data of recording and breeding organizations stored in computers have greatly increased in amount and importance in recent years. Computer-use offers the possibility of much more effective retrieval of information than is possible by hand. However, adequate levels of usage are necessary to pay for the cost of the installation of the computer system and to keep the costs to individual users to reasonable levals.

FAO has successfully followed a policy of collecting and publishing information about animal genetic resources and is continuing to do so. The Commonwealth Agricultural Bureaux of the United Kingdom, and other individual authors have also contributed. Some important publications available are Whitehead (1953) on ancient white cattle of Britain, Joshi and Phillips (1953) on zebu cattle of India and Pakistan, Joshi et al. (1957) on African cattle, French et al. (1966) on European cattle, Mason (1967) on sheep breeds of the Mediterranean, Mason's (1969) World Dictionary of Livestock, Epstein (1969) on domestic animals of China, Epstein (1971) on domestic animals of Africa, Somes (1971) on genetic stocks of poultry in the U.S.A., and Cockrill (1974) on buffaloes.

Somes (1974) discussed the information available on poultry. Although considered desirable for a number of years, no large cooperative effort at conserving poultry stocks has been made because of a lack of funds. However, a 'Catalogue of Poultry Stocks held at Research and Teaching Institutions in Canada' has been published annually since 1967. In 1971 the bulletin 'Registry of *Gallus domesticus* Genetic Stocks in the United States — A Directory of Sources of Specialized Lines and Strains, Mutations, Breeds and Varieties' (Somes, 1971) was published. A second edition planned for 1975 was enlarged to include *Coturnix* quail and turkeys, and the Canadian material. The original directory was in six sections. Section I described and located 87 specialized lines and strains under nine subheadings. Section II gave the inheritance, linkage, characterization, a literature reference and location of 103 mutant traits. Section III contained an up-to-date chromosome linkage map. Section IV described and located 260 breeds and varieties of fowl in an alphabetical list. Section V gave phenotypic descriptions of some of the more complex plumage colour patterns according to their common genotypes. Section VI was an index of breeders and suppliers with their addresses. There was extensive cross-referencing of breeders and suppliers to the other items in the directory. This publication gives an idea of the sort of information that can be assembled for our livestock. The FAO (1973) study group on poultry recommended that a similar directory should be made of world poultry stocks. Somes (1974) agreed that this would be very worthy but that it would be an enormous undertaking.

The importance of information exchange was highlighted also by Bowman (1974) when discussing the formation of the Rare Breeds Survival Trust in the United Kingdom in 1973. This organization, established as a registered charity, aims to carry out the following categories of work:

'(1) Information bank — the building and maintenance of up-to-date information about owners, and animals (their whereabouts, number, breeding habits, environmental needs, etc.) — is the first and foremost objective...

(2) Direct action to preserve — the bringing together of owners through membership of the Trust so that breeding can be encouraged and facilitated. If necessary grants will be given towards the purchase of animals to prevent their slaughter, or the Trust will purchase them.

(3) Research — to see that the gene bank thus created is used for the improvement of breeds currently in use and for the development of new breeds to meet problems brought about by changing economic and environmental circumstances.

(4) Education and conservation — for the benefit of a wider public. Rare breeds of farm animals are just as much part of our heritage as are ancient buildings. There must be opportunities for the public to view these animals and to learn about them. The Trust will encourage the setting-up of farm parks and static exhibitions of high standard. It will promote press, radio and television programmes; it will publish scientific papers and leaflets of general interest.'

The Trust is concerned with the less common breeds of all species of domestic animals. It is initially concentrating on sheep, cattle and pigs. Bowman pointed out that during recent years there has been a marked increase in the interest shown by the general public in many aspects of conservation, and domestic livestock have been encompassed in this interest. In order to cater for the recreational needs of the public and at the same time

Chapter 5 references, p. 104

to give the city dweller an opportunity to see the variety of animals used in agriculture, a number of 'farm-parks' have been opened up by private individuals who collect an entrance free from the public to cover their costs. Upon investigation it became clear that there were many people with suitable areas of land and facilities on which to keep animals, who were willing to maintain breeding nuclei of breeds which were not necessarily of immediate economic importance. From the position about 1960, when the need for breed conservation was considered to involve finance for the purchase and maintenance of animals, the emphasis had changed. The real need was found to be a communication service to monitor the population size and ownership of breeds, to put owners and potential owners in touch with one another and to provide advice. The experience in the United Kingdom is particularly interesting as it shows that the increasing need to develop recreational pursuits and amenity land use has been an important element in the establishment of the Rare Breeds Survival Trust, although the main reasons for breed conservation are more concerned with the long-term need to maintain non-renewable genetic resources.

The work on conservation of animal genetic resources done in France was described by Laurans (1974). The geneticists of the Institute National de la Recherche Agronomique have through their studies defined the problems and laid the bases for sound conservation (e.g. Vissac, 1970). Some historical strains like the Rambouillet flock of Merinos have already been conserved for a long time. The National Museum of Natural History, although primarily concerned with wild animals, has successfully conserved small populations of Poitou donkeys and Soay sheep. Two regional parks (Armorique and Camargue) also have populations of rare breeds (Ouessant sheep and Rove goats, respectively). Since its creation in 1972, the Société d'Ethnozootechnie has placed conservation of domestic breeds prominently among its aims.

The possibility of beneficial mutual interaction between conservation and recreation through farm-parks has already been mentioned. A similar possibility exists to make conservation of Spain's wild galician ponies a tourist attraction (Iglesia-Hernandez, 1974). These ponies usually roam free. Their harvest, and other management operations occur during particular seasons which provide the possibility of exploitation for tourism.

The FAO—UNEP (1975) pilot study on the conservation of animal genetic resources had before it a number of papers setting out the latest position of animal populations needing conservation. Of particular interest was the study by Lauvergne (Annex 2) on disappearing cattle breeds in Europe and the Mediterranean Basin, which has already been discussed as an illustration of the problem of breed displacement. Other valuable data were given in Annex 1 by Lauvergne on the comparative animal production situation in Corsica and Sardinia. Annex 3 by Mason was a preliminary survey of endangered breeds throughout the world. Annex 4 described a study by Mason of the Kuri Cattle of Lake Chad. These cattle are uniquely adapted to subsistence on coarse forage of the shores and to swimming from island to island to reach new pastures. They also have various curious horn forms (buoy-shaped, inflated, very thick, open crescent and lyre-shaped). Mason concluded tentatively that the Kuri breed is not in immediate danger of disappearing. Immediate action to preserve the typical horn forms in the breed was recommended by the pilot study. Another recommendation for immediate action was the evaluation in crossbreeding trials of Brachyceros and Grey Steppe cattle populations typical of the Mediterranean and Balkan countries, before populations decline to dangerously low levels. In the short term the factual situation should be established for Criollo cattle of Latin America, prolific tropical sheep (e.g. Barbados Black-belly), Mediterranean pigs, and West African dwarf shorthorn cattle, which are trypanotolerant. In the longer term a world list should be prepared of herds and flocks of rare and relic breeds, and of genetically interesting lines maintained in zoological gardens, nature reserves, parks (national, regional and private), research stations or under the control of special societies. There should also be a world list of scientific laboratories working in fields that can throw light on taxonomy and phylogeny of domestic species, on the relationships between breeds and on the knowledge of marker genes. These will include laboratories working on chromosomes, blood groups and other biochemical polymorphisms, the metabolism of certain other processes such as melanin formation, and the analysis of gene frequencies. Other recommendations related to the setting-up of experimental genetic herds and a general inventory of breeds. An International Board on Animal Genetic Resources parallel to the International Board on Plant Genetic Resources was suggested as the body to plan and guide a global cooperative programme of animal genetic conservation.

These recommendations, as well as other specific recommendations coming from the FAO study groups (1968, 1971, 1973), show the direction in which steps towards conservation of animal genetic resources have been and are being taken. It is reassuring to note that promising starts have been made.

6. FUTURE OUTLOOK

Maijala (1974) also discussed the arguments against animal conservation. The main one against taking particular actions is the one of cost, especially when contrasted with the possible gains. It is difficult to put a value on simply retaining variation, or on a return that will be made only in a proportion of all possible futures, i.e., only if certain events occur. Another telling argument is that in a conserved stock the general genetic value of the material will have fallen behind the commercial population at the time the special genes are needed. The total effect of using the conserved stock may thus be negative.

These arguments use economic parameters appropriate to short-time periods when we should really be using values appropriate to much longer periods. Conservation of genetic variety is like paying an insurance premium. If a disaster does occur we collect the insurance, i.e. we fall back on conserved stocks. This is very much better than having lost a population completely. It is encouraging to note the obvious concern shown and specific actions taken by so many responsible scientists and organizations in the conservation of variety in domestic animals.

The inertia of present policies will mean that current trends will continue to cause dilution and loss of a significant number of locally adapted populations before the movement towards conservation will be able to prevent this. However, it now seems that many of the useful genetic stocks will be preserved in one way or another.

All higher animals are sexually reproducing diploids. While this makes it more difficult than in many plants to fix and conserve specific genes, it is, on the other hand, an excellent natural mechanism for keeping up variability. As long as some large populations of any domestic species remain, it will be possible by selecting among the continually recurring sexually produced variation to undo most errors of breeding we may have committed. Thus, a responsible mixture of conservation of particularly valuable or interesting genotypes and use of the potential variation guaranteed by large com-

Chapter 5 references, p. 104

mercial populations, should safeguard the future of our domestic animals. The situation will clearly be worse for those species for which we no longer have any use, and thus have no commercial populations. For such species much more careful conservation is necessary, if we wish to preserve them. However, we must again ask the fundamental question: should we conserve such a species?

This chapter closes with one final appeal! — that we should retain the correct balance between looking into the past and looking into the future. It is wise to conserve animal genetic resources, and responsible steps have been and are being taken to increase our efforts in this direction. However, let us not go so far in the increasingly more fashionable direction of conservation and preservation as to tie up all the resources of governments and scientific institutions in massive efforts merely to retain what exists at present. We should remember that we and our animals are part of the ever-changing evolution of life. A certain amount of change will be good. This means that some particular variants must be lost. Any particular loss need not be important in itself. Our aim as scientists and responsible people should be to guide the dynamic processes of evolution for the long-term good of mankind in particular, and of all living things in general.

7. REFERENCES

Barker, J.S.F., 1977. Efficient use of variability in out breeding species. Proc. 3rd Int. Congr. S.A.B.R.A.O., Anim. Breed. Pap., 2c(11), pp. 1—11.

Beilharz, R.G., 1975. Genetic implications of new techniques of reproduction. In: R.L. Reid (Editor), Proc. 3rd World Congr. Anim. Prod. Sydney University Press, Sydney, pp. 644—649.

Bowman, J.C., 1974. Conservation of rare livestock breeds in the United Kingdom. Proc. 1st World Congr. Genet. Appl. Livestock Prod., 2: 23—29.

Cockrill, W.R. (Editor), 1974. The Husbandry and Health of the Domestic Buffalo. FAO, Rome, 993 pp.

Epstein, H., 1969. Domestic Animals of China. C.A.B., Farnham Royal (Bucks), 166 pp.

Epstein, H., 1971. The Origin of the Domestic Animals of Africa. Africana Publishing Corporation, New York, NY, 2 vols., 573 pp. and 719 pp.

Epstein, H., 1974. Vanishing livestock breeds in Africa and Asia. Proc. 1st World Congr. Genet. Appl. Livestock Prod., 2: 31—35.

FAO, 1966. Report of the FAO Study Group on the Evaluation, Utilization and Conservation of Animal Genetic Resources. FAO, Rome, 31 pp.

FAO, 1968. Report of the Second ad hoc Study Group on Animal Genetic Resources. FAO, Rome 26 pp.

FAO, 1971. Report of the Third ad hoc Study Group on Animal Genetic Resources (Pig Breeding). FAO, Rome, 21 pp.

FAO, 1973. Report of the Fourth FAO Expert Consultation on Animal Genetic Resources (Poultry Breeding). FAO, Rome, 19 pp.

FAO, 1977. Mediterranean cattle and sheep in crossbreeding (Report of the first FAO expert consultation on breed evaluation and crossbreeding). FAO Anim. Prod. Health Pap. No. 6, 37 pp.

FAO-UNEP, 1975. Pilot Study on Conservation of Animal Genetic Resources. FAO, Rome, 60 pp.

French, M.H., Johansson, I., Joshi, N.R. and McLaughlin, E.A., 1966. European breeds of cattle, FAO Agricultural Studies No. 67, FAO, Rome, 2 vols., 387 pp. and 424 pp.

Iglesia-Hernandez, P.J., 1974. Conservacion de las Ponies de Galicia (Espana). Proc. 1st World Congr. Genet. Appl. Livestock Prod., 2: 85—93 (in Spanish, English summary).

Jaap, R.G., 1964. Minimum population size and source of stock. Proc. 2nd Eur. Poultry Conf., pp. 429—431.

Joshi, N.R., McLaughlin, E.A. and Phillips, R.W., 1957. Types and breeds of African cattle. FAO Agric. Stud. No. 37, FAO, Rome, 297 pp.

Joshi, N.R. and Phillips, R.W., 1953. Zebu cattle of India and Pakistan. FAO Agric. Stud. No. 19, FAO, Rome 256 pp.

Laurans, R., 1974. Le probleme de la conservation du material genetique en France. Proc. 1st World Congr. Genet. Appl. Livestock Prod., 2: 75—84 (in French, English summary).

Maijala, K., 1970. Need and methods of gene conservation in animal breeding. Ann. Genet. Sel. Anim., 2: 403—415.

Maijala, K., 1974. Conservation of animals in general. Proc. 1st World Congr. Genet. Appl. Livestock Prod., 2: 37—46.

Mason, I.L., 1967. Sheep Breeds of the Mediterranean. FAO, Rome and CAB, Farnham Royal (Bucks.), 215 pp.

Mason, I.L., 1969. A World Dictionary of Livestock Breeds, Types and Varieties. CAB, Farnham Royal (Bucks.), 272 pp.

Mason, I.L., 1974. Introduction to Round Table A: the conservation of animal genetic resources. Proc. 1st World Congr. Genet. Appl. Livestock Prod., 2: 13—21.

Mayr, E., 1970. Populations, Species and Evolution. Belknap Press, Harvard University Press, Cambridge, MA, 453 pp.

Rendel, J., 1975. The utilization and conservation of the world's animal genetic resources. Agric. Environm., 2: 101—119.

Sanchez-Belda, A., 1974. Conservacion de las Razas Ovinas. Proc. 1st World Congr. Genet. Appl. Livestock Prod., 2: 53—59 (in Spanish, English summary).

Somes, R.G., 1971. *Gallus domesticus.* Registry of genetic stocks in the United States. Storrs Agric. Exp. Stn., The University of Connecticut, Storrs, Bull. No. 420, 53 pp.

Somes, R.G., 1974. Conservation of poultry breeds. Proc. 1st World Congr. Genet. Appl. Livestock Prod., 2: 47—51.

Turner, H.N., 1971. Conservation of genetic resources in domestic animals. Outlook Agric., 6: 254—260.

Turner, H.N., 1972. Conservation of genetic resources in domestic animals. Proc. 16th Int. Edison Birthday Celebration, pp. 161—173.

Turton, J.D., 1974. The collection, storage and dissemination of information on breeds of livestock. Proc. 1st World Congr. Genet. Appl. Livestock Prod., 2: 61—74.

Vissac, B., 1970. Etude genetique de la race d'Aubrac. In: L'Aubrac 1. C.N.R.S., Paris, pp. 25—99.

Whitehead, G.K., 1953. The Ancient White Cattle of Britain and Their Descendants. Faber and Faber, London, 174 pp.

Chapter 6

Evolutionary Adaptations and their Significance in Animal Production

J.L. BLACK

1. INTRODUCTION

Throughout evolution, natural selection has led to the retention of those characteristics of an organism which ensure for it a greater reproductive success than other organisms in the same environment. Because of mutations and genetic diversity within a population, this process has resulted in the differential survival of individuals and the continual development of new populations with characteristics that improve their ability to grow and reproduce in a particular environment. These evolutionary adaptations in animals include changes in morphology, biochemistry and behaviour and are passed from one generation to the next. They should be distinguished from physiological adaptations which refer to the capacity of an individual animal to adjust, within its lifetime, to changes in the environment. Even so, the ability of an individual to adapt to environmental change depends upon the environmental extremes to which members of preceding generations were subjected and is therefore influenced by evolution.

Of the several million animal species on earth, few are used by man. Approximately 30 species of mammals and birds and two species of fish have been domesticated for use as sources of food (Blaxter, 1975). Although inclusion of species hunted by man would considerably increase this number, the majority of man's animal products come from cattle (*Bos taurus, Bos indicus, Bos bubalus bubalis*), sheep (*Ovis aries*), goats (*Capra hircus*), pigs (*Sus scrofa*), chickens (*Gallus gallus*), ducks (*Anas platyrhynchos*) and turkeys (*Meleagris gallopavo*). Man undoubtedly has altered the rate and direction of evolutionary adaptations in domestic animals by manipulation of their environment and by genetic selection. Nevertheless, his influence over the past few thousand years has probably been small compared with the adaptations that have occurred during evolution.

Most evolutionary adaptations in animals used by man have some bearing on animal production, and not all are advantageous. Many characteristics selected during evolution differ from those required by man in his quest to optimize the efficiency of conversion of resources into usable animal products. This situation arises particularly through man's desire to transfer his selected species to environments different from, and often less diverse than, those in which they evolved, and also through his aim to optimize production of features not associated with reproductive success. In this chapter, the process of evolution is discussed first and is followed by an examination of the evolution of the main features determining the productive efficiency of animals. The significance of some evolutionary adaptations on animal production is outlined.

Chapter 6 references, p. 130

2. THE EVOLUTIONARY PROCESS

2.1. Expression of genetic information

Current concepts of the way inherited information is expressed in an individual organism and passed to future generations have been reviewed by Davidson (1968) and Watson (1976). Genetic information is carried by the sequence of nucleotide bases which form DNA. The message on DNA is transferred to messenger RNA which is interpreted on polyribosomes and finally transferred into amino acid sequences of structural proteins and enzymes. Each series of three nucleotide bases, or codon, on DNA specifies the incorporation of one amino acid into protein. Each group of codons which specified the amino acid sequence of one protein is called a gene, and each group of genes which specifies the proteins which together control one particular synthesis is known as an operon.

At any particular time, most of the genetic information of an individual cell cannot be transmitted. For example, in higher plants and animals, cells differentiate to perform specialized functions, yet, within an organism, all cells originate from the same initial cell and all contain identical genetic information. Some genes are repressed, leaving active only those required by the cell to perform its specific functions. Thus, a cell may contain active genes, lightly repressed genes and firmly repressed genes. Lightly repressed genes can be activated by appropriate messages and it is by continual activation and repression of genes that an organism adapts to day-by-day changes in the environment. Conversely, firmly repressed genes are not readily activated and, in differentiated animal cells, many are totally unavailable.

The manner by which genes are activated or inactivated (Fig. 6.1) was proposed by Jacob and Monod (1961). The amino acid sequence of each structural protein or enzyme is determined by a structural gene of which there may be several in one operon. It is proposed that each operon contains a region, known as the operator gene, which determines whether the DNA helix will unwind and allow transcription of the operon by messenger RNA. The activity of the operator gene is, in turn, controlled by the product of a regulator gene specific for that operon but situated in some other part of the DNA molecule. It is thought that the protein produced by a regulator gene represses the activity of the operon, but specific inducer substances such as metabolites, hormones or pheromones can inactivate the repressor protein and allow transcription of the operon. In this way, the rate of synthesis of proteins from one operon is controlled by a regulator gene. An operon may contain a regulator gene which determines the activity of the operon producing the regulator protein for its own operator gene. With such feedback mechanisms, the flexibility for control of protein synthesis within an organism is greatly enhanced. Although this kind of regulation of transcription is well known in bacteria, it is not clear that operons exist in eukaryote cells.

2.2. Mechanism of genetic change

Any change in the base sequence of DNA represents a mutation. The most common mutations are substitution of one base for another, but many other types are known including the addition or deletion of bases and the duplication or translocation of existing base sequences (Drake and Baltz, 1976). The probability of a mutation occurring is essentially the same for every amino acid in every protein produced by an organism. However, for any mutation to become fixed into the genetic structure of a population, it must be passed

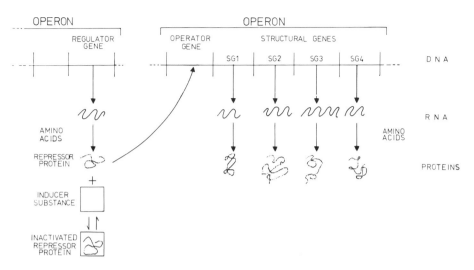

Fig. 6.1. Control of protein synthesis by gene activation and repression. Binding of the repressor protein by an inducer substance allows transcription of the operon controlled by the specific regulator gene.

from the individual in which it first occurred to successive generations until it is predominant in the population. Examination of amino acid sequences in proteins of many organisms suggests that the rate at which mutations are fixed is remarkably constant for each protein and is independent of the generation length of the species (Sarich and Cronin, 1977), but varies greatly between proteins. For example, it has been estimated to take 400 million years for a 1% change to occur in the amino acid sequence of histone 4, whereas the same percentage change is thought to occur in 43 million years for glucagon, 5 million years for growth hormone and only 1 million years for immunoglobulins and snake venom toxins (Wilson et al., 1977). The probability of a mutation becoming fixed depends upon whether the amino acid substitution will be compatible with the biochemical function of the protein and, if not, whether its function can be performed by another existing protein. Thus, the substitution of almost any amino acid in histone 4 is thought to adversely affect its function and without it an organism will not survive, whereas it seems likely that many amino acids in snake venom toxins could be replaced without impairing their functions. Although an amino acid change in an immunoglobulin may have a serious affect on its action, most animals have several related proteins that could perform the same task.

Wilson et al. (1977) contend that the majority of mutations fixed within a population occur in regulator genes which alter the effective concentration of enzymes and the rates of specific biochemical reactions. These regulator gene mutations can facilitate the evolution of metabolic, morphological and behavioural characteristics in a population. By comparison, the evolution of new biochemical functions by alteration of structural genes is thought to be infrequent. Nevertheless, when such changes do occur, as for example through gene duplication, they can have a major influence on the evolution of adaptative mechanisms. In the case of gene duplication, one copy may retain the original function while the other accumulates mutations and may eventually perform a different function. Lactalbumin is thought to have evolved from duplication of the structural gene for lysozyme *c* and has thereby brought new functions to mammals.

Chapter 6 references, p. 130

2.3. Determinants of evolutionary adaptations

Evolution proceeds because the large genetic variation within a population enables differential selection of individuals best adapted to grow and reproduce in a particular environment. The genetic diversity results partly from the large number of gene loci in individual organisms that are heterozygous and partly from the scrambling of genes during the recombination of chromosomal DNA in the formation of gametes during meiosis in sexually reproducing species. Heterozygosity has been estimated to occur in 5—18% of the gene loci in different species (Ayala, 1978) and, in association with the recombination of DNA strands during meiosis, produces such large genetic variation that no two individuals originating from a different fertilization are likely to be the same. Although the reshuffling of genes by recombination produces genetic variation, it does not alter the gene frequency of a population or permit evolution. The ultimate source of genetic variation is from mutations and the frequency of genes in a population is determined by the interaction between mutation and natural selection.

At present, there is controversy concerning the driving force for fixation of mutations (Harris, 1976; King, 1976). One theory contends that gene mutations which become fixed are neutral from the standpoint of natural selection and fulfil the same function as the gene they replaced. The other theory suggests that most mutations are fixed because they provide a positive advantage for the species in natural selection. The theories are not mutually exclusive and both assume that deleterious mutations will be eliminated. Although many neutral mutations probably are fixed, there seems likely to be a greater pressure for the retention of mutations that bestow an advantage in natural selection and a gradual elimination of those characteristics of an organism that have no function.

This notion forms the basis of the economy principle in evolutionary adaptation. The principle postulates that an organism will expend energy on the development or maintenance of a function, whether metabolic, morphological or behavioural, only if it is beneficial for the production of offspring which will themselves survive and reproduce (Curio, 1973). There is little doubt that this principle applies widely and many cases are known where complex functions regress when they become unnecessary for survival. Examples include reduction of eyes and external pigments in underground and cave-dwelling animals and in internal parasites, and the loss of innate recognition of enemies by birds that are found in enemy-free environments (Curio, 1973). The corollary, that functions become more complex because they enhance the ability of a population to survive, also appears true. A striking example is seen in the experiments of Spiegelman (1971) with naked RNA from a bacteria-phage virus grown in vitro under culture conditions devoid of the host bacteria. In one experiment, when the virus was selected for fast replication, the rate of phage turnover greatly increased in only eight generations and, by the 75th generation, the RNA molecule was reduced to 17% of its original size. Because the reproductive success of the selected virus was greatly improved, it could be presumed that the mutant population would have some evolutionary advantage over the wild-type. Although probably true under the artificial experimental conditions, this RNA mutant proved incapable of infecting the bacterial host when returned to natural conditions. By increasing its speed of replication, the mutant had lost the genes responsible for host infection and had become more vulnerable to environmental disturbances. Thus, in evolution, there is a conflict between speed of propagation and survival of the species in different environments. The more complex the environmental influences on a species, the more complex are the evolutionary adaptations required for preservation of the species. Although there appears to be a general relationship between energy flow into progeny and fitness of an animal species to survive (Calow, 1977), enormous variation exists among species depending upon the ecological niche inhabited. Many species have directed energy to body growth and fat deposition at the expense of reproduction, but the advantages of these adaptations outweigh the disadvantages in particular environments and ultimately lead to a greater reproductive success for these species.

3. ANIMAL PRODUCTION IN RELATION TO EVOLUTION

The aim of people involved in animal production is to optimize the efficiency of conversion of available resources into usable animal products. In the broadest sense, this represents optimization of the efficiency of energy flow through an animal system into the desired products. The efficiency of energy utilization in an individual is determinde by the amount and composition of feed eaten, the digestion of that feed and absorption of specific nutrients in relation to requirements. An animal's requirements for nutrients are influenced by the turnover and rate of growth of individual tissues, the amount and composition of secretions and the changes in metabolism induced by environmental stress. The efficiency of utilization of absorbed nutrients depends primarily on the metabolic pathways followed and the coupling of energy released during metabolism to productive functions. In addition to these factors which affect the efficiency of energy flow through an individual, the number of offspring produced by adult females, their age at puberty and the sex ratio of the population influence the efficiency of energy flow through a population. These factors have all been under strong evolutionary pressures and many different adaptations have occurred in animals depending upon the environment in which they lived.

Characteristics of animal morphology, metabolism and behaviour which lead to the most efficient utilization of available feed resources vary, depending upon the products required by man, the nature and availability of feed, and other environmental conditions to which the animals are subjected. Thus, many animal characteristics suitable for meat production differ from those needed for fibre, egg or milk production. Likewise, characteristics desirable in animals raised in the tropics differ from those required in animals raised in deserts, in the subarctic, or at high altitudes. Other desirable characteristics differ for free-grazing animals compared with animals housed indoors. In order to determine the significance of different evolutionary adaptations on animal production, it is necessary to examine the evolution of major functions determining the efficiency of energy flow through an individual animal and through a total animal population.

4. EVOLUTION OF DIGESTIVE SYSTEMS

During digestion, enzymes break down complex organic molecules into simpler units for absorption and utilization. Early in the evolution of life, single-celled organisms like bacteria secreted enzymes into the surrounding environment and digestion occurred outside the cell. This tended to be a wasteful process because some of the digested substrates escaped absorption and single-celled animals evolved with more complex systems, like phagocytosis in protozoa, where the substrate is entirely surrounded by protoplasm during digestion. However, there is an upper limit to the size of feed particles that can be ingested by phagocytosis. Thus, as animals became multi-cellular and cells differentiated, the digestive tract developed with specialized regions for storage of food, secretion of enzymes and absorption of the products of digestion.

4.1. Loss of cellulase

A significant event in the evolution of digestion was the loss, in many higher animals, of the capacity to produce cellulase. This enzyme is secreted by bacteria, protozoa and most phyla of invertebrates (Yokoe and Yasumasu, 1964), but it is apparently not produced by vertebrates. Animals which are unable to secrete cellulase, but eat a significant amount of plant material have overcome the problem in two ways. The most common adaptation has been an increase in the size of the digestive tract to harbour a substantial population of cellulase-secreting bacteria and protozoa. The host animal absorbs and utilizes volatile fatty acids and other substances produced during microbial fermentation of the diet. For this process to function, the rate of passage of feed through the digestive tract must be slowed to permit at least one generation of microbes before passage in the faeces. This has been achieved by an increase in length of the tract and by formation of large, often blind, chambers. These fermentation chambers are most highly developed in insects like termites (*Isoptera*) and in many mammals including ungulates, lagomorphs and rodents. Microbes which produce cellulase are also known in the gut of many species of fish, reptiles and birds, but in most the anatomical complexity of the digestive tract is insufficient to enable digestion of enough plant fibre to supply a substantial proportion of the animal's energy needs (McBee, 1977).

A second, but less common way in which herbivores have overcome the inability to digest plant fibre is seen in a few species of fish and birds. Cellulase has not been found in the gut of either the grass carp (*Ctenopharyngodon idella*) or the goose (*Anser anser*), yet both species can be raised entirely on vegetation. By comparison with related species, the digestive tract of these animals is extremely short and the rate of passage of feed through the tract is fast. The digestive tract of the grass carp is only 2.25 times the length of its body compared with 12 times body length in the silver carp (*Hypophthalmichthys molitrix*) (Cross, 1969). In the goose, grass passes through the digestive tract in 71 min compared with a mean retention time of 6—8 h for particulate matter in another herbivorous bird, the Alaskan rock ptarmigan (*Lagopus mutus*), (McBee, 1977). By comparison, the mean retention time of particulate matter in the rabbit (*Oryctolagus cuniculus*), which is of similar weight to these birds, is about 16 h (Sharkey, 1970). The goose and grass carp both have adapted to a lack of cellulase by being able to ingest vast quantities of feed. With specialized anatomical features for grinding feed as it passes through the digestive tract, they extract the plant cell contents and excrete the fibrous material. However, these animals can survive only on plant material that is low in fibre and high in cell contents.

4.2. Digestive tract anatomy

The anatomy of the digestive tract of animals is closely related to the type of diet. Mouth and teeth structures are especially adapted (Van Gelder, 1969). Carnivores have broad mouths with sharp narrow teeth adapted for seizing prey and tearing it apart, whereas anteaters, which swallow their prey whole, have reduced dentition, long narrow snouts and sticky tongues adapted for penetrating deep into ant nests. Many herbivores lack canine teeth, but have developed broad flat molars for grinding plant material. Browsing herbivores have long flexible tongues for reaching leaves and higher branches, whereas grazing herbivores have developed flat, sharp incisors for cutting grass near ground level.

The digestive tract of carnivorous animals is short and simple. Their diet has a high energy concentration and is readily digested by enzymes secreted into the gut. The stomach is a simple, fully glandular structure, the caecum is either absent or small, and the large intestine is short and narrow compared to animals with other dietary habits. Omnivores have some plant fibre in their diets and the stomach is often enlarged in the cardiac region, and the caecum and large intestine are more capacious than in carnivores. Cellulolytic micro-organisms are found in the gut of these animals, but the amount of fibre digested is generally small. In all herbivorous mammals, the digestive tract is long and complex with specialized chambers adapted to slow the rate of passage of feed and enhance digestion of plant fibre by micro-organisms.

Major modifications to the digestive tract of herbivorous mammals have occurred in both the stomach and the large intestine. In perissodactyls, proboscideans and lagomorphs, the caecum and colon have enlarged and plant fibre is degraded after most soluble components of the diet have been digested and absorbed in the stomach and small intestine. In other herbivorous mammals, typified by ruminants, fermentation chambers have developed in the cardiac region of the stomach. These chambers are generally separated from the acid-secreting pyloric section of the stomach to allow the pH to rise to levels that support vigorous microbial growth. Residual feed particles and a portion of the microbial population pass from the stomach to the small intestine where they are exposed to digestive enzymes of the host. A secondary fermentation also occurs in the caecum and large intestine of these animals.

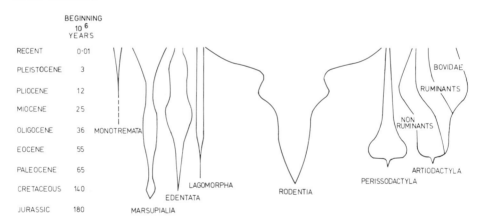

Fig. 6.2. Part of the historical record of mammals. Some orders containing herbivores are represented by a pathway, the width of which is proportional to its known variety during the various periods and epochs. Adapted from Simpson (1949).

Mammals are thought to have evolved from a common ancestry during the Cretaceous period of geological history which began about 140 million years ago. Examination of the number of genera in mammalian orders containing herbivores indicates that perissodactyls were dominant during the Eocene epoch (Fig. 6.2). However, during Miocene and Pliocene their numbers significantly declined and the number of rodent and ruminant-like artiodactyl genera greatly increased. Paleobotanical evidence suggests that, during these drier and cooler epochs, many of the forests of earlier times were

Chapter 6 references, p. 130

TABLE 6.1

Probable lines of evolution of orders in the Class Mammalia showing, for the herbivorous orders, the section of the digestive tract in which feed is fermented, and the families which possess a reticular groove

Geological period	Order and family	Site of fermentation		Reticular groove
		Foregut	Hindgut	
Marsupialia	fam. Macropodidae (kangaroo)	Major	Minor	Present
Insectivora				
Chiroptera				
Dermoptera				
Primates	fam. Cercopithecidae (coloboid and presbytis monkeys)	Major	Minor	Present
Edentata	fam. Bradypodidae (sloth)			
Tubulidentata				
Pholidota				
Cetacea		Nil	Minor	Absent
Lagomorpha		Minor	Major	Absent
Rodentia	fam. Cricetidae (lemming)	Minor	Major	Present
Carnivora				
Proboscidea	fam. Elephantidae (elephant)	Minor	Major	Present
Hyracoidea	fam. Procaviidae (hyrax)	Minor	Major	Present
Sirenia		Nil	Minor	Absent
Perissodactyla		Nil	Major	Absent
Artiodactyla	fam. Tayassuidae (peccary)	Major	Minor	Present
	fam. Hippopotamidae (hippopotamus)	Major	Minor	Present
	fam. Camelidae (camel, alpaca, llama)	Major	Minor	Present
	fam. Cervidae (deer, moose, elk, caribou)	Major	Minor	Present
	fam. Giraffidae (giraffe, okapi)	Major	Minor	Present
	fam. Antilocapridae (pronghorn)	Major	Minor	Present
	fam. Bovidae (sheep, cattle, goat, gazelle, antelope)	Major	Minor	Present
Monotremata				

Geological period (phylogenetic tree labels): Quaternary, Tertiary, Cretaceous, Jurassic; Protoinsectivora, Protungulata, Metatheria, Eutheria, Pantotheria, Prototheria.

Adapted from Black and Sharkey (1970).

replaced by grasslands (Ardrey, 1961). In comparison with forest vegetation, nutritional quality of the grasslands fluctuated widely and often provided only highly fibrous material with little protein and soluble carbohydrate. Animals with fermentation chambers in the stomach are well adapted to survive on such feeds. By recycling blood urea to the stomach, the enhanced supply of nitrogen stimulates microbial growth and digestion of plant fibre. Because much of the microbial protoplasm flowing from the stomach is available for digestion by the host animal, the supply of amino acids and vitamins is also greatly improved. In contrast, perissodactyls with fermentation chambers in the hindgut, digest less plant fibre and fail to absorb most of the protein and vitamins synthesized by microbes. The nutritional significance of this is evident when the hindgut-fermenting rodents and lagomorphs are considered. These orders, which thrived during Pliocene are coprophagic, and by reingesting faecal material, gain much of the benefit from microbial amino acids and vitamins.

Of the two digestive adaptations that appear to have been successful in dealing with low quality grasslands, stomach fermentation has occurred in more genera. Moir (1968) and Bauchop (1977) list the animals in which this adaptation evolved and outline the extent of modifications to the simple stomach. Sacculation of the stomach has occurred in marsupials, primates, edentates, rodents, proboscideans, hyraxes and most frequently in artiodactyls (Table 6.1). Because this adaptation is found in so many animals that have not had a common ancestry since the Cretaceous period, it probably evolved independently on a number of occasions. However, its degree of development appears to be closely related to the habitat in which the animal evolved. For example, stomach sacculation and microbial fermentation of the diet is highly developed in the macropodid marsupial *Macropus eugenii* which inhabits semi-arid environments in Australia, whereas, in a related species, *Thylogale thetis*, which inhabits rain-forest areas, stomach fermentation of the diet appears to be of far less importance (Hume, 1977). In fact, there seems to be a complete gradation in the importance of the site of microbial fermentation in herbivorous mammals from the ruminants, through animals with less complicated stomach sacculations where stomach and hindgut fermentations are about equally important, to the perissodactyls where the stomach is not sacculated and virtually all fermentation occurs in the hindgut.

Although development of fermentation chambers in the stomach of herbivores consuming low quality feeds proved to be a successful adaptation, it created a potential problem for the young. Black and Tribe (1973) showed that when milk flows to the rumen of lambs, it is extensively degraded by micro-organisms and the growth rate and energy retention of these animals is considerably lower than other animals in which milk passes directly to the abomasum. In normal sucking lambs, milk is prevented from entering the rumen by the action of the reticular groove (sulcus reticuli). Black and Sharkey (1970) noted that the reticular groove has evolved in most species in which fermentation of feed occurs in the stomach to any significant extent (Table 6.1). The golden hamster (*Mesocricetus auratus*), in which the oesophagus opens into the acidic pyloric region of the stomach is an exception. Thus, evolution of the reticular groove appears to be an obligatory adaptation for herbivores in which feed passes directly to a stomach fermentation chamber.

4.3. Digestive enzymes

The production of digestive enzymes was an essential step in the evolution of heterotrophic organisms which rely on organic substances for their energy needs. All heterotrophs from bacteria to mammals produce a similar range of digestive enzymes including proteinases, amylases and lipases. There is strong evidence from amino acid sequence studies that, although many enzymes, such as some proteinases, contain the same amino acid residue at the active site, they have evolved from a different origin in bacteria and moulds compared with higher animals (Dixon, 1966). Nevertheless, within a wide range of animals, many digestive enzymes are similar; trypsin and chymotrypsin are found from lower invertebrates to birds and mammals (Neurath et al., 1967). There is also evidence from mammals that the number of digestive enzymes has been increased by gene duplication. Dixon (1966) speculates that trypsin, chymotrypsin A, chymotrypsin B, elastase, thrombin and probably several esterases all evolved from a single origin by this method. The economy principle of evolutionary adaptation is clearly seen with digestive enzymes. Within any species there is a close association between the digestive enzymes present, the concentration and kinetic properties of these enzymes and their need by the animal. For example, the stomachless teleost fish (*Carassius auratus gibelio*) does not secrete pepsin or HCl and the trypsin and chymotrypsin produced are unstable at acidity below pH 6, whereas these enzymes are quite stable at low pH in other animals producing HCl in the stomach (Jany, 1976). Similarly, in ruminants, few carbohydrates reach the small intestine and the range and concentration of intestinal disaccharidases are reduced compared with simple-stomached mammals (Roy and Stobo, 1975). It is interesting to speculate that, as animals became larger, they lost the capacity to synthesize cellulase because it was already produced in quantity by microbes inhabiting the gut.

4.4. Significance of digestive system adaptations

Evolutionary adaptations of the digestive system have a considerable bearing on the diet types that are most suitable for different species of domestic animals. The effect of fibre percentage in the diet on organic matter digestibility is shown for a number of domestic species in Table 6.2. Apparent digestibility of organic matter falls with increasing fibre content in each species, but the depression is less in ruminants than in the pig and the chicken. The results indicate that, whereas all species digest low-fibre diets reasonably efficiently, microbial fermentation in the stomach of ruminants allows these animals to degrade more of the highly fibrous feeds. However, there are other consequences of the microbial fermentation of feed in the stomach of ruminants. At least 10% of the digested energy is lost from the animal as methane and almost as much is released as heat by the microbes during fermentation. In addition, the volatile fatty acids produced during microbial fermentation are used less efficiently than glucose for the production of ATP when absorbed by the animal. The combined effect of these factors on the amount of net energy available to different domestic species from a number of diets is illustrated in Table 6.3. The pig and the chicken with their simple digestive tracts which harbour few micro-organisms utilize the energy from low-fibre, high-starch diets more efficiently than do ruminants. Conversely, highly fibrous diets are used more efficiently by ruminants than non-ruminants although the pig, with some microbial activity in the large intestine, is more suited in this regard than the chicken.

Evolution of fermentation chambers in the stomach of ruminants has a number of other significant effects on the use of these animals. Because many of the rumen microbes can produce protein from non-protein sources of nitrogen, it is possible to feed ruminants low-protein diets supplemented

TABLE 6.2

Apparent digestibility of dietary organic matter by different species in relation to fibre content of the feed

Species	Fibre content (%)			Digestibility depression (per unit change in fibre %)
	0	15	30	
Cattle	86	75	63	0.74
Sheep	89	76	63	0.86
Horses	88	69	51	1.25
Rabbits	97	74	51	1.55
Pigs	94	70	46	1.60
Chickens	86	57	27	1.96

Adapted fom Blaxter (1961).

TABLE 6.3

Effect of diet type on the net energy available to different species

Diet	Energy gained in animal/unit dry matter consumed					
	Absolute values (kJ/g)			Relative values		
	Cattle	Pigs	Chickens	Cattle	Pigs	Chickens
Maize grain	8.5	12.6	12.8	100	148	150
Wheat bran	6.3	7.2	4.4	100	114	71
Dried grass (13% protein)	7.7		2.5	100		32
Wheat chaff	2.6	1.2	0	100	46	0

Adapted from Blaxter (1961).

with cheap sources of nitrogen such as urea. Thus, with non-protein nitrogen, many industrial by-products including straw, citrus pulp and pineapple peelings can be fed economically to ruminants for the production of meat. In addition, intensive beef production from low-protein grains, sugar cane and molasses is possible by replacement of some high-cost protein supplements with urea (Preston and Willis, 1970). Ruminants can also eat many plants which are toxic to other animals because the rumen microbes metabolize and destroy the toxic compounds (James et al., 1975). Conversely, the action of rumen microbes is sometimes disadvantageous for animal production, particularly when ruminants are fed large amounts of diets that are readily degraded in the rumen. Economically, the most important problems resulting from the presence of rumen microbes are bloat in cattle eating legumes and acidosis and bloat in ruminants transferred from high-fibre diets to high-grain diets (Bartley et al., 1975). The rapid breakdown of dietary protein by rumen microbes can also lead to a deficiency in amino acid availability in high producing animals and this is particularly important in sheep used for wool production (Ferguson, 1975). It has been suggested that lambs weighing less than about 25 kg may absorb insufficient amino acid nitrogen to meet their needs when eating immature grasses containing up to 20% crude protein because of degradation of the highly soluble plant protein by rumen micro-organisms (Black et al., 1976; 1979). Under the natural conditions in which these animals evolved, destruction of plant proteins by rumen microbes would not have been a problem because the

Chapter 6 references, p. 130

young are suckled by their mothers beyond this critical weight and the reticular groove evolved to ensure that milk would by-pass the rumen. Another significant evolutionary feature of the ruminant digestive system is the lack of many small intestine disaccharidases which limits the carbohydrates that can be added to milk replacers for calves and lambs (Roy and Stobo, 1975).

5. EVOLUTION OF METABOLISM

5.1. Biochemical aspects

During evolution from the prokaryote cell, basic biochemical pathways of metabolism have remained essentially unchanged. It has been proposed by Sager (1965) that life began with the stabilization of three interacting systems; nucleic acids for replication, either photosynthesis or chemosynthesis for trapping energy, and a protein enzyme system to catalyze the two processes and to form structures. Once this system stabilized, any mutation which strongly modified it would be lethal; hence the extreme conservatism in evolution of biochemical reactions. Rather, evolutionary adaptations have been in the form of additions to the system and modification of its regulation. Hochachka (1976) reviewed the biochemical features of muscle energy metabolism in tissues ranging from obligate-anaerobe to obligate-aerobe in animals from many different environments. He showed that the basic enzymic features are identical in all systems but that each is adapted to its particular circumstances by differences in the concentrations of enzymes and in their regulatory and catalytic properties. For example, diving animals which rely solely on anaerobic metabolism during submersion, achieve this simply by increasing the steady-state concentration of the glycolytic enzymes. A similar example can be seen in ruminants which absorb little glucose, but have available from microbial fermentation acetic and propionic acids. In contrast to non-ruminants, ruminants primarily obtain glucose from propionic acid and amino acids, and synthesize fatty acids from acetate, but the enzyme systems in both groups of animals are similar (Ballard, 1972). The difference lies in the relative importance of the various metabolic pathways and in the concentration of appropriate enzymes.

Likewise, the binding affinities of many ubiquitous enzymes are similar in animals subjected to wide ranges in cell temperature and pressure (Hochachka, 1976). Somero and Low (1976) examined the effects of temperature on the apparent Michaelis constant (K_m) of phosphoenolpyruvate for skeletal muscle pyruvate kinases from four species of fish and one homeotherm, the chicken. The Antarctic teleost, *Trematomus borchgrevinki*, lives at a constant temperature of $-1.86°C$, the warm adapted *Cyprinodon macularius* is found in desert ponds with temperatures near $40°C$ whereas the other two fish, *Gillichthys mirabilis* and *Scorpaena gutatta*, live in habitats where the temperature ranges throughout the year from 10 to $30°C$ and from 7 to $18°C$, respectively. In all species, K_m for the enzyme was almost identical over the range of temperatures to which the cells had adapted. However, outside this range, the binding affinity of the enzyme was substantially reduced. Animals which evolved at constant temperatures showed little stability in K_m with changes in temperature, whereas the fish adapted to wide yearly fluctuations in temperature had stable K_m values over a wide temperature range. This example again demonstrates that many evolutionary adaptations result from regulation of biochemical processes and emphasizes the importance of regulator gene mutations.

During evolution from autotrophic organisms which obtain energy from inorganic sources to the heterotrophic organisms which rely on pre-formed organic material, there was a vivid example of the economy principle of evolutionary adaptation. Many heterotrophs have lost the ability to synthesize compounds essential for survival because they were available from the environment; mutations preventing these syntheses were not disadvantageous and were passed to future generations. Compounds in this category include essential amino acids, essential fatty acids and vitamins. The degree of essentiality of these substances varies among organisms and depends primarily upon the environment to which the organism adapted. Many heterotrophic bacteria require organic matter only for energy and can synthesize all other compounds de novo. Approximately 80% of microbes in the rumen of sheep are thought to be in this category (Bryant and Robinson, 1963), whereas other heterotrophic bacteria require amino acids, vitamins and other substances. *Streptococcus haemolyticus* which has an extremely stable and rich environment within its host animal, requires 19 pre-formed amino acids and a wide range of vitamins and, for *Haemophilus influenzae*, haematin in blood is an essential nutrient (Baldwin, 1949). In mammals, the 10 essential amino acids all require more than three enzymes for their synthesis and all the enzymes needed for the synthesis of non-essential amino acids have other functions in the body (Greenberg, 1961).

Because many animals used for production by man are placed in environments where the feed available is less diverse than in the natural situations they evolved in, loss of the capacity to synthesize amino acids, vitamins and essential fatty acids can have a significant effect on the level of animal production. Deficiencies of lysine in grain-based diets fed to pigs and poultry are particularly important as are deficiencies of sulphur-containing amino acids for wool production in sheep. Before the significance of vitamins was fully realized, diseases induced by vitamin deficiencies were common in pigs and poultry raised indoors.

5.2. Rate of metabolism

Although basic biochemical pathways are similar in all animals, a major change in the rate of metabolism occurred in some species about 70 million years ago (Dawson, 1972). Prior to this, all animals are thought to have been poikilotherms and their body temperature closely followed that of the environment. As already shown, adaptations occurred in poikilotherms which allow individual species to exploit many of the wide range of habitats on earth. Nevertheless, the rate of metabolism and level of activity of many species varies greatly throughout the year in association with environmental temperature. Periods of daily and seasonal torpor are common, and the capacity of most individuals to change their habitat with season to obtain favourable temperatures is limited. It seems probable that animals capable of maintaining a high level of activity throughout the year would have an advantage in survival. Some present-day lizards of the order Squamata, although poikilotherms, have evolved temperature-sensitive neurons in the hypothalamus which enable them to maintain body temperature within narrow limits by behavioural patterns in some environments. However, homeothermy, where body temperature is held constant over a wide range of environmental temperatures, was apparently a more successful method of maintaining activity.

Homeothermy is seen only in mammals and birds, and it probably evolved separately in the two classes (Siegel, 1976). A range in the body temperature is maintained in homeotherms, being lowest in those animals that are judged,

Chapter 6 references, p. 130

from fossil evidence, to be most primitive. In monotremes, body temperature is maintained to about 30°C, whereas in marsupials and primitive eutherian mammals from the orders Insectivora (tenrec, hedgehog, shrew), Pholidota (Pangolin), Hyracoidae (hyrax) and Edentata (sloth, anteater, armadillo), it is maintained around 35—36°C. The highest temperatures are maintained in advanced eutherian mammals and in birds. It can be seen in Table 6.4 that basal metabolic rate is many times faster in homeotherms than in poikilotherms at the same body temperature, and that within the homeotherms, it is fastest in those groups which maintain the highest body temperatures. Coulson et al. (1977) showed that the large differences in metabolic rates between animals cannot be explained by differences in enzyme and substrate concentrations, and suggest that blood flow and hence substrate supply to the tissues is mainly responsible. As a consequence of the higher rate of metabolism in homeotherms than in poikilotherms, obligatory energy requirements and feed intakes are also higher. Because the surface area of an animal increases relative to body mass as mass decreases, the fraction of dietary energy lost as heat is greater in small animals. Thus, feed intake per unit body mass must increase to maintain homeothermy as animals become smaller and the minimum body mass of an animal that can sustain homeothermy appears to be from 3 to 5 g.

Homeothermy, in addition to allowing animals to maintain a high level of activity and survive in a wide range of ambient temperatures, has an important bearing on the functioning of the central nervous system and particularly the cerebrum. The rate of nervous action is closely related to temperature (Zotterman, 1959) and this may have contributed to the success of homeotherms on earth.

Concomitant with the evolution of homeothermy was the development of an array of anatomical, biochemical and behavioural features which enabled animals to either maintain a constant body temperature or to control its

TABLE 6.4

Approximate estimates of body temperatures and basal metabolic rates of poikilotherms and homeotherms

Animal		Body temp. (°C)	Basal metabolic rate ($kJ/W_{kg}^{0.75}/day$)	Ref.
Poikilotherms				
Alligator	(700 kg)	5	1.5	
	(700 kg)	28	10	Coulson et al. (1977)
	(70 kg)	28	12	
	(1 kg)	28	18	
Carp		23	39	Nijkamp et al. (1974)
Lizard		30	31	
Homeotherms				
(a) Mammals				Dawson (1972)
Monotremes		30	142	
Marsupials		35	205	
Primitive eutherian mammals		36	234	
Advanced eutherian mammals		38	293	Kleiber (1961)
(b) Birds				
Passerines		41.5	539	Lasiewski and Dawson (1967)
Non-passerines		40.1	326	Siegel (1976)
Domestic fowl		41.5	375	Hutchinson (1954)

fluctuations in spite of varying environmental conditions. Much has been written on the mechanisms of thermoregulation (e.g., Tromp et al., 1976) and they will not be discussed in this chapter. It is notable that all homeotherms have evolved effective forms of insulation which include hair, feathers and subcutaneous fat deposits as well as effective forms of heat dissipation like panting and sweating. In addition, they have evolved methods of increasing the rate of metabolism in response to cold and can alter blood flow through peripheral arteriovenous anastomoses over at least one order of magnitude to assist in either the dissipation or retention of heat.

5.2.1. Significance of homeothermy

Evolution of homeothermy has had a considerable impact on animal production. Because homeotherms remain active over a wide range of environmental temperatures, they can sustain production in environments widely different from those in which they evolved. Throughout history, this has been important for the development of animal production in new lands. Thus merino sheep are used in the north-west of Australia where maximum temperatures are over 38°C for 100 days each year and also in Central Russia where the temperatures are around 0°C for many months. Nevertheless, within the domestic homeotherms there are species or breeds which are better adapted to particular environments and have higher levels of production. For example, *B. indicus* has a higher level of production in tropical and subtropical environments than *B. taurus* partly because it is better adapted to tolerate heat by the nature of its coat and by its lower basal metabolic rate (Frisch and Vercoe, 1977). The evolution of hair by homeotherms has also been important to man; certain strains of sheep, goats and rabbits are farmed primarily for their fibre.

There is at present controversy concerning the relative efficiency of homeotherms and of poikilotherms in converting dietary energy into animal products (Blaxter, 1975; Calow, 1977). Summarized in Table 6.5 are several estimates of the gross efficiency of conversation of dietary energy to body

TABLE 6.5

Estimates of the gross efficiency of conversion of energy in feed into energy in body gain for different species

Species	Diet	Gross efficiency of energy utilization (Energy gained/ Energy eaten)	Ref.
Poikilotherms			
Carp (*Cyprinus caprio*)	High-protein concentrate	0.27	Nijkamp et al. (1974)
Prawn (*Macrobrachium rosenbergii*)[a]	High-protein concentrate	0.27	Walker (1975)
Alligator (*Alligator mississippiensis*)[a]	Fresh fish	0.25—0.40	Coulson et al. (1973)
Homeotherms			
Chicken (*Gallus gallus*)	Concentrate	0.27—0.33	Nijkamp et al. (1974)
Lamb (*Ovis aries*)	Milk	0.50	Hodge (1974)
	Lucerne chaff	0.07	Sharkey (1970)
Pig (*Sus scrofa*)	Milk	0.52	Hodge (1974)
Rabbit (*Oryctolagus cuniculus*)	Milk	0.27	Sharkey (1970)
	Lucerne chaff	0.06	Sharkey (1970)

[a] Energy efficiencies estimated from feed conversion ratios.

Chapter 6 references, p. 130

gain for poikilotherms and for homeotherms. Such comparisons are not particularly satisfactory because the gross efficiency of utilization of dietary energy is influenced by the level of intake and by the type of diet eaten in relation to the digestive system of the animal (Table 6.3), but they do indicate that there is no major difference between poikilotherms and homeotherms in the efficiency of energy use. However, since the basal metabolic rate of homeotherms is at least six times that of poikilotherms, the rate of production is many times faster in homeotherms when feed is adequate. For example, the carp and chicken examined by Nijkamp et al. (1974) converted dietary energy into body gain with similar efficiency, yet the rate of growth from approximatele 40 g liveweight for 3—4 weeks was only 1.2 g/day in the carp compared with 16 g/day in the chicken. On the other hand, the low metabolic rate of poikilotherms means that many more of them than homeotherms can be maintained on the same amount of feed during periods of feed shortage.

Consequences on animal production of differences in basal metabolic rate are further illustrated by a comparison between *B. indicus* and *B. taurus* cattle. Although evolution of a basal metabolic rate about 10% lower in *B. indicus* than *B. taurus* has been a useful adaptation with regard to heat tolerance, it has apparently an adverse effect on potential growth rate. Because there is a close relationship between voluntary feed intake and basal metabolic rate in cattle, *B. taurus* grow faster than *B. indicus* in a non-stressful environment with adequate good quality pasture (Frisch and Vercoe, 1977). Conversely, when pasture is scarce, *B. taurus* animals lose weight faster than *B. indicus* because of their higher basal metabolic rate. Thus, to improve the efficiency of production from cattle in hot climates with wide seasonal fluctuations in feed supply, an animal with low basal metabolic rate is required, but voluntary feed intake per basal metabolic rate must be higher than has currently evolved. Although the ratio of feed intake to basal metabolic rate is similar for many domestic animals (Kleiber, 1961), it is not universal among homeotherms. The ratio of energy intake to basal metabolic rate in man is about 1.8:1 compared with a value of 4.9:1 for cattle (Kleiber, 1961). The above goal may perhaps be achieved by careful breeding.

5.3. Fluctuations in feed supply

A number of metabolic refinements have evolved to deal with fluctuations in feed supply; the most universal being deposition of body fat. Because selection pressures are towards the improvement of reproductive success, it would seem disadvantageous for animals to direct limited feed resources into fat depots which have no direct bearing on the machinery for fecundity. In fact, within the animal kingdom, there is a tendency for selection of animals which do not deposit body fat. Slobodkin and Richman (1961) measured the energy content/g of tissue in 17 species of animals from six phyla and found that it was skewed towards the lower limit for organic substances; implying that few contain much fat. Conversely, because fat has a high energy content/g, it provides a useful energy store for animals subjected to wide fluctuations in feed supply. Calow (1977) postulates that it is of greater advantage to some species for energy to be stored in the adult for use during periods of nutritional stress than it is for energy to be used to produce young which are more likely to perish during these periods of hardship. An example of this evolutionary trend can be seen in the Platyhelminthes. Some species from this phylum are free-living and suffer wide fluctuations in feed supply, others are endoparasites with unlimited feed, and a third group are ecto-parasites with an availability of feed between the other groups. The energy content of these flatworms is inversely related to the constancy of feed availability, being approximately 21.7 kJ/g for endoparasites, 23.0 kJ/g for ectoparasites, and 26.8 kJ/g for the free-living forms (Calow, 1977). An interesting divergence in this adaptation has occurred in homeotherms. Those animals which inhabit cold climates and are subjected to feed shortage during the winter have developed uniform subcutaneous fat depots which greatly improve their insulation. On the contrary, animals which are subjected to feed shortage in hot climates have localized fat depots, as is seen in the hump of camels (*Camelus* spp.) and in the tail and rump of several breeds of sheep. These depots do not interfere with heat dissipation, and because they generally occur along the back of animals they act as a barrier to the absorption of solar radiation.

The evolutionary adaptation to deposit fat when feed is readily available is of considerable significance in domestic animals used for meat production. Carcase over-fatness is a major problem in many intensive production systems where animals are fed lavishly to promote rapid growth.

There are two other important ways in which animals deal with feed shortage. One is to increase metabolic rate and activity in search of feed. The other is to reduce metabolic rate and activity and wait until feed reappears. Both of these adaptations are highly developed in animals and the one adopted depends upon the endogenous energy reserves of the animal and the probability of finding feed for a given effort as compared to finding it without effort (Calow, 1977). Migratory birds show an extreme example of an adaptive increase in energy expenditure in the search for feed, whereas hibernating animals are well adapted to reducing energy expenditure during feed shortage.

During true hibernation in homeotherms, the rate of metabolism falls to between 1 and 5% of normal and body temperature approximates ambient temperature, but, unlike poikilotherms, these animals are capable of spontaneous or induced arousal to normal levels of metabolism at all times (Hoffman, 1964). True hibernation occurs during winter and is found only in the more primitive mammalian orders of Monotremata, Chiroptera (bats), Insectivora and Rodentia. However, there are other forms of periodic dormancy in homeotherms. Many hibernating animals exhibit a summer dormancy called aestivation, which is induced by high ambient temperatures and water deprivation (Hudson and Bartholomew, 1964). Other animals including bats, rodents and birds from the orders Apodiformes (hummingbirds and swifts), Caprimulgiformes (nightjars and poorwills) and Colliiformes (colys) (Siegel, 1976) can exhibit periods of daily torpor where body temperature falls to levels near ambient. In addition, some members of the order Carnivora, including the black bear (*Ursus americanus*), can exhibit periods of winter sleep. Although, in principle, winter sleep is similar to hibernation in that the rate of metabolism is reduced, body temperature falls only from 38°C to 34°C and then remains constant irrespective of ambient temperature. Black bears do not eat, drink, urinate or defaecate during the 3—5-month period of dormancy, yet the females give birth to and suckle their young (Nelson, 1973). Although non-pregnant bears lose 15—25% of their body weight during winter sleep, it is all fat and there is no net loss in body protein. Metabolic water derived from fat catabolism is sufficient to maintain water balance. Lundberg et al. (1976) showed that the rate of protein turnover was increased 3—5 fold during dormancy and speculate that anabolic pathways for protein synthesis were enhanced and catabolic pathways of amino acid degradation reduced. These special biochemical features appear to be specifically related to winter dormancy, because bears starved during summer are unable to preserve body protein (Nelson et al., 1975).

Chapter 6 references, p. 130

Although periodic dormancy is not seen in domestic animals, a reduction in metabolic rate and level of activity is common in most animals during periods of feed shortage (Graham and Searle, 1975; Calow, 1977). Unfortunately, when domestic species lose weight it includes both fat and protein, and an adaptation similar to that in the black bear during winter sleep could be of considerable advantage in preserving body protein in meat-producing animals subjected to wide seasonal fluctuations in feed supply.

6. EVOLUTION OF ANIMAL SIZE AND RATE OF GROWTH

6.1. Size and growth rate

There is strong evidence that evolutionary pressures have in general operated to increase the size and rate of growth of animals (Boucot, 1976). This trend appears contrary to expections, because during growth limited energy resources are diverted from reproductive output. However, by increasing size and rate of growth, the reproductive success of an individual within a species has apparently been increased. Animals which grow fast and are bigger than their counterparts have a competitive advantage in situations involving direct aggression and tend to be the dominant individuals in a population. This is particularly evident in polygynous species, where the greater competition between males than between females has led to males being the bigger of the two sexes (Jewell, 1976). Size and strength could also confer an advantage in relation to predation. In addition, by growing faster an individual reaches reproductive maturity sooner, decreases generation time, and therefore increases the rate of replication of its genetic characteristics. Other advantages may come to an individual as a direct result of being big, including improved temperature regulation in homeotherms, better water conservation in poikilotherms and a greater choice of feedstuffs (Calow, 1977).

The ultimate size of an individual is determined principally by the final number of cells in each organ, although the maximum size of cells within organs does vary slightly between species (Munro and Gray, 1969); these are both genetically controlled characteristics. Determinants of the rate of growth are not so obvious. For an animal to grow fast, it must be capable of rapidly remodelling existing structures by breaking them down and resynthesizing them again in an expanded form. The gross rate of protein breakdown within a species increases with growth rate (Millward et al., 1976) and there is some evidence that it is faster in sheep than in humans where both species have similar mature weights but vastly different rates of growth (Black, 1976). Robinson (1974) proposes that a major determinant of the rate of protein breakdown is the number and position of glutaminyl and asparaginyl residues in the protein molecule, and that these have been under strong evolutionary pressures. However, fast growth will be achieved only if, in conjunction with rapid protein catabolism, there is a strong impetus for protein anabolism, as is perfected in the black bear during winter sleep, rather than amino acid degradation and urea formation.

Size of domestic animals has a surprisingly small effect on the efficiency of energy utilization. Although basal energy requirement per unit body mass decreases as animals become heavier, voluntary feed intake follows a similar pattern in most domestic animals (Kleiber, 1961). Thus, when feed is freely available, there is little difference between them in the gross efficiency of energy utilization, but the rate of growth is faster in populations of small animals. On the other hand, when feed is scarce, a greater mass of large animals than small animals can be maintained on the same feed supply. Size of animal also affects the convenience of handling the live animal or its component carcase parts.

6.2. Effect of parasites and disease

The rate of growth of an individual is also affected by its susceptibility to parasites and disease. Throughout evolution, both infective agents and host animals have developed elaborate mechanisms to enhance their chances of survival. The wide range of physical and biochemical adaptations that have occurred in host animals to reduce the effects of parasitism are reviewed by Van der Walt and Jansen (1968). These include morphological alterations to the skin and secretion of antimicrobial substances such as lysozyme in tears and lactenin in milk to prevent entry of disease organisms into the body. To combat parasites in the body, physiological adaptations include alterations in blood flow to infected areas, induction of fever to create an environment unsuitable for parasites, and the destruction of infective agents by phagocytes. However, the most important adaptation against disease has been the evolution of plasma immunoglobulins which react with specific foreign substances to render them inert.

Dineen (1978) suggests that the relationships between host and parasite fall into two categories with regard to the immune response. The first category is seen with many micro-organisms which replicate rapidly and reach transmissible levels before the host has had time to develop an effective immune response. A minimum number of organisms is required before the animal can be successfully infected and the severity of the disease is influenced by the number of organisms to which the animal is exposed. If an animal recovers from the infection, it becomes immune to future infection by that strain of organism, but remains susceptible to any random mutation of the parasite that changes its antigenic character. With this type of host—parasite relationship death of the host does not affect the success of the parasite. In natural ecosystems, the level of exposure of animals to disease is generally low and the young develop an immunity following a weak infection. However, in many intensive animal production systems, animals are exposed to high levels of infection and considerable mortality occurs. Nevertheless, evolution of the immune response in animals provides a means by which man can overcome this problem. Through vaccination with the appropriate antigens, immunity to many important animal diseases has been achieved and this has had a significant bearing on the success of intensive production systems. Vaccination is possible because the high level of antigenic disparity between the pathogen and the host means that only a small dose of the antigen is required to produce immunity.

In the second category of host—parasite relationships, parasites have a slower rate of replication and can only produce sufficient progeny to ensure survival of the next generation after the host immune response has had time to develop. Most endoparasites of domestic animals fall into this category. For these parasites to survive, they have evolved characteristics enabling them to elude immunological recognition by the host. However, the antigenic disparity between the host and the parasite is never eliminated and at a certain level of infection the host immune response is provoked and the parasite expelled. Below this threshold level of infection, the parasite survives unrecognized by the host, but extracts nutrients from it. Individual host animals within a population which show the least antigenic disparity with the parasite carry a greater level of infection below the threshold and have a reduced chance of survival. Thus, although this relationship between host

Chapter 6 references, p. 130

and parasite has been successful in nature, the productivity of the host animal is reduced. Recent experiments have shown that elimination of gastrointestinal nematodes by frequent drenching can increase wool production of grazing lambs by as much as 40% (Donald, 1979). The efficiency of animal production, therefore, appears to be adversely affected by this adaptation and no effective method of vaccination has yet been found to reduce the level of infection required to provoke the immune response.

Animal parasites are currently controlled mostly by chemicals. However, evolutionary adaptations within the parasites put the long-term future of this procedure in doubt. Within any population there are generally some organisms with regulator gene mutations which allow them to tolerate the chemical used and a new resistant population quickly evolves. Already 364 species of arthropods are resistant to specific insecticides and several species of gastrointestinal nematodes are resistant to common anthelmintics (Le Jambre, 1978).

7. EVOLUTION OF REPRODUCTIVE STRATEGIES

7.1. Number and size of progeny

As already stated, those individuals of a population which produce the greatest number of surviving offspring in the shortest time have an advantage in natural selection. Thus, reproductive strategies adopted by organisms would have been under strong evolutionary pressures. The main reproductive options open to an organism are concerned with (i) the proportion of available energy directed towards reproduction, and (ii) whether the energy directed to reproduction is incorporated into either a large number of small progeny or a small number of large ones.

The previously discussed bacteriophage experiments of Spiegelman (1971) clearly illustrate that as more available energy is diverted towards replication, the vulnerability of the organism to environmental fluctuations is increased. Calow (1977) suggests that it is always better for an organism to breed rather than to divert energy into other functions unless breeding endangers the adult and the adult has a better chance of survival than smaller progeny. Throughout evolution, there has been a tendency for animals to direct a smaller proportion of resources towards reproduction; progressively more energy has been used for growth, lipogenesis and maintenance of high body temperatures in homeotherms.

With regard to the second option concerning the number and size of offspring, natural selection should, in theory, favour those individuals which give rise to the greatest number of progeny and thereby have the greatest replication of genetic information. However, a smaller number of offspring, each with a greater energy reserve, may have an enhanced chance of survival in harsh environments. Thus, in environments where survival of the young is independent of the number present, the best strategy appears to be the production of a large number of small young, whereas in harsh environments, where survival of the young is inversely proportional to their density, the production of a small number of large individuals appears to be a safer option (Pianka, 1976). Throughout evolution, the trend has been towards producing a few, large progeny. Associated with this has been the development of more complicated reproductive systems, culminating in vivipary, which protect the developing embryo and allow it to grow to a considerable size before being exposed to the outside world (Sharman, 1976). In many species, protection of the young has gone well beyond birth, and in the more intelligent animals, training during this period has been an important aid to survival.

The particular reproductive strategy that has evolved in each domestic species has a large bearing on its productivity. The overall trend throughout evolution for a reduction in the amount of dietary energy flowing to the progeny is seen particularly in grassland-inhabiting mammals where the young move with the adults from birth. In these situations where the risks from predators are particularly high, the evolutionary trends has been for the production of large, strong progeny in numbers small enough to be protected by their mothers. Hence, in the grazing animals that contribute most to animal production — cattle, sheep and goats — single births are common. By comparison, species which during evolution became accustomed to giving birth to and leaving their young in secluded hideouts safe from predators, tend to produce a larger number of small offspring. In domestic animals this is seen particularly in rabbits and pigs. The effect of reproductive strategies on animal productivity is illustrated in Table 6.6. In this example, the gross efficiency of energy utilization by individuals of the four species, rabbit, pig, sheep and cattle, is similar, but the overall productivity of the population is at least 4 fold greater in the species producing a large number of progeny. These estimates of the difference in productivity resulting from the productive strategy that has evolved are conservative because, in intensive livestock systems, the number of young that can be produced by rabbits and pigs is often much greater than the values used in the example.

7.2. Stress and reproduction

The time taken for progeny to reproduce is dependent largely upon their rate of growth relative to the mature size of the species. However, once an individual has reached the size suitable for breeding, its reproductive capacity may be effected by environmental stress. Evolutionary adaptations to stress are different for semelparous animals which reproduce only once in a lifetime than for iteroparous species which reproduce on a number of occasions (Calow, 1977). With the former, reproductive output is reduced only slightly by nutrient restriction but survival of the adult is markedly depressed. By comparison, with iteroparous animals, reproduction may be either prevented or the number of eggs released during ovulation reduced in

TABLE 6.6

Estimates of the gross efficiency of conversion of energy in feed into energy in body gain for populations of domestic animals

	Species			
	Rabbit	Pig	Sheep	Beef cattle
Individual animal				
Energy in gain/energy in feed (relative values)	100	145	116	116
Population				
Average progeny/adult female/year	11.5	7.0	1.15	0.78
Energy gain in progeny/energy intake of population:				
absolute values	9.0	13.2	2.8	2.3
relative values	100	147	32	26

Adapted from Sharkey (1970). Energy cost of males ignored.

Chapter 6 references, p. 130

periods of stress. The course adopted by different species appears to be related to the predictability of the period of stress and to the relative chances of survival of the adult and of the young during the stress. Hence, many animals have evolved periods of anoestrus associated with under-nutrition and lactation. In species which inhabit areas where an annual period of nutritional hardship is predictable, a seasonal anoestrus has evolved. An extreme example is seen in the arctic reindeer (*Rangifer tarandus*) where the female is capable of mating during a 3-week period each year.

These adaptations have a significant effect on the productivity of many animal enterprises where there are large fluctuations in feed availability. It is possible to improve fecundity in many domestic species by feeding them well prior to mating, and 'flushing' has been a common agricultural practice for many years. Nevertheless, in many rangeland environments it may be impracticable to significantly increase the feed intake of breeding females prior to mating. Recent research results suggest that small amounts of high-protein supplements fed a few days before mating significantly increase ovulation rate in under-nourished ewes (Croker et al., 1978) and reduce the length of post-partum anoestrus in beef cattle grazing poor quality pasture (Little, 1975).

7.3. Sex ratio of the population

Another factor that could affect the reproductive capacity of a species is its sex ratio. Theoretically, it would seem to be an advantage to a species if there were more females than males; the latter consume resources but, particularly in polygynous species, contribute little energy towards the progeny. Nevertheless, a 1:1 sex ratio is almost universal in animals whether it is the male heterogamety of XX/XY-type or the female heterogamety of ZW/ZZ-type. It has been proposed that the total expenditure of resources by a population should be equal for both males and females and this results in a sex ratio of 1:1 at the end of the period of parental involvement (Fisher, 1958; Bodmer and Edwards, 1960; Smith, 1978). These authors suggest that some variation in the ratio would occur at fertilization when there is a bias in deaths of one sex during the period of parental expenditure. However, Williams (1978) has questioned the validity of this argument and Ohno (1976) postulates that the equal sex ratio arises because the X(Z) chromosome carries the genes for many ubiquitous enzymes essential for life, such as glucose-6-phosphate dehydrogenase, and must be present in all individuals of a species; hence, a diploid individual can be only XX(ZZ) or XY(ZW). The Y(W) chromosome is thought to carry only genes concerned with sexuality and is therefore not essential for all individuals.

Some asynchronous hermaphrodites have successfully altered the sex ratio by functioning as females when young and converting to males when older. With an age-linked mortality, this results in a predominance of females in the population. It is doubtful whether this has been of much advantage in survival because other asynchronous hermaphrodites function first as males and have a predominance of males in the population (Ohno, 1976). An extremely unusual example of unequal sex ratio has recently been observed by Fredga et al. (1976) in the wood lemming (*Myopus schisticolor*). In this species, there is a definite prevalence of females, some with the normal XX karyo-type and some with an XY karyotype. Both types of females are fertile and give birth to normal young. The authors postulate that, in the XY females, there is an X-linked gene that represses the male-determining characteristics of the Y chromosome. Thus, when a heterozygous XX karyotype mates with a normal male, the sex ratio is 1:3, male:female. When the XY female mates with a normal male, a selective non-disjunction of the chromosomes produces all female progeny.

Agriculturalists have known for a long time that the efficiency of animal production systems is greatly increased by reducing the number of males in the population. The equal sex ratio at birth does not create a large problem in enterprises concerned with the production of meat or fibre because excess males contribute to production. The major consequence in these industries is that the equal sex ratio limits the rate of increase of the population. However, industries concerned with output from reproducing females, such as the egg and milk industries, could obtain a considerable advantage from a sex ratio at birth favouring females. Thus, the viable mutation that has occurred in the wood lemming and allows some XY karyotypes to be normal females producing only female offspring would be extremely useful had it occurred in domestic animals. Such a mutation leads to an extremely rapid rate of increase in the population, and it is interesting to speculate that the lemming has survived as a species only because of the self-destructive potential of some females.

8. CONCLUSIONS

Once the basic biochemical principles of life evolved, gene mutations in combination with natural selection led to the continual development of new organisms with adaptations better suited for survival and reproduction in particular environments. Throughout time, many characteristics of organisms were eliminated as they became unnecessary because of new adaptations and environmental changes. Most adaptations that evolved in animals used by man have an effect on animal production. However, the goals of people involved in animal production to optimize the efficiency of conversion of resources into animal products often differ from the driving forces of evolution.

An examination of the evolution of factors affecting the efficiency of energy utilization in animal populations shows that some adaptations are an advantage in animal production whereas others are a disadvantage. The most useful include evolution of a microbial-based digestive system in the stomach of ruminants which allows effective utilization of much of the low quality grasslands on earth; the evolution of homeothermy which greatly increased the rate of animal growth and led to the development of hair and wool; and evolution of the animal immune mechanism which enables the vaccination of animals against microbial diseases to be successful. Disadvantageous adaptations include loss of the ability of heterotrophic organisms to synthesize essential amino acids, fatty acids and vitamins; low fecundity in animals that evolved where predation and seasonal fluctuations in feed supply were significant; and evolution of the host—parasite relationship which reduces the performance of animals infected with subthreshold levels of endoparasites. In addition, a number of evolutionary adaptations that could be important in animal production occurred in non-domestic animals. These particularly include the predominance of females in wood lemming populations and the ability of the black bear to conserve body protein when losing weight during winter sleep. From an examination of evolutionary adaptations in animals, it is not clear why man selected a few particular species for domestication. Perhaps these animals evolved particular behavioural characteristics which allowed them to co-exist with man.

Chapter 6 references, p. 130

9. REFERENCES

Ardrey, R., 1961. African Genesis. Dell, New York, NY, 416 pp.

Ayala, F.J., 1978. The mechanisms of evolution. Sci. Am., 239(3): 48—61.

Baldwin, E., 1949. An Introduction to Comparative Biochemistry. Cambridge University Press, Cambridge, Great Britain, 164 pp.

Ballard, F.J., 1972. Carbohydrate and lipid metabolism of growing animals. In: W. Lenkeit, K. Breirem and E. Grasemann (Editors), Handbuch der Tierernährung, Vol. 2. Verlag Paul Parey, Hamburg and Berlin, pp. 451—456.

Bartley, E.E., Meyer, R.M. and Fina, L.R., 1975. Feedlot or grain bloat. In: I.W. McDonald and A.C.I. Warner (Editors), Digestion and Metabolism in the Ruminant. University of New England Publishing Unit, Armidale, Australia, pp. 551—562.

Bauchop, T., 1977. Forgut fermentation. In: R.T.J. Clarke and T. Bauchop (Editors), Microbial Ecology of the Gut. Academic Press, London, New York, NY, San Francisco, pp. 223—250.

Black, J.L., 1976. Why is the protein/energy ratio recommended for humans so different from that for domestic animals? Proc. Nutr. Soc. Aust., 1: 34.

Black, J.L. and Sharkey, M.J., 1970. Reticular groove (*sulcus reticuli*) an obligatory adaptation in ruminant-like herbivores? Mammalia, 34: 294—302.

Black, J.L. and Tribe, D.E., 1973. Comparison of ruminal and abomasal administration of feed on the growth and body composition of lambs. Aust. J. Agric. Res., 24: 763—773.

Black, J.L., Graham, N.McC. and Faichney, G.J., 1976. In: T.M. Sutherland, J.R. McWilliam and R.A. Leng (Editors), From Plant to Animal Protein. Reviews in Rural Science 2. University of New England Publishing Unit, Armidale, Australia, pp. 161—166.

Black, J.L., Dawe, S.T., Colebrook, W.F. and James, K.J., 1979. Protein deficiency in young lambs grazing irrigated summer pasture. Proc. Nutr. Soc. Aust., 4: 126.

Blaxter, K.L., 1961. Efficiency of feed conversation by different classes of livestock in relation to food production. Fed. Proc., Fed. Am. Soc. Exp. Biol., 20: 268—274.

Blaxter, K.L., 1975. Conventional and unconventional farmed animals. Proc. Nutr. Soc., 34: 51—56.

Bodmer, W.F. and Edwards, A.W.F., 1960. Natural selection and the sex ratio. Ann. Hum. Genet., 24: 239—244.

Boucot, A.J., 1976. Rates of size increase and of phyletic evolution. Nature (London), 261: 694—695.

Bryant, M.P. and Robinson, I.M., 1963. Apparent incorporation of ammonia and amino acid carbon during growth of selected species of ruminal bacteria. J. Dairy Sci., 46: 150—154.

Calow, P., 1977. Ecology, evolution and energetics: a study in metabolic adaptation. Adv. Ecol. Res., 10: 1—62.

Coulson, R.A., Hernandez, T. and Herbert, J.D., 1977. Metabolic rate, enzyme kinetics in vivo. Comp. Biochem. Physiol. A, 56: 251—262.

Coulson, T.D., Coulson, R.A. and Hernandez, T., 1973. Some observations on the growth of captive alligators. Zoologica, 58: 47—52.

Croker, K.P., Lightfoot, R.J. and Marshall, T., 1978. The fertility of merino ewes fed high protein supplements at joining. Proc. Aust. Soc. Anim. Prod., 12: 250.

Cross, D.G., 1969. Aquatic weed control using grass carp. J. Fish Biol., 1: 27—30.

Curio, E., 1973. Towards a methodology of teleonomy. Experienta, 29: 1045—1058.

Davidson, E.H., 1968. Gene Activity in Early Development. Academic Press, New York, NY, London, 375 pp.

Dawson, T.J., 1972. Primitive mammals and patterns in the evolution of thermoregulation. In: J. Bligh and R.E. Moore (Editors), Essays on Temperature Regulation. North Holland Publishing Co., Amsterdam, London, pp. 1—18.

Dineen, J.K., 1978. The nature and role of immunological control in gastro-intestinal helminthiasis. In: A.D. Donald, W.H. Southcott and J.K. Dineen (Editors), The Epidemiology and Control of Gastrointestinal Parasites of Sheep in Australia. CSIRO, Australia, pp. 121—135.

Dixon, G.H., 1966. Mechanisms of protein evolution. Essays Biochem., 2: 147—204.

Donald, A.D., 1979. Effects of parasites and disease on wool growth. In: J.L. Black and P.J. Reis (Editors), Physiological and Environmental Limitations to Wool Growth. University of New England Publishing Unit, Armidale, Australia, pp. 99—114.

Drake, J.W. and Baltz, R.H., 1976. The biochemistry of mutagenesis. Annu. Rev. Biochem., 45: 11—37.

Ferguson, K.A., 1975. The protection of dietary proteins and amino acids against microbial fermentation in the rumen. In: I.W. McDonald and A.C.I. Warner (Editors), Digestion and Metabolism in the Ruminant. University of New England Publishing Unit, Armidale, Australia, pp. 448—464.

Fisher, R.A., 1958. The Genetical Theory of Natural Selection, 2nd edn. Dover Publications, New York, NY, 291 pp.

Fredga, K., Gropp, A., Winking, H. and Frank, F., 1976. Fertile XX- and XY-type females in the wood lemming *Myopus schisticolor*. Nature (London), 261: 225—227.

Frisch, J.E. and Vercoe, J.E., 1977. Food intake, eating rate, weight gains, metabolic rate and efficiency of feed utilization in *Bos taurus and Bos indicus* crossbred cattle. Anim. Prod., 25: 343—358.

Graham, N.McC. and Searle, T.W., 1975. Studies of weaner sheep during and after periods of weight stasis. I. Energy and nitrogen utilization. Aust. J. Agric. Res., 26: 343—353.

Greenberg, D.M., 1961. Metabolic Pathways, Vol. 2. Academic Press, New York, NY, London, 814 pp.

Harris, H., 1976. Molecular evolution: the neutralist—selectionist controversy. Fed. Proc., Fed. Am. Soc. Exp. Biol., 35: 2079—2082.

Hochachka, P.W., 1976. Design of metabolic and enzymic machinery to fit lifestyle and environment. Biochem. Soc. Symp., 41: 3—31.

Hodge, R.W., 1974. Efficiency of food conversion and body composition of the preruminant lamb and the young pig. Br. J. Nutr., 32: 113—126.

Hoffman, R.A., 1964. Terrestrial animals in cold: hibernators. In: D.B. Dill, E.F. Adolph and C.G. Wilber (Editors), Handbook of Physiology, Section 4, Adaptation to the Environment. American Physiological Society, Washington, pp. 379—403.

Hudson, J.W., and Bartholomew, G.A., 1964. Terrestrial animals in dry heat: estivators. In: D.B. Dill, E.F. Adolph and C.G. Wilber (Editors), Handbook of Physiology, Section 4. Adaptation to the Environment. American Physiological Society, Washington, pp. 541—550.

Hume, I.D., 1977. Maintenance nitrogen requirements of the macropod marsupials *Thylogale thetis*, red-necked pademelon, and *Macropus eugenii*, tammar wallaby. Aust. J. Zool., 25: 407—417.

Hutchinson, J.C.D., 1954. Heat regulation in birds. In: J. Hammond (Editor), Progress in the Physiology of Farm Animals, Vol. 1. Butterworths, London, pp. 299—362.

Jacob, F. and Monod, J., 1961. Genetic regulatory mechanisms in the synthesis of proteins. J. Mol. Biol., 3: 318—356.

James, L.F., Allison, M.J. and Littledike, E.T., 1975. Production and modification of toxic substances in the rumen. In: I.W. McDonald and A.C.I. Warner (Editors), Digestion and Metabolism in the Ruminant. University of New England Publishing Unit, Armidale, Australia, pp. 576—590.

Jany, K.D., 1976. Studies on the digestive enzymes of the stomachless bonefish *Carassius auratus gibelio* (Bloch): endopeptidases. Comp. Biochem. Physiol. B, 53: 31—38.

Jewell, P.A., 1976. Selection for reproductive success. In: C.R. Austin and R.V. Short (Editors), Reproduction in Mammals, Book 6. The Evolution of Reproduction. Cambridge University Press, London, pp. 71—109.

King, J.L., 1976. Progress in the neutral mutation—random drift controversy. Fed. Proc., Fed. Am. Soc. Exp. Biol., 35: 2087—2091.

Kleiber, M., 1961. The Fire of Life. Wiley, New York, NY, London, 454 pp.

Lasiewski, R.C. and Dawson, W.R., 1967. A re-examination of the relation between standard metabolic rate and body weight in birds. Condor, 69: 13—23.

Le Jambre, L.F., 1978. Anthelmintic resistance in gastrointestinal nematodes of sheep. In: A.D. Donald, W.H. Southcott and J.K. Dineen (Editors), The Epidemiology and Control of Gastrointestinal Parasites of Sheep in Australia. CSIRO, Australia, pp. 109—120.

Little, D.A., 1975. Effects of dry season supplements of protein and phosphorus to pregnant cows on the incidence of first post-partum oestrus. Aust. J. Exp. Agric. Anim. Husb., 15: 25—31.

Lundberg, D.A., Nelson, R.A., Wahner, H.W. and Jones, J.D., 1976. Protein metabolism in the black bear before and during hibernation. Mayo Clin. Proc., 51: 716—722.

McBee, R.H., 1977. Fermentation in the hindgut. In: R.T.J. Clarke and T. Bauchop (Editors), Microbial Ecology of the Gut. Academic Press, London, New York, NY, San Francisco, pp. 185—222.

Millward, D.J., Garlick, P.J., Nnanyelugo, D.O. and Waterlow, J.C., 1976. The relative importance of muscle protein synthesis and breakdown in the regulation of muscle mass. Biochem. J., 156: 185—188.

Moir, R.J., 1968. Ruminant digestion and evolution. In: C.F. Code (Editor), Handbook of Physiology, Section 6, Alimentary Canal, Vol. 5. American Physiological Society, Washington, pp. 2673—2694.

Munro, H.N. and Gray, J.A.M., 1969. The nucleic acid content of skeletal muscle and liver in mammals of different body size. Comp. Biochem. Physiol., 28: 897—905.

Nelson, R.A., 1973. Winter sleep in the black bear; a physiologic and metabolic marvel. Mayo Clin. Proc., 48: 733—737.

Nelson, R.A., Jones, J.D., Wahner, H.W., McGill, D.B. and Code, C.F., 1975. Nitrogen metabolism in bears: urea metabolism in summer starvation and in winter sleep and role of urinary bladder in winter nitrogen conservation. Mayo Clin. Proc., 50: 141—146.

Neurath, H., Walsh, K.A. and Winter, W.P., 1967. Evolution of structure and function of proteases. Science, 158: 1638—1644.

Nijkamp, H.J., Van Es, A.J.H. and Huisman, E.A., 1974. Retention of nitrogen, fat, ash, carbon and energy in growing chickens and carp. Publ. Eur. Assoc. Anim. Prod., 14: 277—280.

Ohno, S., 1976. The development of sexual reproduction. In: R.C. Austin and R.V. Short (Editors), Reproduction in Mammals, Book 6. The Evolution of Reproduction. Cambridge University Press, London, pp. 1—31.

Pianka, E.R., 1976. Natural selection of optimal reproductive tactics. Am. Zool., 16: 775—784.

Preston, T.R. and Willis, M.B., 1970. Intensive Beef Production. Pergamon Press, Oxford, 544 pp.

Robinson, A.B., 1974. Evolution and the distribution of glutaminyl and asparaginyl residues in proteins. Proc. Natl. Acad. Sci. U.S.A., 71: 885—888.

Roy, J.H.B. and Stobo, I.J.F., 1975. Nutrition of the pre-ruminant calf. In: I.W. McDonald and A.C.I. Warner (Editors), Digestion and Metabolism in the Ruminant. University of New England Publishing Unit, Armidale, Australia, pp. 30—48.

Sager, R., 1965. Genes outside chromosomes. Sci. Am., 212: 70—79.

Sarich, V.M. and Cronin, J.E., 1977. Generation length and rates of hominoid molecular evolution. Nature (London), 269: 354—355.

Sharkey, M.J., 1970. The digestion and utilization of fibrous feeds by herbivores. Ph. D. Thesis University of Melbourne, Australia, 320 pp.

Sharman, G.B., 1976. Evolution of viviparity in mammals. In: C.R. Austin and R.V. Short (Editors), Reproduction in Mammals, Book 6. The Evolution of Reproduction. Cambridge University Press, London, pp. 32—70.

Siegel, H.S., 1976. Effects of cold on energy metabolism in birds. In: S.W. Tromp, J.J. Bouma, H.D. Johnson (Editors), Progress in Biometerology, Division B, Vol. 1, Part 1, Swets and Zeitlinger, Amsterdam, pp. 259—265.

Simpson, G.G., 1949. The Meaning of Evolution. Yale University Press, New Haven, 364 pp.

Slobodkin, L.B. and Richman, S., 1961. Calories/gm in species of animals. Nature (London), 191: 299.

Smith, J.M., 1978. The Evolution of Sex. Cambridge University Press, Cambridge, 222 pp.

Somero, G.N. and Low, P.S., 1976. Temperature: a 'shaping force' in protein evolution. Biochem. Soc. Symp., 41: 33—42.

Spiegelman, S., 1971. An approach to the experimental analysis of precellular evolution. Q. Rev. Biophys., 4: 213—253.

Tromp, S.W., Bouma, J.J. and Johnson, H.D., (Editors), 1976. Progress in Biometeorology, Division B, Vol. 1, Part 1, Swets and Zeitlinger, Amsterdam, 603 pp.

Van der Walt, K. and Jansen, B.C., 1968. Adaptation to stress and disease. In: E.S.E. Hafez (Editor), Adaptations of Domestic Animals. Lea and Febiger, Philadelphia, pp. 215—230.

Van Gelder, R.G., 1969. Biology of Mammals. Scribner's, New York, NY., 197 pp.

Walker, A., 1975. Crustacean aquaculture. Proc. Nutr. Soc., 34: 65—73.

Watson, J.D., 1976. Molecular Biology of the Gene, 3rd edn. W.A. Benjamin, Inc., New York, NY, 494 pp.

Williams, G.C., 1978. Mysteries of sex and recombination. Q. Rev. Biol., 53: 287—289.

Wilson, A.C., Carlson, S.S. and White, T.J., 1977. Biochemical evolution. Annu. Rev. Biochem., 46: 573—639.

Yokoe, Y. and Yasumasu, I., 1964. The distribution of cellulases in invertebrates. Comp. Biochem. Physiol., 13: 323—338.

Zotterman, Y., 1959. Thermal sensations. In: J. Field, H.W. Magoun and V.E. Hall (Editors), Handbook of Physiology, Section I, Neurophysiology, Vol. 1. American Physiological Society, Washington, pp. 431—458.

10. NOTE ADDED IN PROOF

While this chapter has been in press there has been considerable development in some of the areas discussed. This applies particularly to our understanding of the control of gene expression in eukaryote cells and to the development of techniques which make it possible to incorporate new sequences of DNA into organisms. It is probable that in the future many of the problems in animal production associated with the loss through evolution of enzymes required for the synthesis of essential amino acids and vitamins may be overcome by incorporating the required genes back into our domestic species.

Chapter 7

The Institutionalization of Research
in Animal Science

LYNNETTE J. PEEL

1. INTRODUCTION

Research in animal science has been fostered and motivated by four principal forces:
— the desire of farmers to understand the phenomena they observe, in order to control them and so increase their production and fashion their product;
— the desire of scientists to extend their understanding of nature, to utilize this understanding in a practical way, and to earn a livelihood;
— the desire of industrialists to find a new product and a new market; and
— the desire of governments to help improve the productivity of the farmers and lower the cost of food for local consumption and to increase exports, and of politicians to respond to pressures from farming constituencies.

These forces have been at work over the centuries, sometimes in unison, sometimes in conflict, sometimes with one then another predominating. At all times they have operated within a context of wider political, social, religious and economic change.

2. THE SEVENTEENTH CENTURY

The 17th century has been described as the century of genius (Whitehead, 1975). It was the period which saw the transition from the metaphysical approach to science, with its indebtedness to the works of Aristotle, to the empirical approach and inductive reasoning which developed through the influence of Galileo (1564—1642), Descartes (1596—1650) and Francis Bacon (1561—1626). In this change from deductive to inductive reasoning in science Bacon argued that the intellectual and the utilitarian faculties must be brought together in the one person 'a true and lawful marriage between the empirical and the rational faculty' (Purver, 1967, p. 35). The skill of the artisan should be drawn upon and through experiments utilized to discover the causes of natural phenomena. The advance of knowledge through inductive processes was Bacon's first objective, but secondly this knowledge would allow 'a mastery over nature hitherto undreamed of' (Purver, 1967, p. 49). With knowledge and understanding of nature, power over the functions of nature would be attained, and thereby practical and utilitarian benefits achieved. Bacon proposed a 'Catalogue of Particular Histories' that should be studied. The catalogue covered most of the natural sciences as well as more technical fields. Agriculture was one topic; the processes in the development of animals, and in nutrition 'from the first reception of the food to its complete assimilation' (Purver, 1967, p. 60) were others.

Chapter 7 references, p. 150

The writings of Bacon provided much of the inspiration for the early endeavours of the Royal Society of London, founded in 1660. Among the many initial concerns of the Society was the study of agriculture and a Georgical Committee was set up to oversee endeavours in this field. Its main efforts were directed toward establishing what was known already so that the members could then consider what further improvements might be made, although individual members also undertook their own investigations on their own estates. A questionnaire was drawn up by the Georgical Committee concerning practices in arable land and meadows so that information could be obtained from all parts of the country. The outbreak of plague cut short this activity but a number of descriptions of agriculture were obtained (Lennard, 1932). The attempt illustrates the keen interest of the scientists of the day in practical farming and a wish to apply the knowledge of science to this practice — an interest easily understood given the importance of agricultural matters in the economy and affairs of the nation, and the emphasis placed on the advancement of husbandry by the intellectuals of the Puritan period (Webster, 1975).

The new empirical approach to science was not confined to England. The short-lived Accademia del Cimento (1657—1667) in Florence and the Royal Society were together influential in the development of a third major society, the Académie Royale des Sciences founded in Paris in 1666. The Académie introduced a strong professional element into science in that its members were selected and paid, at least to some extent, by the Crown. In this its organization contrasted with that of the Royal Society which was a voluntary association of gentlemen of private means. The latter was a necessary condition given the lack of any financial patronage. The scientific interests and activities of these institutions were more similar than their organization, and the members of each were well informed of activities within the other (Hall, 1962).

The foundation of the Académie Royale des Sciences and of the Royal Society was important for the long-term evolution of research in animal science for two reasons. First, these societies established patterns of organization for institutional science in their respective countries; second, they provided a professional setting in which experimental findings in chemistry and the biological sciences could be assessed by the best scientists in the land, theories based on these findings propounded, and the results assimilated into the wider knowledge of science.

The most important effect on animal production, of the changed approach to science, was its strong encouragement of the philosophy of experimentation. A few leading landowners were already undertaking their own information gathering, observation and trials, and were demonstrating the benefits of turnips and other roots for winter feeding and introducing into England the legumes and more productive grasses of the Low Countries. The enclosure of open fields and commons made possible the cultivation and management of these fodder species in rotational cropping systems. With more feed available for livestock there was a greater possibility of the animals being fed sufficiently well for them to reach their full genetic potential for the production of manure, wool, milk or meat. Improved feeding and enclosure were necessary precursors of improved breeding in the 18th century (Trow-Smith, 1957; Slicher van Bath, 1963).

3. THE EIGHTEENTH CENTURY

Interest in agricultural improvement and experimentation continued throughout the 18th century and gathered momentum as the century progressed. The main experimenters were to be found amongst the more affluent farmers and landowners for they alone could afford the costs and the risks of loss and failure, inherent in experimenting with different practices. The need for knowledge appropriate to a particular environment and set of conditions was appreciated, and so too was the need to exchange information about more general principles and to compare the results of experiments in one environment with those in another.

Of the improvers in livestock husbandry Robert Bakewell (1726–1795) is probably the best known. He deliberately bred towards a specific type of animal for a specific purpose and is known for his work with longhorn cattle and Leicester sheep. The longhorn was originally used for both milk and meat production but Bakewell, following on from the work of an earlier breeder, bred a beef type which he then fixed through inbreeding. Fecundity, however, was decreased in the process. With sheep Bakewell had more lasting success and his New Leicester sheep were widely accepted.

The demand for meat was strong and Bakewell's aim was to breed animals of quick maturing type for butcher's meat. His own fortunes were built on hiring out his rams and bulls and selling breeding stock amidst wide publicity. Others soon followed his example. The publicity lent impetus to the gradual transition from localized breeds of livestock, well adapted over many years to a local environment and utilized for a variety of purposes, to the development of breeds for a specific purpose within a particular but wider environment. The commercial rewards of breed development then led in the 19th century to a strong growth of the pedigree system and an international exchange of pedigreed animals. As the emphasis on pedigree increased, the specificity of adaptation to environment declined (Trow-Smith, 1959).

The growth of interest in breeding meant that when eventually the science of genetics evolved there was a receptive audience among livestock owners for knowledge of genetics to confirm, explain and extend the practices already developed.

The need for the exchange of information was met through extensive private correspondence and the formation of farmers clubs and societies. Robert Maxwell, Secretary of The Honourable Society of Improvers in the Knowledge of Agriculture (1723–1745), was one secretary who provided in Scotland a focus for the exchange of information on not just a local but a national and international scale (Smith, 1962). Later in the century Arthur Young (1741–1820), in England, became the publicist par excellence for agricultural improvement and experimentation. Although his audience was small it was always influential and extended to America and across Europe to Russia. In 1768, 1770 and 1771 he published accounts of his tours in the south, north and east of England, respectively. These contained first-hand reports of improved methods and trials being carried out on the most progressive estates, written after visits to these estates. It was with new methods that he was concerned rather than with the general system of farming then practised. His considerable service to farming was to make available, in published form, accounts of these latest techniques. The early tours were followed by later travels in Ireland, France, Catelonia and Italy; all were reported upon at length.

In 1784 Young began publication of his *Annals of Agriculture*. This journal eventually ran to 46 volumes until publication ceased in 1815. It carried many articles written by Young but more importantly it provided for the publication of the results of agricultural experiments that previously would only have been communicated by private letter and word of mouth. Circulation of the *Annals* was small, no more than about 350 copies regularly in 1791. But its circulation and influence were international (Mingay, 1975; Rossiter, 1976).

Young was a prolific writer and a prolific experimenter on his own farm. He was in contact with the English scientists of his day, visited Lavoisier in Paris, and was elected to a fellowship of the Royal Society. He was, however, a publicist rather than a scientist, and his writings reflected and did much to encourage the spirit of improvement then prevalent (Gazley, 1973).

The leading landowners of the 18th century were anxious to improve their production by the only way they thought feasible — trials on their own estates at their own expense and the exchange of information with like-minded people. They were in contact with the scientists of the day, some of whom were certainly counted among their number, but apart from the general philosophy of experimental enquiry, laboratory science still had little specifically to offer. Indeed there was some frustration that agricultural experimentation was so unamenable to the controlled situation of the laboratory and required such a long time span before results were obtained, in contrast to the rapid turnover of laboratory experiments. The costs in relation to the information gained were correspondingly greater.

Towards the end of the 18th century in both France and England the state sought to be advised on, and to promote farming practices. In France the government established the Comité d'Agriculture (Assemblée de l'Administration d'Agriculture) in 1775. Various members of the Académie des Sciences were appointed to it and Antoine Lavoisier, an active member of the Comité, undertook to carry out experiments for it on his own farms. He had been experimenting with farming practices on his farms for some years. The Comité was also instrumental in disseminating information on improved practices throughout the country. However, by 1787 changes had occurred in the government and it showed only indifference to the Comité; in that year the Comité ceased to function and some of its duties were transferred to the Société d'Agriculture (Smeaton, 1956).

In England government funds were used to establish a Board of Agriculture in 1793 and Arthur Young was appointed as its secretary. It is best known for its published surveys of agriculture in varous parts of the Kingdom. The Board met regularly but in general it was not particularly effective in promoting agricultural improvement. The grant to the Board ceased in 1821 and the Board faded out of existence (Mitchison, 1962). There had been a certain will in government in both England and France for the promotion of agricultural improvement but the ways and the means were illusive. From the Royal Society in the sixteen-sixties to the Board of Agriculture at the end of the 18th century corporate attempts at improvement in agriculture did not advance beyond gathering information on the most recent developments and disseminating it. Experimentation remained the province of the individual landholder and this was not questioned.

4. THE NINETEENTH CENTURY

Antoine Lavoisier (1743—1794), scientist, economist, social reformer and agricultural experimenter, conducted his agricultural experiments with precision. He deliberately selected the poorest land on his estate and, over a period of 10 years, through experiments, careful measurement and recording, sought to improve the productivity of his integrated livestock and cropping husbandry (McKie, 1952). At the same time he was working on his laboratory experiments, again giving close attention to accurate measurement and building on the work of his predecessors and contemporaries in France,

Britain, Germany and Sweden. In 1789, a few months before the storming of the Bastille, he published his *Traité Elémentaire de Chimie.* This described the new system of chemistry previously outlined by Lavoisier and his colleagues to the Académe Royale des Sciences. It was approved, also, prior to publication at Lavoisier's request, by the Société d'Agriculture (McKie, 1952; Smeaton, 1956). The *Traité Elémentaire de Chimie* marked the final transition from the phlogiston theory to an understanding of oxidation and chemical combination, and to the formulation of a new system of chemical nomenclature (Hall, 1962). By the early 19th century, his new approach had been developed into a powerful tool for chemical analysis. This then opened the way for the study of the composition of plants, animal substances, soils and fertilizing material. From the macro-effect on crop growth of three loads of sheep manure/acre compared with two loads, as recorded by Lavoisier, there was now the possibility of studying at the micro-level what constituents of sheep manure were important in increasing the growth of plants — a type of information long desired but hitherto difficult to acquire because of the limited progress in chemical knowledge (Russell, 1966). Lavoisier, himself, did not live to bring together his laboratory and field experimentation, but his new system of chemistry prepared the way for his successors to do this in the 19th century. Antoine de Fourcroy (1755—1809), a collaborator with Lavoisier, lectured in 1798 'on chemistry applied to the physics of vegetables and animals' and its application to agriculture and medicine (Smeaton, 1962). But it was Humphry Davy (1778—1829) in England who was to popularize the application of chemistry to agriculture. Lavoisier, meanwhile, had met an untimely death at the guillotine in 1794.

Davy began his study of chemistry by reading Lavoisier's *Traité Elémentaire de Chimie* in 1797 (Hartley, 1966) and in 1801 he obtained a professional scientific appointment as Assistant Lecturer in Chemistry and Director of the Laboratory at the Royal Institution in London. The following year he was appointed Professor. The Royal Institution had been established in 1799 to bring scientific improvements and teaching to the benefit of the practical arts. Some half of the founding members were improving landowners and many were also members of the Board of Agriculture. Thus the importance of science to agriculture was a strong theme of the Royal Institution in its early years, although by the eighteen-thirties the rural interest had waned and the Institution was proceeding in other directions under the leadership of Michael Faraday (1791—1867) and an urban-based membership (Berman, 1978).

In a course of lectures on chemistry in 1802 Davy included a section on agriculture, and later that year the Board of Agriculture arranged with the Royal Institution for Davy to give a series of lectures on the application of chemistry to agriculture in 1803. To prepare himself for this he visited farms and undertook experiments on land made available by Sir Thomas Bernard. Davy was already required to do soil analyses for the Royal Institution Proprietors (Berman, 1978). Davy repeated his lectures annually until 1812, and in 1813 published them under the title *Elements of Agricultural Chemistry, in a Course of Lectures for the Board of Agriculture.* This work went through a number of editions and translations throughout the remainder of the first half of the century. In it Davy brought together existing knowledge on the composition and analysis of soils, plants and manures and the nutrition of plants in agriculture. Reference to animals was mainly restricted to the analysis of animal manures and their value for plant growth. He did attempt to assess the feeding value of grasses by determining the percentage of water-soluble matter in them but this was not a determination of value (Russell, 1966).

Chapter 7 references, p. 150

Davy's work prepared the way for a wide, if at times sceptical, acceptance of agricultural chemistry. Progress was slow for the three decades after its publication until Justus von Liebig (1803—1873), Professor of Chemistry at Giessen University, published his book *Organic Chemistry in its Applications to Agriculture and Physiology* in 1840. This brought a stimulus to activity in agricultural chemistry — an activity most frequently thought of and written of in terms of plant nutrition, fertilizers and the eventual development of a fertilizer industry. The heroes are Justus von Liebig in his chemical laboratory at Giessen in Germany and John Bennet Lawes (1814—1900) with Joseph Henry Gilbert (1817—1901) at Rothamsted in England. Liebig was the laboratory analyst and theoretician, and above all the teacher and inspirer of professional chemists. Rossiter (1975) lists 182 foreign students from France, Switzerland, Great Britain, America, Russia and other countries studying under him at various times between 1829 and 1852. Lawes and Gilbert were the practical experimenters who utilized experiments and analyses in their chemical laboratory in association with practical, long-term trials in the fields of Rothamsted farm owned by Lawes. Gilbert, whom Lawes engaged in 1843, was a trained chemist who had spent a session studying under Liebig. Lawes, the more practically inclined and largely self-taught chemist, already had not only developed superphosphate and shown its worth in field trails but also had set up a factory to produce it.

The controversies between Liebig and his followers, on the one hand, and Lawes and Gilbert, on the other, during the mid-part of the century are well known. They centred among other things on Liebig's mineral theory of plant nutrition, based on his analyses of the ash of plants, and divergent views on the source of nitrogen for plant growth. Liebig erroneously concluded that sufficient nitrogen for plant growth would be washed out of the atmosphere. Eventually it was shown that this was not the case and that nitrogen, phosphorus and potassium were not only important elements for plant growth but that they could be supplied to plants in the form of simple compounds added to the soil (Russell, 1966).

As the century progressed agricultural chemistry became almost synonymous with agricultural science. In the eighteen-thirties and eighteen-forties in Great Britain innovative landowners were anxious to gain benefit from an association of science with agriculture. 'Practice with science' was adopted as the motto of the Royal Agricultural Society of England, founded in 1839 — a society which had a profound influence through the auspices of its journal, in a similar way to that of Arthur Young through his *Annals of Agriculture*. The Society explored the application of science to agriculture through chemistry, geology, botany and entomology, and papers by Liebig, Lawes and the entomologist John Curtis appeared in its journal. It was an uneasy exploration. Eagerness for new knowledge of practical significance outstripped the ability of science to provide it. Farmers rightly had become wary of scientific theories of plant growth too widely promulgated before they had been tested in the field (Peel, 1976). The Society appointed its own consulting chemist and financed successful research on absorption in soils, but by the late eighteen-fifties farmers were increasingly buying the new fertilizing materials and their needs were for practical information rather than theoretical. Science could tell the farmer if his fertilizer or feeding stuff had been adulterated, and it was this practical service to agriculture that consulting chemists, either privately or through attachment to agricultural societies, were able to provide for the rest of the century.

Meanwhile the farm-scale, macro-agricultural experiment of the improving landlord was ceasing to be fashionable. The agricultural revolution of the 17th and 18th centuries and the high farming, with land drainage and heavy fertilizer applications, of the later 19th century had proved the worth of experiments on the macro-scale on private estates. But experiments were very costly and time consuming and new avenues of improvement were no longer so obvious. The rewards had to be considerable for a private land-owner to afford the cost of experiments and no longer were they so. Furthermore the importance of agriculture in national affairs was now less dominant — with the advent of the industrial revolution there were other matters of interest competing for the attention of wealthy landlords. The example of Rothamsted had required a rare combination of chemical and agricultural understanding, and the financial resources to bring the two together. Few could meet such requirements. The agricultural societies debated such possibilities but were only partially convinced of the usefulness of experimental facilities established by a society rather than under the control of an experienced landowner, and tended to be deterred by the cost despite the series of investigations undertaken by the Royal Agricultural Society at Woburn (Ramsay, 1879; Russell, 1966).

By the second half of the 19th century, therefore, the old experimental ethos of the 17th century scientific revolution, so willingly adopted and exploited by the improving landowners, had wained in its influence with them. To take its place was a new growing awareness of the usefulness of dis-ciplinary science — a usefulness not wholly to be trusted but certainly worthy of further investigation.

The genesis of this new ethos was undoubtedly the application of chemistry to the understanding of plant nutrition; a field that would dominate agricul-tural science and research well into the 20th century. But throughout the years of rapid increase in knowledge of agricultural chemistry as applied to plant growth there were parallel investigations of the respiration, nutrition and growth of animals. Here, though, was a subject perhaps more formidable and less easily rendered immediately useful to the practising farmer.

Lavoisier and Pierre-Simon Laplace (1749—1827) in the seventeen-eighties had investigated the nature of respiration and concluded that:

> When an animal is in a constant state and at rest; when it can live for a considerable time without inconvenience in the medium surrounding it; in general, when the conditions in which it is placed do not in any way sensibly change its blood and humours, so that after several hours the animal system shows no appreciable variation; the conservation of its heat is due, at least in great part, to the heat produced by the combination of the pure air breathed by the animal with the base of fixed air supplied to it by the blood (McKie, 1952, p. 108).

From this basis, as improvements in analytical chemistry allowed, the practitioners of chemistry and physiology from their respective points of view proceeded to investigate animal respiration, digestion and nutrition. The Parisian chemists who followed on from Lavoisier early in the 19th century were particularly active in this field. By 1841 Jean-Baptiste Dumas (1800—1884) and Jean-Baptiste Boussingault (1801—1887) were able to summarize their conclusions thus far in terms that: 'The vegetable kingdom constitutes an immense reduction apparatus'; whereas animals 'constitute, from the chemical point of view, combustion apparatuses' (Holmes, 1974, p. 23). In this, in association with the atmosphere with its constant proportions of nitrogen and oxygen, Dumas saw the cycle of organic life.

Shortly afterwards Dumas and Liebig were engaged in sharp controversy on the nature of the nutrients of the animal body and particularly the sources of animal fat. Liebig was involved in various controversies on matters of science. Scientists were outspoken in the mid-19th century and rivalry, particularly international rivalry, had developed. Liebig believed that animals transformed substances in their bodies and suggested that on chemical

Chapter 7 references, p. 150

grounds sugar or starch could be converted to fat in the animal body. Dumas, on the other hand, believed that fat in the body of the animal was derived from the fat in the food. Both supported their theories by interpretation of limited experimental evidence. It was at this time, in 1842, that Liebig published his book *Animal Chemistry* summarizing his chemical approach to digestion, nutrition and respiration. The debate continued, knowledge of organic chemistry was extended and experimental evidence accrued — it tended to support the views of Liebig rather than of Dumas. The debate was finally resolved by the careful experiments of Boussingault on his farm at Bechelbronn in Alsace; experiments of a type, incidentally, that Lyon Playfair (1818—1898) in England would have liked to have tried but could not afford. Boussingault experimented with cows, pigs, geese and ducks. He carefully determined and analyzed the food eaten, and excrements and milk produced. The animals and birds were weighed frequently and some were dissected to determine the amount of fat produced in the body. Eventually he had to conclude that the animals could produce more fat in their bodies than there had been fat in their feed (Holmes, 1974).

Boussingault worked with great care and precision and his combination of agricultural knowledge and practical experiment on his farm with analysis in his laboratory may be compared with the similar approach of Lawes and Gilbert at Rothamsted. The outcome of his work, however, did not have such immediately practical application and furthermore, unlike Lawes and Gilbert, he was seen to be the loser in his debate with Liebig. But for this the name of Bechelbronn might have become as well known as that of Rothamsted. During this intensive activity in the application of chemistry to animal nutrition in the eighteen-forties it was gradually realized that animal digestion was far more complex than previously supposed. As already appreciated far more by the physiologists, the search for understanding had to proceed to the internal functions of the animal body and this is the path that later was taken. In due course the animal became a part of the laboratory scene; an object of physiological and chemical functions quite removed from its former agricultural context. Pursuit of these laboratory studies was seen mainly in terms of the furtherance of medical science rather than of agricultural science. This is not the place to trace the subsequent unfolding of physiological and biochemical knowledge. It needs only to be remarked that it was a development that greatly aided a subsequent more detailed understanding of nutrition, digestion and respiration in their relation to livestock production.

The livestock farmer was not, however, totally ignored, for parallel with these developments in agricultural chemistry and physiology leading to the study of animal functions at the micro-level was another more practical development. The comparative nutritive value of different feedingstuffs for livestock had for long been of interest to farmers. The essential question was if a certain amount of milk, manure and meat was produced by an animal from a given quantity of one feedingstuff, how much of another feedingstuff was required to give the same level of production. Tyler (1975) quotes a study by Bergen published in 1781 in which three different groups of six pigs were fed on measured quantities of half-and-half crushed barley and crushed vetch, crushed barley and crushed vetch plus potatoes, or potatoes. From the gain in weight over 14 weeks it was concluded that about 2 bushels of potatoes were equivalent to about 1 bushel of barley and vetch for fattening pigs.

This type of experiment could be carried out readily in practical terms but financially it was very expensive because of the careful weighing and control of feedingstuffs and weighing of animals that was needed over an extended period. By the end of the 18th century various equivalences of feedingstuffs for livestock production had been published but it was not always clear which were the result of careful experiment, which were an experienced stockman's approximation, or, by the early nineteen-hundreds, which were the result of theoretical calculations based on chemical analyses. Works on animal nutrition have for a long time attributed the concept of hay equivalents of feedingstuffs to Albrecht Thaer (1725—1828) who published tables of weights of various feedingstuffs considered equivalent in nutritive value to a standard weight of hay. These figures were republished by various authors throughout the 19th century. But as Tyler (1975) has recently shown, a great deal of confusion has been introduced into the literature by inaccuracy of quotation and misunderstanding of the original work. He indicates that many of Thaer's hay equivalents were probably based on analyses carried out by his colleague Heinrich Einhof (d. 1808) in which nutritive material was separated from fibre by maceration and water extraction. Although Einhof used acid, alkali and alcohol in subsidiary extractions, Tyler (1975) emphasizes that extraction with these reagents almost certainly did not contribute to the evaluation of the hay equivalents published by Thaer despite the claims of later writers. Davy's work in assessing the nutritive value of individual species of grasses by determining the water-soluble matter in them would support this assertion (Davy, 1813).

The notion of hay equivalents was particularly attractive because of its practical application, and it thus remained of considerable interest to agricultural chemists and progressive farmers. It also provided a point of contact between the laboratory chemist and the practical farmer. Eventually in 1859 Wilhelm Henneberg (1825—1890) showed by experiment that the hay equivalents thus far published were erroneous in that different rations with very different hay equivalents were each capable of maintaining an ox. Henneberg and Friedrich Stohmann (1832—1897) at Weende in Germany then laid the basis for the technique of proximate analysis of feed and faeces which, with the addition of the Kjeldahl determination for nitrogen in 1883, has continued in use ever since and is only now being challenged by automatic analytical techniques and more refined concepts. In the technique of proximate analysis the material is analyzed into the components of moisture, crude protein, crude fibre, ether extract, nitrogen-free extract and ash using wet chemical methods and heating ovens. The virtue of the technique was that when tested against the results from digestibility trials with the animals it was shown to have predictive value. Henneberg and Stohmann carried out a number of digestibility trials and determined the maintenance requirements of an ox in terms of digestible nutrients, i.e., digestible protein, digestible fat and digestible carbohydrate. They began to make respiration experiments and this work was carried on by Gustav Kühn (1840—1892), a student of Henneberg, who went to the Royal Agricultural Experimental Station at Möckern as director. He died suddenly in 1892 and was succeeded as director by Oscar Kellner (1851—1911) who carried on the work and developed carbon and nitrogen balance studies, and eventually his system of starch equivalents of feedingstuffs based on their fat-producing powers relative to starch. This system was widely adopted in Europe and countries of the British Commonwealth (Tyler, 1959, 1975; McDonald et al., 1973).

Germany, in the second half of the 19th century, provided a favourable setting for the type of work just described. The post-Napoleonic period had witnessed a rejuvenation of the numerous German universities which was accompanied by a rapid growth of research in the natural sciences and the birth of the university research school, particularly in chemistry. Liebig established the pattern in his analytical laboratory at Giessen and it was

Chapter 7 references, p. 150

quickly duplicated in other German universities during the eighteen-fifties. For the remainder of the century Germany was an unrivalled centre for training in chemistry (Farrar, 1975).

In such an environment chemical analysis was promoted, and young men trained in analytical techniques were leaving the universities anxious to use their new knowledge to earn a living. Unlike the situation in Great Britain, the leading farmers in the German states became involved in the establishment of experimental stations. True, much of the work at these stations was routine analysis of soils, feedingstuffs and fertilizers, the same type of work that consulting chemists were performing in Great Britain and other countries, but there was also the added dimension of at least some experimental work in a number of instances. For example, the Animal Physiological Experimental Station at Weende where Henneberg did his work was established in 1852 by the Royal Agricultural Society of Hanover; the Royal Agricultural Experimental Station at Möckern, Saxony, where Kühn and then Kellner were directors, was founded in 1851 by the Economic Society of Leipzig upon its estate in Möckern and brought under the control of the State of Saxony in 1879. The Empire of Germany with its confederation of states, each with its own separate government, did not encourage the growth of centralized institutions. Instead, organizations were state-based and within the states it was organizations of farmers and other patrons who established and provided for the experimental stations. By the beginning of the 20th century the experimental stations still received more than two-thirds of their funds from these sources although by then most also received subsidies from their respective governments. Management of the stations took a variety of forms depending on origin and subsequent fortunes. In 1888, at Weimar, the Association of Agricultural Experimental Stations in the German Empire was formed and thereafter met annually to consider matters of mutual interest concerning analysis and questions of policy (True and Crosby, 1904).

The influence of the German university teaching and research departments, and experimental stations was far reaching. The Japanese, for example, after the Meiji restoration in 1868, had at first looked to England as a source of guidance in advanced agriculture, but with the spread of the influence of Liebig's work and a closer affinity with German farming systems, the Japanese invited German agricultural scientists, including Oscar Kellner, to take up posts in Japan and help and guide them in developing their own experimental work. Kellner returned from an appointment in Tokyo to take up the directorship at Möckern. The Japanese Government also looked to the United States for agricultural guidance and the Americans were influential in developing the Sapporo Agricultural College in the northern island of Hokkaido and in the general development of agriculture in that more sparsely settled, frontier land of Japan (Japan FAO Association, 1959).

The development of agricultural experimentation in the United States was largely a reflection of developments in this area in England during the 18th and early part of the 19th centuries and in Germany in the later part of the 19th century. Young Americans embarking on a career in chemistry went to Germany to seek training and returned to establish the German-type analytical teaching and research laboratory. From this basis a distinctive American form of agricultural research gradually emerged. Federal support for agricultural colleges became a reality with the passing of the Morrill Act in 1862. In the same year the United States Department of Agriculture was established. A powerful lobby had emerged that believed that through education in the new science of agriculture, farming could be improved, and the principle of federal aid for agricultural advancement was thus established (Gates,

1962; Rossiter, 1976). Aid for agricultural experimentation was also a recurring request in various states but it was not satisfied until 1875 when the State of Connecticut established an agricultural experiment station, the culmination of much advocacy by German-trained scientists among the farming community over many years, an advocacy that stressed the application and usefulness of scientific knowledge to the farmer. An advocacy which, however, at times ran into difficulties when confidence outstripped science and farmers were invited to take a too simplistic view of science and were then disappointed when the results from the test tube did not always produce the expected profits. A delicate balance was required. Potential patrons had to be persuaded and yet the limitations and uncertainties of science had to be allowed for in persuading people who tended to want straight answers and had little time for subtleties and uncertainties.

As with the experimental stations in Germany and elsewhere the new stations in the United States had to meet the needs of their direct or indirect sponsors, the local farmers, and a great deal of time was spent in carrying out routine analyses of fertilizers, feedingstuffs and soils. Support for unspecified 'research' was still a novel concept. By the eighteen-eighties a new generation of locally-trained scientists was being appointed to staff the growing number of experimental stations and as their numbers grew, specialization increased and professional associations were established. Pressure for support for the means to carry out research intensified and in 1887 Congress gave its support and the Hatch Act was passed which provided for a grant of $15 000/year to each state for the support of an experiment station and set up the Office of Experiment Stations within the United States Department of Agriculture. Subsequently in 1906 the Adams Act increased the appropriation to $30 000/year (Rosenberg, 1976; Rossiter, 1976).

Such funding was liberal indeed in relation to other countries and also opportune for agricultural science in a context reaching far beyond the shores of the United States. Agricultural chemistry, particularly German agricultural chemistry, had created an analytical science of demonstrable use to the farmer. Chemical analysis of soils, feedingstuffs and fertilizers had become routine and mundane, particularly to test for the adulteration of feedingstuffs and fertilizers. Agricultural experimental stations had been established on the basis of this demonstrated usefulness of science, but by the eighteen-eighties a new lead and a new impetus was needed if agricultural science was to go beyond routine analysis. On the world scene it was the Americans who were able to provide this lead. Not solely, the nutrition work in Germany has already been noted, bacteriological work by Louis Pasteur (1822—1895) in France was a critical forerunner, and vitamin work by F. Gowland Hopkins (1861—1947) in England, and others elsewhere, was a parallel development. But the sheer resources of the American research system at the critical time in research development ensured that from the eighteen-eighties to the nineteen-twenties momentum in agricultural research on the world scale was maintained until other countries, too, under the influence of the American example, contributed state funds to agricultural research.

Three developments were particularly significant in the promotion of animal research in America during this period — the Babcock test for butterfat, the work on protein chemistry leading to the discovery of vitamin A at the Wisconsin and Connecticut Agricultural Experiment Stations, and the work of Henry P. Armsby (1853—1921) in nutrition studies similar to those of Kellner. The work of Stephen M. Babcock (1843—1931) and his colleagues in dairy chemistry and bacteriology, at the University of Wisconsin Agricultural Experiment Station, was part of a deliberate programme to promote dairying in the state and yielded practical results that could be shown to be relevant to the farmers' needs. With a test for butterfat available the value of a milk sample for butter production could be measured and a fair basis for payment obtained. Babcock began his work at Wisconsin in 1887 (Rosenberg, 1976).

In 1883 in Denmark the government provided the equivalent of U.S. $33 000 to establish an Agricultural Experimental Laboratory at the Royal Veterinary and Agricultural College at Copenhagen. Comparative feeding experiments with milking cows were begun and work on tuberculosis in dairy cows and the development of dairying equipment and machinery were added (True and Crosby, 1904). The knowledge gained from this type of work in Denmark, America and other countries enabled a quite rapid change from a cottage industry to a farm-production, factory-processing industry to occur in dairying during the decades just before and after the turn of the century. This success ensured an expanding role for research in dairy science in various countries.

By the end of the 19th century the American agricultural experiment station scientists were engaged in routine analyses and regulation work, field demonstrations and extension activities, lecturing at the associated land grant colleges, and research. The more gifted of the experiment station directors had developed their institutions by assiduously building up contacts with the leading farmers and sympathetic politicians in their respective states. In this way they had gained political support for their activities and additional financial support. The costs of these achievements they also recognized: the ever increasing demands for advice and extension activities which left all too little time for the investigations and research needed to back up the advice; the increased involvement of the paymasters in the affairs of the station, with their desire for immediate results and lack of interest in, or support for, more basic long-term investigations; the discouragement of rigorously trained scientists of the German tradition when they found themselves left with minimal time, and few resources, to devote to research when all the other demands on their time had been met. It required leadership ability of a high order and sound judgment of men, as well as scientific understanding, to keep these conflicting forces in balance (Rosenberg, 1976).

The early years of the new century found the agricultural experiment stations, as institutions, well established. The pioneering phase was over. At the same time the agricultural scientists in the stations were increasing in number and devolving their interests more and more into different specializations. A new generation was replacing the founding scientists, a generation more confident in its role and yet firmly attached to the academic and increasingly professional ethos of science — an attitude now fostered by the growing opportunities for professional employment outside the sphere of the experiment stations and land grant colleges. For the continued prosperity of their experiment stations the scientists were dependent on the good will of the farmers and politicians; yet for their own personal advancement they were dependent on the judgment of their scientific peers within the context of research publication and scientific professionalism.

The scientists were growing more skilful in their lobbying activities for the promotion of research, and with the passing of the Adams Act in 1906 the federal appropriation for the experiment stations was doubled, with the additional funds being designated for original scientific research only. The old dichotomy had perhaps become more even, certainly it was more firmly established. The scientists had gained in their opportunity to undertake research. The farmers, on the other hand, were adamant that an experiment station or land grant college should carry out the practical experiments that they could not afford and they expected to find on it better crops and better livestock than they themselves produced, as testimony to the superior skills and knowledge derived from the state-supported learning and experimentation. Indeed, such were the pressures in these directions that great effort was expended at stations and colleges in preparing animals for competition at local shows. Much was made of the publicity value of success in these endeavours until eventually there were moves to exclude the college professors from competing at local shows following a run of successes (Rosenberg, 1976, 1977; Rossiter, 1979).

5. THE TWENTIETH CENTURY

Three years after the United States Government had voted extra funds to support agricultural research, the United Kingdom Parliament approved its first substantial vote destined for the same purpose, under the terms of the Development and Road Improvement Funds Acts of 1909 and 1910. The aim was to promote the economic development of the United Kingdom, and £2.9 million was set aside for this purpose to be spent over 5 years. It was a bold and unexpected move by the Chancellor, Mr. Lloyd George, in what proved to be an exceedingly turbulent parliamentary session. It was a move in response to the economic circumstances of the time rather than as a result of specific advocacy by farming or scientific interests. There was a general awareness in the community of the superiority of the Germans in science and technology and of the superiority of German and American universities in these fields. The need for state support for scientific research had been raised from time to time in organizations such as the British Association for the Advancement of Science, but with little result. In 1905 a pressure group, the British Science Guild, was formed by the President of the British Association to promote state support for science but its formation reflected more the lack of real zeal for such developments within the larger scientific body than any substantial pressure for government aid to scientific research. Many held that the freedom necessary for good research would be lost if research was done at the behest of government paymasters. It was only the advent of war in 1914 that revealed how far behind the United Kingdom was in the development of science and its application to industry (Poole and Andrews, 1972).

The Development Fund was intended to promote economic development by aiding the development of agriculture, including the organization of co-operation, forestry and fisheries. The fund was to be administered by eight commissioners, all but one of whom were part-time. They did not have executive powers and had to work through other bodies. As the activities of the Commission evolved, the main emphasis became centred on agricultural education and research as being the most promising fields for improvement. Only one of the Commissioners, Alfred Daniel Hall (1864–1942), was experienced in agricultural research and it was natural that he should take a leading role in the consideration of agricultural matters and particularly the devising of a scheme for agricultural research. Hall had succeeded J.B. Lawes and J.H. Gilbert (who had briefly taken over from Lawes after his death) as Director of the privately endowed Rothamsted Experimental Station in 1902 (Dale, 1956).

The Commission began planning its future activities in 1910 and decided on three initial lines of action: the development of agricultural research and the promotion of agricultural knowledge amongst farmers; investigation of

Chapter 7 references, p. 150

the economic possibilities of establishing crops such as flax, hemp, tobacco and sugar beet; and the encouragement of the organization of agricultural co-operation. When the Commissioners came to consider agricultural research Hall proposed the establishment of a series of research institutes. Each institute, he suggested, should be linked to a university or college and should specialize in a particular branch of agricultural science. This basic concept was accepted. Thus, firstly, as in the United States, the agricultural research institutes were to be funded by the state but were to be associated with seats of learning; carrying out the research was not seen as a function appropriate to the state. Secondly, and in contrast to the United States, agricultural science was to be subdivided, for the purposes of research, largely according to its disciplinary science basis rather than on any production or regional basis — although dairying and fruit growing already posed a challenge to this scheme of things. In the proposal submitted to the Treasury in 1911 agricultural science was divided into the following 11 subjects: plant physiology, plant pathology, plant breeding, fruit growing (including the practical treatment of plant diseases), plant nutrition and soil problems, animal nutrition, animal breeding, animal pathology, dairying, agricultural zoology, and the economics of agriculture. These subject divisions have provided the framework for the system of agricultural research institutes in the United Kingdom ever since.

The Development Commission began establishing institutes according to this scheme as quickly as possible, making use of existing people and institutions with appropriate training and interests as points of expansion. The Animal Nutrition Research Institute was established at Cambridge University in 1911, the National Institute for Research in Dairying at University College, Reading, in 1912, and funding was provided for research in animal pathology at the Royal Veterinary College and the Veterinary Laboratory. A total sum of £ 30 000/annum was allocated for the support of all the proposed research institutes combined. In addition two small funds were established, one to provide special grants to aid meritorious individual workers at universities and elsewhere outside the institutes, the other to provide for scholarships to train scientists for work in the institutes (Dale, 1931, 1956; Russell, 1966; Whetham, 1978).

The outbreak of war in 1914 brought a temporary halt to the establishment of new institutes and the expansion of those already in existence. In the post-war period creation of new institutes was resumed but it was not an easy period. Funds were short and were soon coming from various sources but only in small amounts and often for specific projects. Qualified staff in both the universities and the associated institutes were few and, despite the efforts of the Development Commissioners, extension activities were limited. Farmers, meanwhile, were increasingly seeking analyses of soils, fertilizers and feedingstuffs and turning to some of the new institutions and university departments to provide them. Institute directors for their part had to seek for funds wherever they could to augment the state allocation. Economic uncertainty, recovery from war and some pressure from farmers dictated a gradual and uncertain progress.

During the nineteen-twenties some badly needed funds and impetus came from two unexpected sources, the £1 million fund from the Corn Production Acts (Repeal) Act of 1921 and the Empire Marketing Board of 1926–1933. During the war the government had promoted the ploughing up of grassland for corn production to replace the loss of grain imports. This policy had been supported by offering a guaranteed price to farmers for wheat and oats. The policy was continued by a further Act in 1920 when prices were particularly high; by 1921 it had become obvious, as prices for grain fell steeply, that the government might find itself liable to pay the farmers up to £30 million in support of the coming harvest. It therefore quickly repealed the 1920 Act but, as part of the deal to pacify the farmers, and at the insistence of the National Farmers' Union, a sum of £1 million was provided for the benefit of agricultural education and research. Gratifyingly for the research institutes, agricultural colleges and university departments, the farmers had come to champion their cause (Dale, 1956; Whetham, 1978).

The Empire Marketing Board was similarly created as a result of the government going back on its commitments. In 1923 the government had given the Dominions a pledge concerning preferential treatment for Empire produce on the British market. The political balance had then shifted towards free trade and in order to honour the pledge while pursuing a free-trade policy it was decided to set aside £1 million/year to promote the marketing of Empire produce in the United Kingdom. The Empire Marketing Board was established to carry out this function. It interpreted its remit in the widest possible terms and developed not only a major promotion campaign for Empire products on the United Kingdom market but embarked on the support of agricultural research so that these products might be more readily produced. This support was extended to research institutes throughout the Empire and included the appointment, in 1931, of a plant-breeding expert for the study of Empire grassland problems in association with the Welsh Plant Breeding Research Station, Aberystwyth. A senior member of the station, William Davies (1899–1968) was appointed after his return from a period spent on loan to New Zealand. The advent of the economic depression of the early nineteen-thirties soon brought an end to government financial support for the Empire Marketing Board and it was abolished. It had, however, played an important part in helping fledgling research institutes through a difficult period and its reports convey a picture of important research being carried out enthusiastically with limited means in various parts of the British Empire (Huxley, 1970; Barnes and Nicholson, 1980).

During these formative decades for research institutes and the university departments of agriculture, up to the second world war, much still depended on the leadership abilities of a few people, just as it had done during the early years of the experiment stations in the United States. In the United Kingdom, in animal and pasture science, the activities of John Hammond Snr (1889–1964) at the University of Cambridge in developing his work on artificial insemination and the growth and physiology of farm livestock; of John Boyd Orr (1880–1971) as director of the Rowett Research Institute at Aberdeen, which became a leading centre for studies in nutrition; of Reginald George Stapledon (1882–1960) director of the Welsh Plant Breeding Station, Aberystwyth, which he developed into a major centre for grassland research; and of Robert Stenhouse Williams (1871–1932) who established the National Institute for Research in Dairying at Reading, were particularly important. Their influence was felt throughout the British Empire (Russell, 1966).

As previously remarked, it was the advent of war in 1914 which highlighted the paucity of scientific research for the benefit of manufacturing and chemical industry in the United Kingdom. In response to the needs of wartime a Committee of the Privy Council for Scientific and Industrial Research was formed in 1915 and subsequently, in 1916, the Department of Scientific and Industrial Research was established. On the other side of the world in Australia there had been a number of limited moves to promote the idea of nationally funded scientific research but until 1915 these had not met with any success. There was little pressure from scientists themselves apart from the activities of a branch of the British Science Guild. In agriculture, the state departments of agriculture were slowly expanding and providing a limited investigation and advisory service to the farmers. In 1915, however, a copy of the British Government White Paper (Cd 8005) describing the scheme to establish the Committee for Scientific and Industrial Research reached Australia and proved there to be the catalyst that provoked a similar development in Australia. In 1916 the Commonwealth Government set up an Advisory Council of Science and Industry which was required to initiate research through existing institutions pending the establishment of an institute; in 1920 the Institute of Science and Industry was established, and in 1926 this became the Commonwealth Council for Scientific and Industrial Research.

The arguments put forward in favour of establishing what became the CSIR were similar to those used in advocating the establishment of the DSIR in the United Kingdom. In particular, the example of the advancement of German industry through the application of science was cited, but also there was a rapidly growing belief in the power of science to further general economic development and a desire to be part of a wider British Empire development of the application of science to industry. Other Dominions of the British Empire were similarly influenced. In the United Kingdom provision had already been made for research in agriculture, and the DSIR was established to further research in industry. In Australia a comparable agricultural research system did not exist and, furthermore, the urban industrial base was comparatively small. Industry meant largely, in terms of national output, rural industry and mining, and in Australian terminology rural industry meant predominantly the cropping and pastoral industries. Thus through this chain of circumstances a major organization was established. In 1930, having considered the limited resources available to it, the CSIR decided to concentrate on research for the rural industries and to this end established Divisions of Animal Nutrition, Animal Health, Economic Entomology, Plant Industry, Soil Science and Forest Products. A similar development occurred also in New Zealand (Currie and Graham, 1966; Peel, 1973).

In the nineteen-twenties the transformation of the British Empire into the British Commonwealth was proceeding; nevertheless a very strong feeling for the bond of Empire persisted and this found expression in many ways, including through the organization of agricultural research. In 1927 the first Imperial Agricultural Research Conference was held in London. An Imperial Bureau of Entomology and Mycology had already been established and the conference led to the establishment of eight new Imperial Bureaux (now Commonwealth Agricultural Bureaux) for the interchange of information among research workers throughout the Empire. The Bureaux were established in association with United Kingdom agricultural research institutes and reflected their division of the subject matter of agricultural science. Meanwhile the Empire Marketing Board, through its various activities, was further strengthening and promoting the Empire concept of agricultural research (Dale, 1931).

By 1931, that scientific research was a matter for state support, was no longer questioned in the United Kingdom and the Agricultural Research Council was established on the pattern of the Medical Research Council and the Advisory Council of the Department of Scientific and Industrial Research. It was made responsible to the Privy Council for promoting and co-ordinating agricultural research in Great Britain. A third of its members were appointed for their experience in agriculture and the remainder for their qualifications in one of the basic sciences. With the establishment of the ARC the general structure of the agricultural research system was confirmed. The period of the nineteen-thirties was not favourable for the development of research because of economic depression and then the onset of war, but with the end of war in 1945 a framework for the rapid expansion of research was available (Political and Economic Planning, 1938).

The structure of agricultural research in the United Kingdom provided institutions which, when adequately funded by the state, gave singular protection from direct pressures from the farmers and biased the weight of influence in favour of the agricultural scientists and the pure scientists.

The establishment of the ARC in the image of the other research councils, and the predominance of basic scientists on it, reinforced this bias. Indeed, so favourable was the organizational structure to science it probably accounts for the lack of development of a strong corporate identity among agricultural scientists as such in the United Kingdom as compared with other scientists. Throughout the life of the research institutes there have been notable instances of close contact with the farming community, but the institutes were not dependent on the approval and backing of the farmers for their prosperity as their predecessors had been in Germany and the United States. Scientists were not forced to learn about the ways of the farmer. As Whetham (1978, p. 280) has commented, on the period to 1939, 'the reports of the ARC leave the impression that there remained a gap between the work of the scientists and the practice of ordinary farmers, a gap created partly by the lack of finance for the advisory services, and partly by the lack of accurate knowledge on many aspects of agriculture among the scientists'. If the pendulum had swung too far in one direction in the United States it had possibly swung too far in the opposite direction in Great Britain.

The post-war years after 1945 witnessed a gradual recovery of research activity in animal science as nations emerged from the aftermath of war in the late nineteen-forties and early nineteen-fifties. This was followed by a period of very rapid expansion in the late nineteen-fifties and the nineteen-sixties when there was enormous confidence in the power of agricultural science to overcome obstacles and point a rather simple and, above all, logical way forward. Industry became very active in research as it developed new chemicals for pest and disease control, bred special lines of poultry for intensive production, and formulated feedingstuffs for bulk manufacture. Gradually the confidence was shattered, particularly amongst non-scientists, as unwelcome effects from the excessive use of pesticides and other chemicals emerged. Science, too, felt the repercussions of the world-wide wave of political and moral liberalization which swept through many countries in the late nineteen-sixties and early nineteen-seventies. Cries for accountability were heard in many spheres, research in animal science, and in the agricultural sciences in general, was not immune. In Great Britain, the United States, Australia and elsewhere, these political forces led to enactments which curtailed the freedom of agricultural scientists to pursue their research in the directions they themselves chose, and attempted instead to establish means to direct their activities toward what might be more generally perceived as the needs of the farming community. As formerly high-spending national governments were forced by changing economic circumstances to reduce their spending, agricultural research became one of many activities which was left less well funded than previously. The long-term effects of these wider political and economic changes on research in animal science still remain to be seen (Passmore, 1974; Mayer and Mayer, 1976).

Within the subject matter of research there has also been unease. The more important gains from the application of research findings in practice in the immediate post-war years owed much to research carried out immediately before the war. Since then, as research expanded, it tended to be directed toward the smaller and smaller — from the animal to the tissue to the cell to

Chapter 7 references, p. 150

the subcellular level. The organizational structure of research in some countries, as has been discussed, directed effort within one institute towards breeding, at another towards nutrition. It was difficult to arrange that an animal be treated as a whole in all its aspects, even more difficult to bring together studies on all the aspects of plants and animals interacting together as on a farm. To meet this challenge agricultural systems studies have emerged but the established organizational framework for research frequently does not assist this approach. Yet in the poorer countries and less developed parts of the world, as highly financed livestock development projects are tried and found to fail, not just the animal—plant factors are implicated but the much broader factors of animal, plant, environment, man and his politics, economics and society. The design of agricultural research systems in these countries must surely be an open question.

In agricultural research there now exists an essential tension. When scientist and farmer were the one man, when Boussingault or Lawes could treat as one the affairs of their fields and the affairs of their laboratories, science and agriculture were truly brought together. But as science became professionalized and increasingly answerable to governments, and as basic science was accorded prestige over applied science, the scientist and the farmer were drawn apart. The opportunity for an amalgam of knowledge decreased. Yet the challenge of the sheer complexity of the science of animal production in its widest context is immense. What will the imaginative approach be that will solve this tension, meet this challenge and carry with it the support of community and statesman?

6. REFERENCES

Barnes, J. and Nicholson, D., 1980. The Leo Amery Diaries, Vol 1: 1896—1929. Hutchinson, London, 653 pp.

Berman, M., 1978. Social Change and Scientific Organization, The Royal Institution, 1799—1844. Heinemann Educational Books, London, 224 pp.

Currie, G. and Graham, J., 1966. The Origins of CSIRO. Science and the Commonwealth Government 1901—1926. Commonwealth Scientific and Industrial Research Organization, Melbourne, 203 pp.

Dale, H.E., 1931. Agricultural research and education. In: H. Hunter (Editor), Baillière's Encyclopaedia of Scientific Agriculture, Vol. 1. Baillière, Tindall and Cox, London, pp. 4—14.

Dale, H.E., 1956. Daniel Hall, Pioneer in Scientific Agriculture. John Murray, London, 241 pp.

Davy, H., 1813. Elements of Agricultural Chemistry, in a Course of Lectures for the Board of Agriculture. Longman, Hurst, Rees, Orme and Brown, London, Appendix, pp. 1xi—1xiii.

Farrar, W.V., 1975. Science and the German university system, 1790—1850. In: M. Crosland (Editor), The Emergence of Science in Western Europe. Macmillan, London, pp. 179—192.

Gates, W., 1962. The Morrill Act and early agricultural science. Mich. Hist., 46: 289—302.

Gazley, J.G., 1973. The Life of Arthur Young 1741—1820. American Philosophical Society, Philadelphia, 727 pp.

Hall, A.R., 1962. The Scientific Revolution 1500—1800, 2nd edn. Longmans, London, 394 pp.

Hartley, H., 1966. Humphry Davy. Nelson, London, 160 pp.

Holmes, F.L., 1974. Claude Bernard and Animal Chemistry, the Emergence of a Scientist. Harvard University Press, Cambridge, MA, 541 pp.

Huxley, G., 1970. Both Hands, An Autobiography. Chatto and Windus, London, 254 pp.

Japan FAO Association, 1959. A Century of Technical Development in Japanese Agriculture. Japan FAO Association, Tokyo, 128 pp.

Lennard, R., 1932. English agriculture under Charles II: the evidence of the Royal Society's 'enquiries'. Econ. Hist. Rev., 4: 23—45.

Mayer, A. and Mayer, J., 1976. Agriculture, the island empire. In: G. Holton and W.

Blampied (Editors), Science and Its Public: The Changing Relationship. Boston Studies in the Philosophy of Science, Vol. xxxiii. D. Reidel Publishing Company, Dordrecht, pp. 83—95.

McDonald, P., Edwards R.A. and Greenhalgh, J.F.D., 1973. Animal Nutrition, 2nd edn. Longman, London, 479 pp.

McKie, D., 1952. Antoine Lavoisier, Scientist, Economist, Social Reformer. Constable, London, 335 pp.

Mingay, G.E. (Editor), 1975. Arthur Young and His Times. Macmillan, London, 264 pp.

Mitchison, R., 1962. Agricultural Sir John. The Life of Sir John Sinclair of Ulbster 1754—1835. Geoffrey Bles, London, 291 pp.

Passmore, J., 1974. Man's Responsibility for Nature. Duckworth, London, 213 pp.

Peel, L.J., 1973. History of the Australian pastoral industries to 1960. In: G. Alexander and O.B. Williams (Editors), The Pastoral Industries of Australia, Practice and Technology of Sheep and Cattle Production. Sydney University Press, Sydney, pp. 41—75.

Peel, L.J., 1976. Practice with science: the first twenty years. J. R. Agric. Soc. of Engl., 137: 9—18.

Political and Economic Planning, 1938. Report on Agricultural Research in Great Britain. PEP, London, 146 pp.

Poole, J.B. and Andrews, K. (Editors), 1972. The Government of Science in Britain. Weidenfeld and Nicolson, London, 358 pp.

Purver, M., 1967. The Royal Society: Concept and Creation. Routledge and Kegan Paul, London, 246 pp.

Ramsay, A., 1879. History of the Highland and Agricultural Society of Scotland with Notices of Anterior Societies for the Promotion of Agriculture in Scotland. William Blackwood and Sons, Edinburgh, 592 pp.

Rosenberg, C.E., 1976. No Other Gods. On Science and American Social Thought. The Johns Hopkins University Press, Baltimore, 273 pp.

Rosenberg, C.E., 1977. Rationalization and reality in the shaping of American agricultural research, 1875—1914. Soc. Stud. Sci., 7: 401—422.

Rossiter, M.W., 1975. The Emergence of Agricultural Science, Justus Liebig and the Americans, 1840—1880. Yale University Press, New Haven, 275 pp.

Rossiter, M.W., 1976. The organization of agricultural improvement in the United States, 1785—1865. In: A. Oleson and S.C. Brown (Editors), The Pursuit of Knowledge in the Early American Republic. The Johns Hopkins University Press, Baltimore, pp. 279—298.

Rossiter, M.W., 1979. The organization of the agricultural sciences. In: A. Oleson and J. Voss (Editors), The Organization of Knowledge in Modern America, 1860—1920. The Johns Hopkins University Press, Baltimore, pp. 211—248.

Russell, E.J., 1966. A History of Agricultural Science in Great Britain 1620—1954. George Allen and Unwin Ltd., London, 493 pp.

Slicher van Bath, B.H., 1963. The Agrarian History of Western Europe A.D. 500—1850. Edward Arnold (Publishers) Ltd., London, 364 pp.

Smeaton, W.A., 1956. Lavoisier's membership of the Société Royale d'Agriculture and the Comité d'Agriculture. Ann. Sci., 12: 267—277.

Smeaton, W.A., 1962. Fourcroy, Chemist and Revolutionary 1755—1809. W. Heffer and Sons Ltd. for W.A. Smeaton, London, 288 pp.

Smith, J.H., 1962. The Gordon's Mill Farming Club 1758—1764. Oliver and Boyd, Edinburgh, 156 pp.

Trow-Smith, R., 1957. A History of British Livestock Husbandry to 1700. Routledge and Kegan Paul, London, 286 pp.

Trow-Smith, R., 1959. A History of British Livestock Husbandry 1700—1900. Routledge and Kegan Paul, London, 351 pp.

True, A.C. and Crosby, D.J., 1904. Agricultural Experiment Stations in Foreign Countries, revised edn. Government Printing Office, Washington, 276 pp. U.S. Department of Agriculture Office of Experiment Stations Bulletin No. 112 (Revised).

Tyler, C., 1959. The historical development of feeding standards. In: D.P. Cuthbertson (Editor), Scientific Principles of Feeding Farm Livestock. Farmer and Stock-Breeder Publications Ltd., London, pp. 8—18.

Tyler, C., 1975. Albrecht Thaer's hay equivalents: fact or fiction? Nutr. Abstr. Rev., 45: 1—11.

Webster, C., 1975. The Great Instauration: Science, Medicine and Reform 1626—1660. Duckworth, London, 630 pp.

Whetham, E.H., 1978. The Agrarian History of England and Wales, Vol. VIII 1914—39. Cambridge University Press, Cambridge, 353 pp.

Whitehead, A.N., 1975. Science and the Modern World. Collins, Fontana Books, Glasgow, 252 pp. (first published 1926).

Chapter 8

The Ethics of Animal Use

PETER SINGER

1. INTRODUCTION

Human beings use non-human animals in many ways. The animal sciences are built upon the use of animals. They use animals as the subjects of their experiments. Some of the major branches of the animal sciences have as their principal aim the more efficient utilization of non-human animals for human purposes, particularly for the purpose of providing food for humans. So widespread is the use of non-human animals, and so long is its history, that we seldom question its ethics. The fact that a practice is ancient and widespread is, however, no guarantee that it is ethically sound — slavery, for example, is probably almost as ancient and was once almost as widespread — and there are serious doubts that can be raised about the ethics of the uses we make of non-human animals.

In this chapter some of the ways in which humans have, in the past, regarded animals and sought to justify using them are outlined. (Emphasis is placed on the historical tradition of the West, since that has now become the dominant influence on attitudes to animals throughout the world.) Some recent and radical challenges to the ancient assumption that we have the right to use animals in the ways in which we do, and on which many of the animal sciences are based, are then examined.

2. ANIMALS FOR HUMAN USE

Karl Marx was not the first to draw attention to the influence that the production of food has on the production of ideas. In searching for food, humans develop categories and classifications. For instance, they learn to classify things into edibles and inedibles. Although enthusiastic vegetarians sometimes try to argue that humans are natural herbivores, most of the evidence suggests that our ancestors killed and ate other animals on occasion, even if their diet was largely vegetarian. Thus other animals fell into the category of edibles. No doubt, too, when the weather grew cold, early humans found that the furs of the animals they killed would keep them warm. So an attitude to other species was there from the very beginning: they could be used for human purposes. They were means for avoiding hunger and cold rather than ends in themselves. The idea that other species are means to human ends has been and generally still is the foundation of human attitudes towards them.

Chapter 8 references, p. 164

2.1. Genesis

In describing these attitudes it is difficult to know where to begin, but the creation of the universe seems a fitting date. The biblical story of the creation may not be the earliest document of the Western tradition, but it is certainly one of the early ones, and it sets out very clearly the nature of the relationship between man and animal, as the Hebrew people at the time conceived it to be. It is a superb example of myth echoing reality:

> And God said, Let the earth bring forth the living creature after his kind, cattle and creeping thing, and beast of the earth after his kind: and it was so.
> And God made the beast of the earth after his kind, and cattle after their kind, and everything that creepeth upon the earth after his kind: and God saw that it was good.
> And God said, Let us make man in our image, after our likeness: and let them have dominion over the fish of the sea, and over the fowl of the air, and over the earth, and over every creeping thing that creepeth upon the earth.
> So God created man in his own image, in the image of God created he him; male and female created he them.
> And God blessed them, and God said unto them, Be fruitful, and multiply, and replenish the earth, and subdue it; and have dominion over the fish of the sea, and over the fowl of the air, and over every living thing that moveth upon the earth (*Genesis*, I, 24—28).

The Bible tells us that God made man in His own image. We may regard this as man making God in his own image. Either way, it allots man a special position in the universe, as a being that, alone of all living things, is God-like. One might think this enough to ensure human supremacy over the rest of creation; but in case man should ever wonder if his conduct towards the other creatures was justified merely because of his God-like nature, there were also God's own words to fall back on. God explicitly gave man dominion over every living thing. It is true that, in the Garden of Eden, this dominion may not have involved killing other animals for food, but certainly after the Flood, God is recorded as having explicitly said to Noah: 'Every moving thing that liveth shall be meat for you' (*Genesis*, IX, 1—3).

This gives the basic stance taken in the ancient Hebrew writings towards non-humans. It is true that the prophet Isaiah condemned animal sacrifices, and the book of Isaiah does contain a lovely vision of the time when the wolf will dwell with the lamb, the lion will eat straw like the ox, and 'they shall not hurt nor destroy in all my holy mountain'. This, however, is obviously a utopian vision, not a command for immediate adoption. There are other scattered passages in the Old Testament encouraging some degree of kindliness toward non-human animals, although some of these appear to be motivated by concern for animals as the property of humans, and as instruments for labour, rather than for their own sake. Most importantly, there is nothing to challenge the overall view, set down in *Genesis*, that man is the pinnacle of creation, and all the other creatures have been delivered into his hands.

2.2. Aristotle and Aquinas

The second of the two ancient traditions that make up Western thought has its origins in Greece. Although there were a number of conflicting schools of philosophy in Ancient Greece, it is the views of Aristotle which have had the greatest influence on Western thought. Aristotle regarded nature as a hierarchy, in which the function of the less rational and hence less perfect beings was to serve the more rational and more perfect. So, he wrote:

> ... Plants exist for the sake of animals, and brute beasts for the sake of man —
> domestic animals for his use and food, wild ones (or at any rate most of them) for
> food and other accessories of life, such as clothing and various tools.
>
> Since nature makes nothing purposeless or in vain, it is undeniably true that she
> has made all animals for the sake of man (Aristotle, *Politics*, 1256b).

Aristotle's ideas came to the West via Christianity. The early Christian writers were no more ready than Aristotle to give moral weight to the lives of non-human animals. When St. Paul, in interpreting the old Mosaic law against putting a muzzle on the ox that treads out the corn, asked: 'Doth God care for oxen?' it is clear that he was asking a rhetorical question, to which the answer was 'No'; the law must have somehow been meant 'altogether for our sakes' (Corinthians, IX, 9—10). Augustine agreed, using as evidence that there are no common rights between humans and lesser living things, the incidents in the gospels when Jesus sent devils into a herd of swine, causing them to hurl themselves into the sea, and with a curse withered a fig tree on which he had found no fruit (Augustine, *The Catholic and Manichean Ways of Life*, p. 102).

It was Thomas Aquinas who blended Aristotle and the Christian writings. Echoing Aristotle, he maintained that plants exist for the sake of animals, and animals for the sake of man. Sins can only be against God, one's human neighbours, or against oneself. Even charity does not extend to 'irrational creatures', for, among other things, they are not included in 'the fellowship of everlasting happiness'. We can love animals only 'if we regard them as the good things that we desire for others', that is, 'to God's honor and man's use.' Yet if this was the correct view, as Aquinas thought, there was one problem that needed explaining: why does the Old Testament have a few scattered injunctions against cruelty to animals, such as 'The just man regardeth the life of his beast, but the bowels of the wicked are cruel'? Aquinas did not overlook such passages, but he did deny that their intention was to spare animals pain. Instead, he wrote, 'it is evident that if a man practices a pitiable affection for animals, he is all the more disposed to take pity of his fellow men ...' So, for Aquinas, the only sound reason for avoiding cruelty to animals was that it could lead to cruelty to humans. (See Aquinas, *Summa Theologica*, I, II, Q 72, art. 4; II, I, Q 102 art. 6; II, II, Q 25 art. 3; II, II Q 64 art. 1; II, II, Q 159 art. 2; and *Summa Contra Gentiles* III, II, 112.)

The influence of Aquinas has been strong in the Roman Catholic Church. Not even that oft-quoted exception to the standard Christian view of nature, Francis of Assisi, really broke away from the orthodox theology of his co-religionists. Despite his legendary kindness to animals, Francis could still write: 'every creature proclaims: "God made me for your sake, O man!" [*St. Francis of Assisi, His Life and Writings as Recorded by His Contemporaries*, translated by L. Sherley-Price; see also Passmore (1974, p. 112)]. As late as the 19th century Pope Pius IX gave evidence of the continuing hold of the views of Paul, Augustine and Aquinas by refusing to allow a Society for the Prevention of Cruelty to Animals to be established in Rome because to do so would imply that humans have duties toward animals (Turner, 1964, p. 163).

2.3. Kant

It is not, however, only among Roman Catholics that a view like that of Aquinas has found adherents. Calvin, for instance, had no doubt that all of nature was created specifically for its usefulness to man (see Calvin, the *Institutes of Religion*, Book 1, chapters 14, 22; Vol. 1, p. 182 and elsewhere. This ref. is taken from Passmore, 1974, p. 13) and in the late 18th century

Chapter 8 references, p. 164

Immanuel Kant, in lecturing on ethics, considered the question of our duties to animals, and told his students:

> So far as animals are concerned, we have no direct duties. Animals are not self-conscious and are there merely as a means to an end. That end is man (Kant, *Lectures on Ethics*, pp. 239—240).

So Kant also believed that animals do not count, and we may do exactly as we please with them, provided our conduct has no bad consequences for humans.

Why could Kant not admit that the non-human animals are ends in themselves? One might try to explain this by reference to what is, for him, the ultimate ethical commandment, namely, to respect all persons as ends. By 'person' Kant means a rational, self-directing being; and it is the ability to think and act rationally that, according to Kant, gives a human being moral worth, and makes humans ends in themselves. All other aspects of human nature are of no moral significance for Kant. Virtues based on feelings, like sympathy, benevolence, pity, and compassion contribute nothing to a person's moral worth; they are feelings which one either happens to have, or happens not to have; they are not the outcome of one's free will (though precisely how it is possible to act freely remains mysterious in Kant's philosophy).

Kant's belief that there is something of unique moral value about a being that can act freely and rationally may partially explain his refusal to recognize non-rational animals as ends, but it is not the whole story. If it were, he would, in consistency, have excluded infants, lunatics, and the senile from what he called 'the kingdom of ends' as well; yet he clearly meant to include all human beings in this kingdom. Why, then, include non-rational beings in human form, but exclude non-rational (or more accurately, non-fully rational) beings in the form of dogs and chickens? Is this anything other than a prejudice in favour of the species to which Kant belongs?

In the period in which Kant was writing — the period known as 'The Enlightenment' — it is possible to find signs of the development of a different attitude to animals, one which today is probably more commonly voiced than the view of Aquinas or Kant. The Enlightenment was a period in which compassion came to be important. People recognized that other animals are capable of suffering, and that this entitles them to some consideration. This did not mean that animals had any rights; their interests were still to be sacrificed whenever they clashed with human interests. The Scottish philosopher David Hume expressed this attitude well when he said that we are 'bound by the laws of humanity to give gentle usage to these creatures' (Hume, 1777). 'Gentle usage' nicely sums up the attitude to animals which now prevails in countries like the United States and England. We are entitled, according to this view, to use animals, but we ought to do so gently. So we may kill a steer in order to eat it, but we should not torment it in the bull ring. Those who defend this kind of attitude within a Judeo-Christian frame-work assert that the dominion God gave us over the animals is a kind of stewardship, for which we are answerable to God, and not a license for tyranny and unlimited exploitation.

These ideas are certainly distinct from those of Aristotle, Aquinas or Kant, but there is another perspective on the ethical status of animals, from which the differences appear less striking than similarities.

3. ANIMALS AND HUMANS AS EQUALS

3.1. The case for equality

The belief that human beings are entitled to regard animals as having an inferior moral status, and as objects available for human use, has sporadically been challenged in the past. Pythagoras, Buddha, Plutarch, Porphry, Montaigne, Voltaire, Bentham, Schopenhauer, Lewis Gompertz and Henry Salt all dissented strongly from the attitudes which prevailed in their times and cultures, and which to a large extent still prevail in ours. In recent years this conventional attitude has once again been questioned, this time by a mixed group of academic philosophers and animal welfare activists who seek to extend such notions as equality and the possession of moral rights to non-human animals. Since this position is still a relatively unfamiliar one, it requires careful exposition if we are to make clear the basis for the claim that human and non-human animals are, in some sense, moral equals.

It will be helpful to begin with the more familiar claim that all human beings are equal. When we say that all human beings, whatever their race, creed or sex are equal, what is it that we are asserting? Those who wish to defend a hierarchical, inegalitarian society have often pointed out that by whatever test we choose, it simply is not true that all humans are equal. Like it or not, we must face the fact that humans come in different shapes and sizes; they come with differing moral capacities, differing intellectual abilities, differing amounts of benevolent feeling and sensitivity to the needs of others, differing abilities to communicate effectively, and different capacities to experience pleasure and pain. In short, if the demand for equality were based on the actual equality of all human beings, we would have to stop demanding equality. It would be an unjustifiable demand.

Fortunately the case for upholding the equality of human beings does not depend on equality of intelligence, moral capacity, physical strength, or any other matters of fact of this kind. Equality is a moral ideal, not a simple assertion of fact. There is no logically compelling reason for assuming that a factual difference in ability between two people justifies any difference in the amount of consideration we give to satisfying their needs and interests. The principle of the equality of human beings is not a description of an alleged actual equality among humans; it is a prescription of how we should treat humans.

Jeremy Bentham (1789) incorporated the essential basis of moral equality into his utilitarian system of ethics in the formula: 'Each to count for one and none for more than one.' In other words, the interests of every being affected by an action are to be taken into account and given the same weight as the like interests of any other being. A later utilitarian, Henry Sidgwick (1901), put the point this way: 'The good of any one individual is of no more importance, from the point of view (if I may say so) of the Universe, than the good of any other.' More recently, the leading figures in contemporary moral philosophy have shown a great deal of agreement in specifying as a fundamental pre-supposition of their moral theories some similar requirement which operates so as to give everyone's interests equal consideration — although they cannot agree on how this requirement is best formulated.

It is an implication of this principle of equality that our concern for others ought not to depend on what they are like, or what abilities they possess — although precisely what this concern requires us to do may vary according to the characteristics of those affected by what we do. It is on this basis that the case against racism and the case against sexism must both ultimately rest;

Chapter 8 references, p. 164

and it is in accordance with this principle that speciesism is also to be condemned. If possessing a higher degree of intelligence does not entitle one human to use another for his own ends, how can it entitle humans to exploit non-humans?

Many philosophers have proposed the principle of equal consideration of interests, in some form or another, as a basic moral principle; but not many of them have recognized that this principle applies to members of other species as well as to our own. Bentham was one of the few who did realize this. In a forward-looking passage, written at a time when black slaves in the British dominions were still being treated much as we now treat non-human animals, Bentham wrote:

> The day *may* come when the rest of the animal creation may acquire those rights which never could have been withholden from them but by the hand of tyranny. The French have already discovered that the blackness of the skin is no reason why a human being should be abandoned without redress to the caprice of a tormentor. It may one day come to be recognized that the number of the legs, the villosity of the skin, or the termination of the *os sacrum*, are reasons equally insufficient for abandoning a sensitive being to the same fate. What else is it that should trace the insuperable line? Is it the faculty of reason, or perhaps the faculty of discourse? But a full-grown horse or dog is beyond comparison a more rational, as well as a more conversable animal, than an infant of a day, or a week, or even a month, old. But suppose they were otherwise, what would it avail? The question is not, Can they reason? nor Can they *talk*? but, *Can they suffer*? (Bentham, 1789).

In this passage Bentham points to the capacity for suffering as the vital characteristic that gives a being the right to equal consideration. The capacity for suffering — or more strictly, for suffering and/or enjoyment or happiness — is not just another characteristic like the capacity for language, or for higher mathematics. Bentham is not saying that those who try to mark 'the insuperable line' that determines whether the interests of a being should be considered happen to have selected the wrong characteristic. The capacity for suffering and enjoying things is a pre-requisite for having interests at all, a condition that must be satisfied before we can speak of interests in any meaningful way. It would be nonsense to say that it was not in the interests of a stone to be kicked along the road by a schoolboy. A stone does not have interests because it cannot suffer. Nothing that we can do to it could possibly make any difference to its welfare. A mouse, on the other hand, does have an interest in not being tormented, because it will suffer if it is.

If a being suffers, there can be no moral justification for refusing to take that suffering into consideration. No matter what the nature of the being, the principle of equality requires that its suffering be counted equally with the like suffering — insofar as rough comparisons can be made — of any other being. If a being is not capable of suffering, or of experiencing enjoyment or happiness, there is nothing to be taken into account. This is why the limit of sentience (using the term as a convenient, if not strictly accurate, shorthand for the capacity to suffer or experience enjoyment or happiness) is the only defensible boundary of concern for the interests of others. To mark this boundary by some characteristic like intelligence or rationality would be to mark it in an arbitrary way. Why not choose some other characteristic, like skin colour?

The racist violates the principle of equality by giving greater weight to the interests of members of his own race, when there is a clash between their interests and the interests of those of another race. Similarly the 'speciesist' allows the interests of his own species to override the greater interests of members of other species.

3.2. Equal consideration of interests

If the case for animal equality is sound, what follows from it? It does not follow, of course, that animals ought to have all of the rights that we think humans ought to have — including, for instance, the right to vote. It is equality of consideration of interests, not equality of rights, that the case for animal equality seeks to establish. But what exactly does this mean, in practical terms?

If I give a horse a hard slap across its rump with my open hand the horse may start, but it presumably feels little pain; its skin is thick enough to protect it against a mere slap. If I slap a baby in the same way, however, the baby will cry and presumably does feel pain, for its skin is more sensitive. So it is worse to slap a baby than a horse, if both slaps are administered with equal force. But there must be some kind of blow — I do not know exactly what it would be, but perhaps a blow with a heavy stick — that would cause the horse as much pain as we cause a baby by slapping it with our hand. That is what I mean by 'the same amount of pain', and if we consider it wrong to inflict that much pain on a baby for no good reason then we must, unless we are speciesists, consider it equally wrong to inflict the same amount of pain on a horse for no good reason.

There are other differences between humans and animals that cause other complications. Normal adult human beings have mental capacities which will, in certain circumstances, lead them to suffer more than animals would in the same circumstances. If, for instance, we decided to perform extremely painful or lethal scientific experiments on normal adult humans, kidnapped at random from public parks for this purpose, every adult who entered a park would become fearful that he would be kidnapped. The resultant terror would be a form of suffering additional to the pain of the experiment. The same experiments performed on non-human animals would cause less suffering since the animals would not have the anticipatory dread of being kidnapped and experimented upon. This does not mean, of course, that it would be right to perform the experiment on animals, but only that there is a reason, which is not speciesist, for preferring to use animals rather than normal adult humans, if the experiment is to be done at all. It should be noted, however, that this same argument gives us a reason for preferring to use human infants — orphans perhaps — or retarded humans for experiments, rather than adults, since infants and retarded humans would also have no idea of what was going to happen to them. So far as this argument is concerned non-human animals, infants and retarded humans are in the same category, and if we use this argument to justify experiments on non-human animals we have to ask ourselves whether we are also prepared to allow experiments on human infants and retarded adults; if we make a distinction between animals and these humans, on what basis can we do it, other than a bare-faced — and morally indefensible — preference for members of our own species?

There are many areas in which the superior mental powers of normal adult humans make a difference: anticipation, more detailed memory, greater knowledge of what is happening, and so on. Yet these differences do not all point to greater suffering on the part of the normal human being. Sometimes an animal may suffer more because of his more limited understanding. If, for instance, we are taking prisoners in war time we can explain to them that while they must submit to capture, search, and confinement they will not otherwise be harmed and will be set free at the conclusion of hostilities. If we capture a wild animal, however, we cannot explain that we are not threatening its life. A wild animal cannot distinguish an attempt to overpower and confine from an attempt to kill; the one causes as much terror as the other.

Chapter 8 references, p. 164

It may be objected that comparisons of the sufferings of different species are impossible to make, and that for this reason when the interests of animals and humans clash the principle of equality gives no guidance. It is probably true that comparisons of suffering between members of different species cannot be made precisely, but precision is not essential. Even if we were to prevent the infliction of suffering on animals only when it is quite certain that the interests of humans will not be affected to anything like the extent that animals are affected, we would be forced to make radical changes in our treatment of animals that would involve our diet, the farming methods we use, experimental procedures in many fields of science, our approach to wildlife and hunting, trapping and the wearing of furs, and areas of entertainment like circuses, rodeos, and zoos. As a result, a vast amount of suffering would be avoided.

So far a lot has been said about the infliction of suffering on animals, but nothing about killing them. This omission has been deliberate. The application of the principle of equality to the infliction of suffering is, in theory at least, fairly straightforward. Pain and suffering are bad and should be prevented or minimized, irrespective of the race, sex, or species of the being that suffers. How bad a pain is depends on how intense it is and how long it lasts, but pains of the same intensity and duration are equally bad, whether felt by humans or animals.

While self-awareness, intelligence, the capacity for meaningful relations with others, etc., are not relevant to the question of inflicting pain — since pain is pain, whatever other capacities, beyond the capacity to feel pain, the being may have — these capacities may be relevant to the question of taking life. It is not arbitrary to hold that the life of a self-aware being, capable of abstract thought, planning for the future, complex acts of communication, etc., is more valuable than the life of a being without these capacities. To see the difference between the issues of inflicting pain and taking life, consider how we would choose within our own species. If we had to choose to save the life of a normal human or a mentally defective human, we would probably choose to save the life of the normal human; but if we had to choose between preventing pain in the normal human or the mental defective — imagine that both have received painful but superficial injuries, and we only have enough painkiller for one of them — the choice is not so clear. The same is true when we consider other species. The evil of pain is, in itself, unaffected by the other characteristics of the being that feels the pain; the value of life is affected by these other characteristics.

Normally this will mean that if we have to choose between the life of a human being and the life of another animal we should choose to save the life of the human; but there may be special cases in which the reverse holds true, because the human being in question does not have the capacities of a normal human being. Therefore this view is not speciesist, although it may appear so at first glance. The preference, in normal cases, for saving a human life over the life of an animal when a choice has to be made is a preference based on the characteristics that normal humans have, and not on the mere fact that they are members of our own species. This is why when we consider members of our own species who lack the characteristics of normal humans we can no longer say that their lives are always to be preferred to those of other animals. This issue comes up in a practical way in the following section. In general, the question of when it is wrong to kill (painlessly) an animal is one to which we need give no precise answer. As long as we remember that we should give the same respect to the lives of animals as we give to the lives of those humans at a similar mental level, we shall not go far wrong.

4. THE CONSEQUENCES OF EQUALITY

The case for animal equality has far-reaching implications. The most important of these, particularly for those concerned with animal sciences, are the areas of experimentation on animals, and the use of animals for food.

4.1. Animal experimentation

Many experimenters today operate in accordance with the ethical attitudes of Aristotle or Aquinas rather than those who hold that animals are entitled to equal consideration. The pages of several scientific journals bear testimony to the fact that experiments are being performed which involve the deaths and suffering of countless numbers of animals, with no real prospect of compensating benefits for humans or any other animals. If the interests of animals are to be considered as we consider the interests of humans, such experiments are indefensible.

Does the rejection of speciesism rule out all experiments which take the life of an animal, or inflict suffering upon it? Not necessarily. It is compatible with the principle of equal consideration of interests that we sacrifice the lesser interest of one being to the greater interest of another. If, for the reasons discussed in the preceding section, we believe that the life of a normal adult human being is of more value than that of, say, a rabbit, it may be permissible to take the life of the rabbit if this is the only way in which the human life can be saved.

I say that this 'may' be permissible because on some ethical views we should not sacrifice the interests of one human being to benefit another. Thus, most experimenters would reject the idea of taking the life of a severely and irreparably retarded human being even if this was the only way to save the life of a normal human being. Those who hold this view cannot, consistently with the principle of equal consideration, believe that we should do to a rabbit what they would not do to the retarded human. The rabbit may well be more aware of what is happening, more self-directing, and at least as sensitive to pain, as the retarded human. There seems to be no relevant characteristic which retarded humans possess that an adult rabbit does not have to the same or a higher degree. The experimenter, then, shows a bias in favor of his own species whenever he carries out an experiment on a non-human for a purpose that he would not think justified using a human being at an equal or lower level of awareness, sentience, etc. (To avoid complications I have assumed that there are no parents or other relatives involved who have special feelings for the retarded human — this would be the case if the retarded human were an orphan, or had been abandoned in an institution.)

This comparison provides a basis for answering the question: what experiments on animals can be justified without speciesism? The short answer is: only those experiments which would also be justified if performed on an orphaned, irreparably retarded human being at a comparable level of sentience, awareness, etc.

Exactly which experiments this criterion justifies depends on the extent to which one believes that it is permissible to sacrifice the interests of one human being to benefit another. Since I take a broadly utilitarian stance on these issues, I believe that the criterion does justify some experiments, although I also believe that there are very many experiments which it does not justify.

Chapter 8 references, p. 164

4.2. Animals as food

For most people in modern, urbanized societies, the principal form of contact with non-human animals is at meal times, when animal flesh or animal products are taken as food. The use of animals for food is probably the oldest, and remains the most widespread, form of animal use. There is also a sense in which it is the most basic form of animal use, the foundation stone on which rests the belief that animals exist for our pleasure and convenience.

It is obvious that if animals count in their own right, our use of animals for food becomes questionable. This is especially apparent when animal flesh is a luxury rather than a necessity. The eskimo, living in an environment where he must kill animals for food or starve, might be justified in claiming that his interest in surviving overrides that of the animals he kills. Most of us cannot defend our diet in this way. Citizens of industrialized societies have available to them a wide range of foods from which an adequate diet can be obtained without the use of animal flesh. The overwhelming weight of medical evidence indicates that animal flesh is not necessary for good health or longevity. When the animals to be eaten are fed on grains or other feedstuffs which are also suitable for direct human consumption, the end result is that less, rather than more, food is available to humans. Hence animal flesh is consumed neither for health, nor to increase our food supply; it must therefore be considered a luxury, consumed only because people like its taste.

In considering the ethics of the use of animal flesh for human food in industrialized societies, we are considering a situation in which a relatively minor human interest must be balanced against the lives and welfare of the animals involved. The principle of equal consideration of interests does not allow the major interests of one being to be sacrificed for the minor interests of another. Therefore the eating of animal flesh would appear to be incompatible with this principle.

The case against using animals for food is at its strongest when animals are made to lead miserable lives so that their flesh can be made available to humans at the lowest possible cost. Modern forms of intensive farming are the application of science and technology to the attitude that animals are objects for us to use. The animal is regarded as a kind of machine for converting relatively cheap fodder into more expensive flesh. No thought is given to the welfare of the animal, except in so far as ill-health may affect production. As Ruth Harrison (1964, p. 3) has said, 'cruelty is acknowledged only where profitability ceases'. Unfortunately, profitability may be enhanced by conditions that cause stress and discomfort to the animals involved. Even where output per animal is demonstrably lower under intensive conditions than where alternative methods are used, overall profits can be higher, since output per animal may be a less important economic factor than capital or labour costs. As a result, hens are crowded eight or nine to a 45 × 60 cm wire cage, veal calves are denied roughage and kept on an iron-deficient diet to ensure that their flesh remains pale and tender, and breeding sows are confined in stalls that do not permit them to turn around. There is ample evidence that under these conditions the animals lead miserable lives (Harrison, 1964; Brambell, 1965; Singer, 1975, chapter 3). If this is correct, intensive animal farming is ethically indefensible.

It is sometimes urged that the ethical position is different when animals lead pleasant lives, as they may sometimes do under free-range conditions. Then, even though the animals are slaughtered in the end, their lives have been worthwhile on the whole; and if the animals had not been destined for slaughter, they would never have lived at all.

This argument is interesting because it attempts to reconcile human use of animals with equal consideration for the interests of the animals themselves — it claims that by eating pork, we benefit pigs. The difficulty with the argument is that it is not clear which pig we benefit. We certainly do not benefit the pig whose flesh we are eating, for that pig would have been better off if it had been allowed to go on living. Do we then benefit pigs who are yet to be born, and who would not be born if it were not for people who ate pork? Perhaps, but the idea of benefiting a non-existent animal is peculiar, to say the least. If the pig had not been born, there would have been no pig at all to gain or lose by our actions.

The problems involved in comparing existence and non-existence are too complex to go into (see Singer, 1975, pp. 264—266). Even if the argument is valid, however, it will only justify some forms of animal farming — those in which the animals involved lead a life which is, on balance, a desirable one, and then are killed swiftly and painlessly. If animals are to be caged or confined in ways that place them under the stress of crowding, maternal deprivation, and a restriction of their natural activities, it is hard to believe that we do them any good in bringing them into existence.

Not all uses of animals for food involve killing animals and eating their flesh. When we keep dairy cattle, or laying hens, we are using animals in ways that do less violence to the animals themselves. Are these uses of animals easier to justify?

The fact that a given use of an animal does not involve killing the animal does not, of course, eliminate all problems of justification. The principle of equal consideration of interests still requires us to ask whether what we are doing is contrary to the interests of the animal, and if it is, whether our own interests would be so greatly harmed by changing our present practices that an impartial observer would regard our loss as greater than the animals' gain. Now we do not, strictly speaking, need eggs and dairy produce, any more than we need meat. People can, and do live entirely on plant foods, although greater care is needed with such a diet than with one that merely contains no meat. Some believe that if no animal products at all are taken, vitamin B12 supplementation is required, but others deny this (see Ellis and Montegriffo, 1971). Thus there is no vital human interest that requires us to use animals to obtain eggs or milk, and it will only be defensible to use animals in this way if it can be done without thwarting important interests of the animals.

The nature of farming practices is again relevant at this point. The confinement of birds in wire cages is no more defensible when the birds are kept for their eggs than it would be if the birds were being raised for the table. This means that the methods of egg production that prevail in industrialized nations are incompatible with the equal consideration of the interests of hens. Where hens are allowed to graze freely, however, taking their eggs does not appear to violate any of their major interests. The hens do not show signs of great distress when the eggs are removed, and hens with free access to outside runs appear to live contented lives. So there do not seem to be any major ethical objections to using free-ranging hens for egg production.

When we turn to dairy farming there is an additional point to consider, apart from the distinction between intensive and free-range dairying. This is the fact that to prevent the cow's milk from drying up, it is necessary to make the cow pregnant at regular intervals and then — if the milk is to be available for humans — take the calf away. The removal of the calf causes distress to both calf and mother, and the remainder of the calf's lifetime is likely to be a most unhappy one, whether it is taken to market for sale and immediate slaughter, or raised for a few months to produce veal. This raises a serious ethical question about dairy farming.

Chapter 8 references, p. 164

5. CONCLUSION

We have seen that the principle of equal consideration of interests requires us to reject ways of using animals which are fundamental to the animal sciences and the animal farming industry. In the light of its implications, it may be thought that the principle of equal consideration is therefore either unsound, or else unrealistic, in its application to non-human animals. I do not believe that the principle is unsound. If it applies to all human beings — and there is very wide agreement that it does — it is quite arbitrary to deny that it applies to non-human animals. There are, as we have seen, some human beings who are inferior, by any test of rationality, autonomy, or any other morally relevant capacity, to some non-humans. If we are not prepared to degrade such humans to the status of means for our ends, to use as we please as experimental subjects or objects for our dinner table, we have no basis for treating sentient non-humans in this manner.

Whether the principle is unrealistic is a more serious issue. What does 'unrealistic' mean in this context? The principle is not unrealistic in the sense that it would be impossible to put into practice. There are many people who do abstain from eating animal flesh, and even from eating any animal products; and we could, of course, exist without experimenting upon animals. But is it unrealistic in the sense that it is unlikely ever to be put into practice, on a large scale? That may be. We cannot, however, be sure of this at the present time. The extent to which human beings can be moved by the interests of other species is still unknown. There are signs of increasing public concern over experimentation and intensive farming, and it is not altogether beyond the bounds of possibility that the view that animals are things for us to use will, in a few hundred years, be thought of with the same repugnance that we now have for the idea that some humans are, because of their race or lowly birth, things for other humans to use.

6. REFERENCES AND FURTHER READING

Aquinas, T. Summa Theologica, translated by the Dominican Fathers. Benziger Bros., London. 1918.

Aquinas, T., Summa Contra Gentiles, translated by the Dominican Fathers. Benziger Bros., London. 1928.

Aristotle. Politics, translated by E. Barker, Clarendon Press, Oxford. 1946.

Augustine. The Catholic and Manichean Ways of Life, translated by D.A. Gallagher and I.J. Gallagher. The Catholic University Press, Boston. 1966.

Bentham, J., 1789. An Introduction to the Principles of Morals and Legislation. Hafner, New York, NY, 1948.

Brambell Report, 1965. The Report of the Technical Committee to Enquire into the Welfare of Animals kept under Intensive Livestock Husbandry Systems, Command Paper 2836. Her Majesty's Stationery Office, London.

Calvin, J., Institutes of Religion, translated by F.C. Battles. London. 1961.

Clark, S., 1977. The Moral Status of Animals. Clarendon Press, Oxford.

Ellis, F.R. and Montegriffo, W.M.E., 1971. The health of vegans. In: Plant Foods for Human Nutrition, Vol. 2, pp. 93–101.

Francis of Assisi, His Life and Writings as Recorded by His Contemporaries, translated by L. Sherley-Price. Mowbray, London. 1959.

Godlovitch, S., Godlovitch, R. and Harris, J. (Editors), 1972. Animals, Men and Morals. Gollancz, London.

Gompertz, L., 1824. Moral Inquiries on The Situation of Man and of Brutes. Published by the author, London.

Harrison, R., 1964. Animal Machines. Stuart, London.

Hume, D., 1777. An Enquiry Concerning the Principles of Morals. Collier, New York, NY, 1965.

Kant, I., Lectures on Ethics, translated by L. Infield. Harper and Row, New York, NY, 1963.

Linzey, A., 1976. Animal Rights. SCM Press, London.

Passmore, J., 1974. Man's Responsibility for Nature. Scribner, NY.

Regan, T. and Singer, P, (Editors), 1976. Animal Rights and Human Obligations. Prentice-Hall, Englewood Cliffs, NJ.

Ryder, R., 1975. Victims of Science. Davis-Poynter, London.

Salt, H., 1892. Animals' Rights. G. Bell and Sons, London.

Sidgwick, H., 1901. The Methods of Ethics, 7th edn. Macmillan, London.

Singer, P., 1975. Animal Liberation. A New York Review Book, New York, NY.

Turner, E.S., 1964. All Heaven in a Rage. Michael Joseph, London.

Chapter 9

The Rational Use of Wild Animals

H.P. LEDGER

1. INTRODUCTION

The study of the role of animals in our complex ecosystem is a part of the relatively new scientific discipline of Ecology. There is still much to learn of the manner in which animals may be both conserved and used rationally for the long-term benefit of man and there is considerable anxiety lest some species may become extinct before their true value is recognized. From a utilitarian point of view animals may be considered in terms of their ability to convert organic products, mainly vegetation, into animal products for human use, viz:

(a) foods (meat, milk, eggs and blood);

(b) utilities (draught power, wool, hair, hides, skins and feathers); and

(c) means for relaxation (racing, riding, hunting, photography, tourism, pets for company, etc.).

The more animal species which are utilized the wider the range of organic substances that can be usefully converted and the wider the range of habitats and environmental situations that can be used effectively without long-term ecological degradation. For example, the relative independence of free water supplies which is shown by some wild species allows them to utilize areas of land remote from watering points while other species are well adapted to extreme environmental temperatures. Similarly, disease resistance enables others to utilize grazing and forest lands denied to those stock which lack the protection of natural immunity or expensive prophylactic drugs. For example, the disease trypanosomiasis, the vector of which is the tsetse fly (*Glossinia* sp.), at present precludes the use by domesticated stock of some 60 million km² of African bush grazing. Some markets demand animal products which can only be met by unconventional livestock.

Wild animals thus provide an opportunity to add to the very limited range of species so far domesticated, but whether domestication would necessarily increase the productivity of wild species remains debatable. In one capacity or another many species can be used to increase the return per unit area of land either by the direct utilization of a wider range of herbage or, indirectly, by their control of undesirable grass or bush growth, thereby improving the habitats for other animals.

However, wildlife is not always an asset; in certain circumstances wild animals compete with domesticated stock for feed which the latter may use to better advantage. Some wildlife may also constitute a reservoir of disease which kills or debilitates farm animals, and others may damage fences and other farm installations, ruin crops or give rise to unacceptable levels of expense for crop protection. It has often been claimed that wildlife are less

Chapter 9 references, p. 188

destructive of their habitat than are domesticated species but, as Pratt (1968) points out, 'although it is true that wild species have an inherent ecological advantage over some domesticated species — especially in those areas that are poorly watered and infested with tsetse fly — it cannot therefore be concluded that all wild species are paragons of ecological virtue, incapable of causing damage to their habitat. Any species, wild or domestic, will damage its habitat if it is present in uncontrolled abundance.' The examples of rabbit populations in Australia, elephants in Kenya (Glover, 1963) and hippopotamus in Uganda (Thornton, 1971) have been cited in support of this observation.

Man's dependence upon animals for food, clothing and power is widely appreciated but little is understood about his dependence upon them for company and relaxation. This aspect of wild animal utilization is likely to become progressively more important in those parts of the world where urbanization is increasing and where rising living standards are providing more time and money to spend on leisure.

2. DEVELOPED AND UNDER-DEVELOPED REGIONS OF THE WORLD

The use that is made of wild animals depends partly upon biological considerations and partly upon economic and sociological factors which vary markedly from country to country, and frequently within a single country.

It is customary to divide the world into two zones: the 'developed' zone comprising the United States of America, Canada, Western Europe, Eastern Europe (including the U.S.S.R.) and Oceania (Australia and New Zealand); and the 'under-developed' zone encompassing the Middle East, Asia, Africa and Latin America. Poorer under-developed countries generally have the highest densities and annual increases of human populations and the lowest levels of education. Moreover, most of them cannot solve their population problems by emigration. Some 'under-developed' countries, particularly in the Middle East, are fortunate in having substantial resources of fossil fuels or minerals. However, although these resources contribute, at least temporarily, to national wealth of an unusually high order, these countries remain, for the most part, under-developed in terms of their biological and human resources. Under-developed countries cannot rely on cheap imports of raw materials from colonial empires and most of them lack the necessary resources to increase food production at a rate sufficient to keep pace with population growth. Thus the circumstances which dictated animal usage in the past in the developed zones may provide irrelevant or erroneous solutions to the present-day problems of under-developed countries.

What part can wild animals play in the increase and redistribution of the world's food resources? Can they help to alleviate the need for grain imports to under-developed countries by converting crop residues or otherwise unusable herbage into animal products to meet the growing need for food, or can they provide clothing, power, or income, particularly foreign currency, from tourism? Is it because of accident, thoughtlessness, bad judgement or necessity that developed countries now contain less variety of wildlife than is found in many under-developed regions? When and how do wild animals supplement or compete with other forms of land use? What are the essential factors which together determine that a particular animal species is desired, tolerated or classified as a pest? To what extent are man's aesthetic and material needs inevitably competitive? Can a hungry or rapacious man be persuaded to preserve animals that are worth more to him dead than alive? How can a rich nation like the U.S.A. which, for example, reduced its buffalo herd from an estimated 60 million in 1700 to a few dozen by 1900 (Taylor, 1970), explain to an impoverished neighbour their need to preserve their wild animals?

3. CONSERVATION OF ANIMAL RESOURCES

Inevitably technological innovations will continue to involve changes in animal use and the direction and extent of these changes will partly be determined by the sociological circumstances of a community. In many developed countries the ox has been replaced as a source of power by the horse which, in turn, has been replaced by the tractor, whilst in most under-developed countries the ox or buffalo is still the prime mover in agricultural practice. This has meant that in many developed economies the cattle breeder, free of the need to produce draught animals, has been able to concentrate his efforts on selecting superior livestock for beef or milk production. Similarly, nowadays, the horse breeder is almost solely concerned with the breeding of animals for their recreational value. The growth of horse-breeding enterprises under these changed circumstances reflects the growing needs of affluent people for animals for entertainment and recreation. The role of the dog in society has also changed. It is now seldom kept for the original purpose of helping man to hunt his prey but is trained and bred for the herding of sheep and cattle, the protection of property, the provision of 'eyes' for the blind, or as a companion. Inevitably, as increasing human populations and their domesticated livestock make larger demands on limited natural resources, the role of wild animals must change. They have to be contained on smaller areas of land, much of it relatively infertile and ecologically fragile. It may become necessary to domesticate some wild species completely, other species may be contained within boundaries for periodic 'harvesting', whilst others may be left relatively free to range their habitat without excessive restriction. In this process of continuing adjustment the danger is that many animals, including some whole species, may be exterminated. Even if we understand our present needs, those of the future are not easily predicted and species extermination is a horribly permanent process. The replacement of the horse by the tractor serves to illustrate the point. A few years ago the breeding of cart horses had decreased in popularity to the point where it survived only in the hands of a few enthusiasts. Now, as a result of the rising cost of fossil fuels, there are signs of a resurgence of horse breeding for draught purposes at highly remunerative prices. Whether these animals will ever be used again in large numbers is not so important as the fact that there are some left from which to breed should such a need arise.

Arguments concerning the present or the potential usefulness of wild animals to man have been proposed particularly by those conservationists who hope that evidence of the material worth of wildlife may help to ensure their survival. These efforts have caused some animal husbandrymen, agricultural scientists and veterinarians to re-evaluate the usefulness of animals as a natural resource by extending their horizons beyond the boundaries of domesticated species. The majority of stockmen still view the presence of wild animals on their farms with suspicion. Undoubtedly, in certain situations, wildlife compete with domesticated stock for food and water, threaten damage to crops, fences and other farm installations, or harbour diseases or parasites which may kill, debilitate or devalue farm animals. With a few notable exceptions it remains to be shown that, on balance, additional monetary profit is likely to result from the substitution of wild for domesticated animal species, or even the integrated use of both.

The exaggeration of the possible advantages of wild animal use by some over-enthusiastic conservationists has served to bolster the prejudices and scepticism of many in the farming community. The over-reactions of some conservationists and farmers have been counter-productive and have stemmed from a failure to understand the complexity of the ecosystem in which we

Chapter 9 references, p. 188

live and a lack of appreciation of the management difficulties in any enterprise, new or old, which seeks both to conserve and to use rationally the
animal resources that are available.

The pendulum has swung from the euphoria of the early nineteen-sixties,
which envisaged the use of wild ungulates as a major source of meat for the
protein-starved populations in under-developed countries, to the disillusion-
ment of the mid-nineteen-seventies, which resulted from the failure of efforts
to achieve this aim in practice. The time is now ripe to review existing knowledge and examine afresh the possibilities and practicalities of wild
animal usage.

4. INTERNATIONAL CONCERN FOR WILD ANIMAL PROTECTION

There has never before been so widespread and great an interest in the
preservation of wild animals and their habitats. It is no coincidence that this
concern has occurred at a time of unprecedented increase in the world's
human population.

Those who have been privileged to enjoy the beauty and wonder of wildlife have been conscious of the threat to its existence by the pressures imposed by expanding human populations who have to feed and entertain
themselves on decreasing per capita areas of land. This threat is made all the
more serious because of the possible environmental effects of the intensive
technologies (such as irrigation, inorganic fertilizers, chemical sprays, herbicides, etc.) now used to stimulate food production.

Since the turn of the present century, the successful efforts of relatively
few people to encourage the conservation and preservation of wild animals
is reflected by the number and growth of national organizations which have
been established for this purpose throughout the world. In the majority of
cases these societies have been financed by private contributions, but they
are increasingly being encouraged by public funds, statutory recognition and
government legislation. Enormous impetus was given to these conservation
movements by the birth of world organizations such as the International
Union for the Conservation of Nature and Natural Resources (IUCN), the
International Biological Programme (IBP) and the World Wildlife Fund
(WWF), together with such other voluntary institutions as the Washington-
based African Wildlife Leadership Foundation.

Because of the fear, and even likelihood, of an escalation in the rate of
species extermination, conservationists have had to concentrate mainly on
animal preservation, often through the development of appropriate reserves,
rather than on animal use. A detailed description of the development of
nature reserves, national parks and conservation organizations throughout
the world, together with a reference to species exterminated since 1950 and
those now threatened with a similar fate, has been provided by H.R.H. Prince
Philip and James Fisher (1970). Simon and Géroudet (1970) also described
48 animal species currently in danger of extinction.

The Survival Service Commission of the IUCN was formed to keep records
of endangered and extinct species and to assess the reasons for their plight.
Lists of these species are updated from time to time in the *Red Data Book*,
published by the IUCN.

The present trends of the relative rates of species emergence and disappearance point to a world of decreasing variety and, enthusiastic though the
conservation organizations have been, their efforts only provide a breathing
space in which to gather evidence of the usefulness of wild animals and,
therefore, of the need to conserve them.

There is no reason to assume that man, as the world's dominant species, cannot live without the presence of wild animals. Already many people are virtually unaware of their existence and it is conceivable that man's inventiveness could find a more efficient way of converting vegetation and crop residues to his various needs than by processing them through animals. The extent to which non-human animals are desirable or essential in man's ecosystem is debatable. It is a debate for which we are ill-prepared because of the lack of factual evidence to support or refute the high levels of emotionalism that such a discussion inevitably invokes.

5. THE NEW SCIENCE: ECOLOGY

That we lack such information is partly due to the disproportionate, if understandable, emphasis that has been placed on the intensified systems of agriculture, and partly due to the comparative youthfulness of the science 'Ecology'. This discipline has been defined as 'the branch of biology dealing with the relations between organisms and their environments' (*New Elizabethan Dictionary*). H.R.H. Prince Philip and Fisher (1970) have indicated the relative immaturity of this branch of science by reference to the dates on which many of the current definitive words and technical publications first appeared:

> '... the word *ecology* itself did not come into general use until 1868. Most of the other words we use in general biology are also young. Even *comparative anatomy* (1617) and *cell* (in the sense of an animal or plant cell, 1665) are fairly young. *Embryology* dates from as late as 1772, *taxonomy* from 1813, *chlorophyll* from 1817, *nucleus* from 1823, and the first use of *evolution* in the modern sense from 1831. *Palaeontology* dates from 1836, *protoplasm* from 1839, *biocenosis* from 1877, *enzyme* from 1878, *symbiosis* from 1879, *plankton* from 1888, *mutation* from 1894, *autecology* and *synecology* from 1896.
>
> The first ecological work was surely T.R. Malthus's *Essay on the Principle of Population* in 1798. Many works of the nineteenth century, notably Darwin's, are full of implied ecology. But as far as I can find the first book with "ecology" in its title was E. Warning's *Oecological Plant Geography*, as late as 1896. F.E. Clements's pioneer *Research Methods in Ecology* dates from 1905, and the first book on strict animal ecology was C.C. Adams's *Guide to the Study of Animal Ecology* of 1913. The *Journal of Ecology* was established in the same year; *Ecology*, another fine journal, in 1920, the *Journal of Animal Ecology* in 1932.'

During the last 50 years, publications on ecology have been numerous and diverse.

6. SOCIO-ECONOMIC ATTITUDES AND CONSTRAINTS

Too little attention is often given by the livestock developer to the likely impact of local traditions, religious taboos and cultural habits on the attitudes of people towards wild animal use. These influences are variable in their importance and are sometimes highly localized, but they are by no means specific to 'under-developed' societies. For example, Wilson (1973) drew attention to the fact that social as well as biological factors limit the use of intensive forms of animal production in the United Kingdom. The revulsion that some people feel towards 'factory farming' such as battery-house poultry production and intensive methods of pig husbandry are instances of ethical standards imposing a limit on what others claim to be technological progress (see Chapter 8).

Chapter 9 references, p. 188

Just as man's consideration for others has grown from the suppression of the slave trade to the provision of welfare-state conditions to protect the less-able and under-privileged members of society, so has concern grown for the preservation and well-being of non-human animals. The pressures and successes of organizations such as the anti-vivisectionists, anti-blood sports associations and the Societies for the Prevention of Cruelty to Animals, confirm these changes in the outlook of society.

When they reviewed the changes of man's attitude to animal life over the centuries, Regan and Singer (1976) pointed out that:

> 'The environmental movement has made millions aware of what we have done to wild animals. When whole species disappear forever, we can hardly fail to think about what we have done. New discoveries about the abilities of nonhuman animals, including the ability of chimpanzees to learn a complex sign language, have made us realize how closely related we are to the other animals. The threat of global famine has led to a spate of articles pointing out that modern methods of rearing animals for food waste more protein than they produce, and this in turn leads some people to ask, If the mass rearing and slaughter of animals does not help to keep us fed, how is this practice to be justified?'

The social constraints to which Wilson (1973) referred are also to be found among those conservationists who object to management practices designed to preserve wild animals so that they can be used for commercial gain. For example, some protest against organized shooting to control wild animal populations in national parks in order to avoid habitat damage caused by over-grazing. There is also a vociferous minority who would close zoos and ban circuses, even though animals in them may be fit, well-treated, comfortably housed and adequately fed. It can be argued that excessive concern for the maintenance of the status quo of wild animals is not only impracticable but may well be to their eventual disadvantage. Species preservation is as much a concern of those who would 'use' wild animals as those whose aim it is to conserve them. Yet much investigation to this end has been blocked by conservationists who are reluctant to have animals killed or captured on their territory even for purposes of research. Little research has so far been reported on the physiological and metabolic responses of wild animals to normal or imposed circumstances of environmental stress and, until much more fundamental information of this sort is available, we will not understand how best wildlife can be managed to their own and man's advantage. The preservationist who would deny to many access to large tracts of 'wilderness' country in the interests of the few (animals and humans) is no less open to criticism from other claimants to his territory than is the industrialist who denies others enjoyment of the countryside by the pollution which results from his enterprise.

Nothing increases the value of a product more than its rarity and nothing puts it in greater danger than man's rapacity. Thus the increasing values of rare birds' eggs, leopard skins, crocodile skins, elephant tusks and rhino horn constitute a one-way ticket on the journey to extinction for the species concerned.

There are at least three courses open to slow down and, hopefully, reverse this process of extinction before it is too late. The first is to try and persuade a world public not to purchase wildlife trophies' products and to legislate against their sale. This has been attempted in some countries but in view of the acquisitiveness of man, who is even prepared to spend large sums of money on stolen art treasures only to keep them unseen in a vault, this approach cannot be viewed with optimism.

The second means for ensuring the survival of a species is for man to protect it within its own habitat. The success of this will depend upon local pressures for possible alternative land use of the site chosen and the ability of local management to withstand corruption, obtain adequate finance, and combat the predatory incursions of the poacher or collector. It has been pointed out frequently that species survival is rarely challenged by the killing of animals for meat and skins by local people for their own consumption and use.

The third way is to recognize public demand and to farm, harvest or otherwise use commercially the animals which are to be conserved. If, as has been suggested, the leopard is in danger of becoming extinct because of the high prices paid for its skin and claws, one way of ensuring its survival might be to breed leopards to satisfy and even encourage the commercial demand. This type of suggestion tends to invoke considerable emotional opposition and indignation from many people although these feelings are often less strong and, therefore, more inconsistent, when the breeding and use of crocodiles for their skins is suggested. However, species survival aside, why should there be any greater objection to the wearing of a leopard skin coat than to one made of mink or sheepskin, or to a pair of shoes made from elephant ear than to those made of cowhide? Why should not such products be paid for by a kid-gloved hand from a well-filled crocodile or pig-skin wallet? Surely people able to earn a living associated with the use of wild animals should be considered at least as respectable as a farmer, butcher or tanner. If such a development could ensure a species gene pool from which conservation areas could be replenished or revitalized, so much the better.

That there may be a possibility of 'farming' the more exotic carnivores is suggested by the extent to which the breeding of lions in captivity, even in temperate zones, has become an embarrassment. Although their productivity has outstripped the demand for replacements in zoos, circuses and game parks, pressures exerted by preservationist groups oppose the development of such projects. Recently there was a public outcry in Europe lest surplus lions might be sold for the skins and claws they produced. One reason for protest was that such a practice might render more difficult the control of the traffic in wild lion skins and claws. However, it could be argued that lions in the wild would be better served by the selling of trophies and skins from their 'farmed' counterparts. The quality of the products from the latter would certainly be superior, as few hunted trophies are free from some form of damage.

However, it is not always possible to utilize animals in captivity even though their products are known to have a high commercial value. For example, there is the case of the vicuna (*Vicugna vicugna*), a close relative of the llama and alpaca, which is noted for the extreme fineness of its fleece. According to Simon and Géroudet (1970):

> 'Various attempts have been made to domesticate the vicuna, but with little success. The rearing of the young presents no great difficulty, provided they are captured at a very early age. [Adults are] less tractable, and captive animals refuse to mate. . The only recent success was achieved by Francisco Paredes, owner of the Hacienda Cala Cala in the Azangaro Province of the Department of Puno, who spent many years attempting to domesticate the vicuna. He started with a nucleus of ten hand-reared animals, accustomed from the earliest age to being handled, which were kept in large enclosures under semi-wild conditions: but eighteen years elapsed before the first breeding success was obtained. Since then the herd has gradually increased to about 500 to 600 animals, all ranch born.
>
> A further difficulty is that in spite of vicuna wool being the most valuable of all natural fibres ... the annual yield of about a third to half a pound per animal is so low that, paradoxically, domestication is uneconomic.
>
> Attempts have therefore been made to cross the vicuna with the alpaca ... While some success has been achieved, the resultant progeny are invariably sterile...'

Chapter 9 references, p. 188

7. DOMESTICATION

Any discussion of management systems for wild animals must consider the advantages and disadvantages of domestication. Definition of 'domestication' is difficult. The popular concept is an animal which can be handled with ease, readily be trained, and which is amenable to management for the purpose for which it is kept. However, are animals kept in zoos and circuses more or less domesticated than free-ranging cattle which are only rounded up annually? There is nothing absolute about the concept of a 'domesticated animal' and, in many circumstances, the degree and manner of domestication determine the feasibility and success of a wildlife utilization project.

Ratner and Boice (1975) noted that 'domestication' is an ambiguous word and that its definition largely depends upon the discipline of the person using it. Scientists engaged in behavioural studies associated domestication with the degeneration of traits of the wild animal from which the domesticates were derived. This view may be disturbing and unacceptable to the breeder of highly productive farm animals! However, numerous records have shown that initially the process may result in the development of generations that are smaller than their wild counterparts, reproductively less efficient and maternally less able to rear their young. Therefore, those who would domesticate wild animals must be prepared for initial disappointments and they may well have to contend with breeding problems. The length of time taken to overcome these problems may be extensive.

For example, Berry (1969) noted the results of King's experiments on the domestication of the rat (King and Donaldson, 1929, 1939). Early generations of King's 'captive greys' showed a high degree of female infertility and infant mortality, and it was necessary to foster the first generation on to 'well domesticated' mothers. It was suggested that the infertility of wild females and the relatively high proportions of sterile animals was due to the disturbance of the nervous system brought about by fear and confinement in cages. The incidence of female sterility declined from 37 to 6% in the first 10 generations and all females chosen for breeding in the 13th generation onwards were fertile. Subsequently, females became pregnant at an increasingly early age and had a longer breeding life.

Such results do not finally answer all questions about the biological effects of domestication. Was the original management of the newly captured rats responsible for the infertility? If the diet had been different, the cages larger, the daylight hours altered or individual escape chambers provided, etc., would the results have been different? However, published results indicate that problems of considerable magnitude frequently occur at the outset of many domestication programmes.

Smellie (1838) believed servitude, degradation and disfigurement to be characteristic of domestication. Darwin (1868) regarded abnormal modification to the benefit of man as a distinctive attribute of domestication, a view hardly compatible with that of Smellie. More recently, ethologists such as Eibl-Eibesfeldt (1970) have drawn attention to the shortening of the muzzle and extremities as degenerative effects of domestication.

Obesity is another characteristic that has been cited as a degenerative product of domestication, though where a desirable degree of increased fatness ends and obesity begins is difficult to define. Is any amount of fat-cover more than that normally found in wild animals an index of the degree of domestication or of an animal's suitability for domestication?

A better concept would seem to be that of Boice (1973) who considered domesticated animals to be 'adaptive' rather than 'inferior'. This observation serves well to describe the domestic potential of captive animals, some of which adapt readily and breed easily under their new circumstances. It is interesting to note that an animal's disposition in the wild is a poor criterion of its suitability for domestication. For example, the aggressive lion breeds well in captivity whilst the timid vicuna does not.

The main advantage of domestication is that it enables animal products to be obtained when required and with relative ease. The more tame or docile the animal the better suited it is for intensive forms of animal production. Conversely, the wilder the animal the more difficult it becomes to utilize its products and this disadvantage is only partly offset by an animal's greater ability to fend for itself and survive under extensive forms of management.

Of some 4237 species of Mammalia (Morris, 1965) only 15—16 species are used 'extensively' (Wright, 1954) while 26 species are said to be of 'substantial importance' (Huntington, 1925). Hale (1969) listed 23 species of Mammalia and 11 species of Aves which are of major importance to man (Table 9.1). Early civilizations domesticated a few more species of animals and birds than are now currently in use. According to Ucko and Dimbleby (1969) most of these were used in religious rites, on ceremonial occasions and as emblems of ancient deities. Others were used as aids for hunting, fishing and for sport. There have also been attempts from time to time by pioneer emigrants to domesticate some indigenous animals found in their new homelands. Possibly the most successful of these attempts has been that of domesticating the ostrich in South Africa, the plumes of which found a ready market in Europe during the Victorian era. This enterprise is of particular interest because it is one of the very few examples where the possible use of an animal in its own right was recognized and a market was specifically created for the purpose of exploiting its natural potential.

Attempts to domesticate indigenous animals for use in conventional agriculture have largely been unsuccessful, not necessarily because the wild animals were not amenable, but usually because the demand for their products was not sufficient to justify the efforts involved in domestication. The pioneer settler moving into new territory found it easier to hunt existing game animals for the meat, hides and skins he required and to use the domesticated stock he had brought with him for draught purposes and milk. Once established in his new home he found it easier, or more profitable, either to select from his exotic domesticates those best adapted to their new environments or to ameliorate the environments to suit the exotic animals. The limited success of attempts to domesticate additional species of wildlife indicates that man's main requirements are still being satisfied adequately by the original domesticates.

The initial selection of domesticates from among herbivorous animals living in open or lightly wooded terrain or forest clearings was logical because this was the habitat in which ancestral man could hunt most easily and best defend himself against predatory animals. The basis of the original selections is obscure and why they were so successful remains an intriguing conundrum. Zeuner (1963) has suggested that domestication had little to do with conscious selection for particular traits but resulted from 'the social relationships of animal species of which man was one'. Undoubtedly this association of man and animal must have been important because man would necessarily have focussed his attention on those animals which were least aggressive towards him. They in turn lived alongside him without undue fear and had a limited symbiotic relationship with him. Herbivorous animals no doubt benefited by the regenerated growth promoted by man-made bush fires, and carnivores and man jointly benefited from the prey that each of them killed on those limited areas of short, young herbage.

Chapter 9 references, p. 188

TABLE 9.1

Generic distribution of major domestic species

Class Mammalia	Class Aves
Order Perissodactyla	**Order Anseriformes**
Family Equidae	Family Anatidae
Equus caballus — horse	*Anas platyrhynchos* — duck
Equus asinus — ass or donkey	*Cairina moschata* — muscovy duck
	Anser anser — goose
Order Artiodactyla	*Branta canadensis* — Canada goose
Family Suidae	
Sus domesticus — swine	**Order Galliformes**
Family Camelidae	Family Phasianidae
Camelus bactrianus — Bactrian camel	*Gallus gallus* — chicken
Camelus dromedarius — Arabian camel	*Coturnix coturnix* — Japanese quail
Lama pacos — alpaca	*Phasianus colchicus* — ring-necked
Lama glama — llama	pheasant
Family Cervidae	*Pavo cristatus* — peafowl
Rangifer tarandus — reindeer	Family Numididae
Family Bovidae	*Numida meleagris* — guinea fowl
Bos taurus — European cattle	Family Meleagrididae
Bos indicus — Brahman (Zebu),	*Meleagris gallopavo* — turkey
Indian and Afrikaner cattle	
Bos grunniens — yak	**Order Columbiformes**
Bibos sondaicus — banteng	Family Columbidae
Bibos frontalis — gayal	*Columba livia* — pigeon
Bos bubalus bubalis — Indian buffalo	
Ovibos moschatus — musk ox	
Ovis aries — sheep	
Capra hircus — goat	
Order Carnivora	
Family Canidae	
Canis familiaris — dog	
Family Felidae	
Felis catus — cat	
Order Proboscidae	
Family Elephantidae	
Elephas maximus — Asiatic elephant	
Order Rodentia	
Family Muridae	
Rattus norvegicus — rat	
Family Caviidae	
Cavia porcellus — guinea pig	
Order Lagomorpha	
Family Leporidae	
Oryctolagus cuniculus — rabbit	

From Hale (1969).

Even today such symbiosis exists; contemporary observers, including the author in Uganda, have seen Africans drive lions from their kill, remove the hindquarters of the quarry for human consumption and leave the remains for the lions to consume on their return. The lions have then been followed on the 'kill' by the lesser carnivores and scavengers, such as hyena and jackal, whilst the vultures awaited their turn for the final pickings.

Early man used animal fats for fuel, nourishment and as a dressing to soften the hides and skins he wore, as well as to smear over his body as a further protection against the cold. Compared with domesticated species, many wild ungulates, particularly those inhabiting the tropics, are singularly lacking in dissectable body fat. One wonders whether man consciously selected animals for a high fat content or whether those animals which were the most docile happened also to be those with most depot fat. If this was an accidental selection, it was a most fortunate one because man's dependence upon animal fats grew as he moved into harsher environments and his physical work-load increased. Until comparatively recently meat-producing animals in developed countries were highly prized for the amount of fat they carried and they are still much valued for this reason in many developing societies today.

8. CHANGING OBJECTIVES FOR ANIMAL PRODUCTION

The uses to which wild animals can be put have to be considered against the following circumstances of our time: the exponential growth of human populations increasingly located in urbanized communities (World Bank, 1972); increased mechanization with its resultant reduction of man's physical work-load and his greater mobility; improved standards of living with associated changes in tastes and values; rising costs of high-energy fuels, fertilizers and foods necessary to sustain intensive methods of agriculture and animal husbandry; decreasing per capita land areas for food production; and increasing demands from urban populations for a variety of forms of rural relaxation — some of which involve the viewing or hunting of wild animals.

These changing circumstances call for a reappraisal of our animal resources (Bowman, 1977), particularly in view of the probability that some of our domesticated species may be nearing the peak of their biological potential (Wilson, 1973), or are as near to it as our attitudes towards animal welfare will permit (Singer, 1975). Such a reappraisal indicates that land of high fertility will progressively be used primarily for crop production for direct human consumption and that animals will be used to make better use of less fertile lands, crop residues, urban by-products and wastes.

Increasing efforts are now being made to improve the productivity of marginal lands both in terms of meeting man's material requirements and his aesthetic needs for relaxation. To make marginal lands more productive it may be necessary to select those animals which make best use of the existing herbage in the prevailing climate rather than attempt to use those animals best able to respond to improved management and ameliorated environments.

Particular care needs to be taken when choosing criteria by which the usefulness of different species is compared. Such comparisons need to be set in stated socio-economic and ecological contexts, and should involve critical, relevant and precise methodologies. For example, comparisons of growth rates and body composition changes between wild and domesticated species have too often been made using entire male wild animals and domesticated castrates. This is because, inevitably, the male of the wild species is always entire whilst most research results relating to domesticated species refer to castrates ('steers' in the case of cattle, 'wethers' for sheep and 'barrows' for pigs). Such comparisons are frequently misleading because the effect of castration is to make the growth pattern of the male more similar to that of a non-pregnant female, i.e. the castrated male deposits more fat and is less able to develop neck and shoulder muscles with approaching maturity.

From the results of Laflamme and Burgess (1973) it is possible to calculate that their bulls grew 18% faster than their steers, ate 21% more food, and achieved a 14% greater efficiency of food conversion. Similarly, Moore and Brown (1977) recorded that red deer stags were 16% lighter in weight at 22 months of age if they were castrated at 5 months old.

Another comparison that can be misleading is that of relating the productivity of wild, free-ranging animals to that of herded, domesticated species. Too often such studies in fact compare management systems rather than the productive capacities of species.

Knowledge of the productivity of wild animals is an essential ingredient in the formation of a land-use policy but, in the final analysis, it may well be that a lower return per unit area of land of easily obtainable produce (i.e., from domesticated stock) is more profitable than a higher unit return of less easily harvested material. This particularly applies to the production of meat from domesticated and wild animal species, where ease of collection not only helps to regulate market supplies but also ensures hygienic inspection and control of this highly perishable product.

9. WILD ANIMALS AS A SOURCE OF LEAN MEAT

An increasing demand for lean meat has been one of the prominent changes in taste and appetite which have been brought about by changes in living conditions. The more people are engaged in sedentary occupations, as a result of the mechanization of industry, the less fat they seek in their diets. This trend is accentuated when the range of available dietary ingredients becomes more extensive.

Public resistance to the purchase of fat meats, aided and abetted by their rising costs, has resulted in efforts by geneticists and nutritionists to produce lean meat, preferentially deposited in the expensive part of a carcase, namely the hindquarters. Because domesticated animals have long been selected for the production of meat with a high fat content, it has proved difficult to reduce the total amount of fat in carcases except by sending younger animals for slaughter at lighter weights. However, the more extensive use of larger, leaner breeds of cattle, such as the Charolais, Simental or Cianina, which were originally bred for draught purposes, for crossbreeding with the fatter, conventional meat breeds such as the Angus or Hereford, has become increasingly popular. Nevertheless, attempts to improve the preferential distribution of muscular tissue within the carcase have been unsuccessful and there appears to be little genetic scope for such improvement within a species (Berg and Butterfield, 1976).

The type of animal now required for lean meat production is well represented by several species of wild ungulates, particularly those found in the tropics. Many of these species have a higher carcase lean weight content per unit of liveweight (lean constant) than do cattle, and several also exhibit better developed hindquarters (Tables 9.2 and 9.6).

The chemical composition, as well as the amount, of fat present in meat has recently become an important factor in human nutrition. The amount of fat affects the texture and flavour of meat but not necessarily its tenderness, whilst the composition of the fat is becoming increasingly important on medical grounds.

Fats can be broadly divided into two types. Those that are fully hydrogenated, i.e., those which contain the maximum number of hydrogen ions per molecule, are the saturated fats, such as triglycerides which occur in the visible tissues of animals. Unsaturated fats, which are incompletely hydrogenated, are mainly present in vegetable oils. Consumption of saturated fats seems to be associated with high cholesterol levels in humans and has been

TABLE 9.2

Comparative meat production ability of some mature East African ruminants

Species	Sex	n	$\bar{x}$ Liveweight		Carcase wt. as % of liveweight		Carcase lean as % of liveweight	
			kg	±S.D.	%	±S.D.	%	±S.D.
Hippopotamus	M	4	1489.8	224.9	43.0	2.4	32.3	1.8
	F	4	1277.2	101.5	41.9	2.4	29.6	1.8
Buffalo	M	8	753.0	69.6	50.5	2.3	40.6	2.4
Eland	M	5	508.1	63.2	59.1	3.6	46.7	3.7
Zebu bulls	M	10	483.9	65.1	58.0	3.0	39.8	2.1
Zebu steers	M	70	469.8	66.8	57.6	2.4	31.6	2.4
Zebu fat cows	F	4	394.8	41.1	59.4	1.0	31.9	1.9
Zebu thin cows	F	9	298.4	48.6	46.8	1.4	30.3	1.5
Wildebeest (N)	M	10	243.3	14.6	55.7	1.4	43.8	1.4
	F	10	192.0	9.3	53.2	1.6	40.9	1.1
Waterbuck	M	10	237.7	18.3	58.6	1.4	48.5	1.8
	F	10	181.0	11.2	58.9	1.8	46.5	2.2
Wildebeest (S)	M	10	203.0	11.4	50.0	3.2	39.2	3.2
	F	10	160.3	12.6	51.4	2.2	38.6	3.5
Oryx	M	10	176.4	12.1	57.0	1.7	45.9	2.5
	F	10	161.5	20.3	58.9	2.5	45.5	1.6
Kongoni (Hartebeest)	M	5	142.5	11.1	57.2	1.4	46.1	1.5
	F	5	126.2	7.7	58.1	2.0	45.9	1.6
Topi	M	10	130.8	9.1	54.2	1.9	43.6	1.4
	F	10	103.9	8.0	54.0	2.1	44.0	1.5
Kob	M	10	96.7	5.8	57.7	1.9	47.8	1.7
	F	10	62.1	4.2	58.3	2.9	47.1	2.3
Lesser kudu	M	10	92.1	14.4	62.1	1.5	50.0	2.2
Warthog	M	10	87.8	7.5	54.7	2.5	45.4	3.1
	F	10	60.2	7.9	55.7	1.9	46.7	2.1
Grant's gazelle	M	6	60.1	6.2	60.5	2.2	48.2	1.7
	F	5	41.3	1.5	59.0	3.3	45.7	2.4
Impala	M	10	56.7	2.6	58.1	0.9	47.3	1.7
	F	10	42.0	2.6	58.3	3.0	47.1	2.2
Gerenuk	M	5	31.2	2.1	65.0	2.1	52.4	2.4
Thomson's gazelle (N)	M	10	25.3	1.6	58.6	2.1	48.1	2.6
	F	10	18.4	1.2	57.1	2.1	45.5	1.8
Thomson's gazelle (S)	M	10	20.3	1.7	54.2	1.7	43.3	1.9
	F	10	16.9	1.3	53.6	3.7	40.8	3.2

From Ledger (1968). N = Nairobi Game Park; S = Serengetti Game Park.

implicated as a causal factor in cardiac failure and arterial disease (Table 9.3). Unsaturated fats are thought to have a beneficial effect in preventing arterial deposits. Thus, when converting poor quality protein in herbage into high quality animal protein, animals convert some of the unsaturated vegetable oils into less desirable saturated fats. Although controversy concerning the degree to which animal fats should be incriminated as a cause of human vascular disease still exists, there is ample evidence to show that many wild ungulates produce greater amounts of edible protein per unit of liveweight than do domesticated stock and achieve maturity without the deposition of large quantities of fat (Tribe and Peel, 1963; Ledger et al., 1967; Coop and

Chapter 9 references, p. 188

TABLE 9.3

Cardiovascular risk profile
Standardized regression coefficients for incidence of cardiovascular disease for specified
risk factors: men and women aged 35—64 years (Framingham Study: 18-year follow up)

Risk factors	Coronary disease	Brain infarction	Intermittent claudication	Hypertensive heart failure[a]	Total cardio-vascular disease
Men					
Systolic blood pressure	0.245	0.587	0.205	0.542	0.326
Cigarette smoking	0.214	0.326	0.602	0.204	0.272
ECG-LVH[c]	0.054	0.018	0.093	0.130	0.110
Glucose intolerance	0.051	0.158	0.221	0.287	0.133
Serum cholesterol[b]	0.441	0.412	0.424	0.099	0.403
Women					
Systolic blood pressure	0.329	0.478	0.040	0.504	0.360
Cigarette smoking	−0.087	−0.005	0.209	0.392	0.019
ECG-LVH	0.068	0.074	0.158	0.093	0.083
Glucose intolerance	0.102	0.064	0.275	0.196	0.124
Serum cholesterol[b]	0.391	0.574	0.207	0.062	0.301

From Kannel et al. (1976).
[a]Congestive heart failure in the absence of coronary or rheumatic heart disease.
[b]Regression coefficients for cholesterol at age 45.
[c]ECG-LVH = Electrocardiographic evidence of left ventricular hypertrophy.

Lamming, 1976; Forss, 1976). Domesticated ruminants store saturated fats in visible sites, such as in the subcutaneous layers and intermuscular and perinephric deposits, but much of the little fat of wild animals is accounted for by invisible 'structural' fats high in phospholipid content (Forss, 1976).

Crawford (1968) has related differences in the amounts of saturated fats present in an animal to variations in diet. Animals feeding in woodlands have lower contents of saturated fats than those grazing in open grasslands (Table 9.4), and Forss (1976) has suggested that the greater proportion of unsaturated fat in the browsing animal is because some of its foods, such as seeds, pass through the rumen unaffected by its hydrogenating enzymes.

The difference between the body compositions of domesticated and wild ruminants may be of dietary and/or hereditary origin. For example, Forss reported that 'deer put on only half to one third as much fat as sheep given identical food and treatment; on poor restricted rations or on very rich ad lib rations, they are much leaner'. This leanness may be associated

TABLE 9.4

Ratio of polyunsaturated fat to saturated and monounsaturated fats in meat from animals on differing feeding regimes

Animal	Environment	Ratio of polyunsaturated to saturated and mono-unsaturated fats
Buffalo	Park	0.11
Buffalo	Woodland	0.43
Topi	Grassland	0.087
Topi	Woodland	0.30
Giraffe	Zoo	0.042
Giraffe	Woodland	0.64

From Crawford (1968), cited by Forss (1976).

with a higher metabolic rate (about 10—20% higher than sheep on a metabolic liveweight basis) since protein is more costly to maintain than fat (R. Kay, personal communication, 1970). The higher maintenance requirements of wild animals was also reported by Rogerson (1968) who recorded that cattle, wildebeest and eland derived a similar quantity of metabolizable energy from similar amounts of the same diet and utilized it with a similar degree of efficiency for maintenance purposes (82% for wildebeest and 80% for eland compared to the generally accepted range 80—85% for cattle and sheep). However, the metabolizable energy required per unit of metabolic weight varied considerably, the eland and wildebeest requiring from 20 to 30% more metabolizable energy for maintenance than did cattle. Such results, indicating that per unit of liveweight these wild animals required greater amounts of food for maintenance than did cattle, must be considered when comparisons of wildlife and domesticated stock are made on the basis of their efficiencies of feed utilization.

Moreover, Rogerson's findings did not support the claim that wild animals predominantly utilize their food surpluses to lay down protein. His comment on this aspect of his experiment was as follows:

'In view of the lean nature of most game carcasses, it would be convenient to suggest that the wildebeest was laying down body protein rather than fat; in support of this theory the results obtained by Ritzman and Colovos (1943), who showed that ruminants utilized energy more efficiently for growth than for fattening, could be quoted. Unfortunately, the carbon nitrogen balance data invalidate this possibility as about 90% of the retained energy is in the form of fat, and it would seem that a satisfactory explanation of what appears to be an appreciable difference in efficiency of utilization must await a more detailed study.'

A useful guide to the composition of domestic carcase meat was provided by Callow (1947) who showed that fat-free carcase tissue has a stable water content, close to 77%, irrespective of the type or condition of animal from which it is taken (Table 9.5). A similar result has been obtained for a wide

TABLE 9.5

Chemical composition of American boneless meat

Meat	Grade	Fat %	Protein %	Water %	Water on fat-free basis (calcd.)
Beef	Thin	14.0	18.8	66.0	76.7
	Medium	22.0	17.5	60.0	76.9
	Fat	28.0	16.3	55.0	76.4
	Very fat	39.0	13.7	47.0	77.0
Lamb	Thin	14.8	17.1	66.3	77.8
	Intermediate	27.7	15.7	55.8	77.2
	Fat	39.8	13.0	46.2	76.7
Pork	Packer's thin	35.0	14.1	50.0	76.9
	Packer's medium	45.0	11.9	42.0	76.4
	Packer's fat	55.0	9.8	35.0	77.8
Veal, including kidney and kidney fat	Thin	10.0	19.4	70.0	77.8
	Medium	14.0	18.8	66.0	76.7
	Fat	19.0	18.0	62.0	76.5

$\bar{x}$ 77.0 ± 0.5

From Callow (1947).

Chapter 9 references, p. 188

range of wild ruminants in East Africa (Ledger, unpublished data, 1970). Callow (1944) also acknowledged an observation by Watson that the carcase muscular tissue of a steer (*Bos taurus*) is always approximately a third of its liveweight, irrespective of the animal's age or condition. This is equally true for *Bos indicus* steers but not for bulls, neither is it true for many of the wild ungulates whose productivity of animal protein per unit of liveweight considerably exceeds this value (Table 9.2).

With regard to carcase 'balance', it is rare to find heifers or steers with hindquarters which exceed 53% of carcase weight when carcases are divided between the 10th and 11th ribs, but in several game species the females exceed a hindquarter balance of 55%. Modern market requirements for meat carcases can be better satisfied by some of the wild animals than by many domesticated animals (Table 9.6). One can only wonder what advances would have been made towards providing today's 'ideal carcase' if those efforts to change the composition of domesticated animals had been expended on the selection for 'manageability' or docility of some of today's wild ungulates.

10. INTENSIVE MANAGEMENT OF WILD ANIMALS

Animal production systems can be broadly classified as 'intensive' or 'extensive' and it is convenient to consider the feasibility of wild animal use at these two levels.

Intensive forms of animal husbandry are predominantly concerned with animals that are essentially tame and highly productive, often when kept in conditions of close confinement, e.g., broiler-house methods of producing poultry meats, battery-housed hens for egg production and large pig units for the production of bacon. The keeping of wild animals in zoos and circuses is also an intensive form of animal production designed to provide entertainment and education instead of food. Dairying and the fattening of cattle in feedlots may be considered as second-order systems of intensive husbandry, with fenced grazing animals representing the point where intensive systems merge into the extensive.

The process of increasing animal production by intensive methods has involved raising levels of nutrition, health and management, and the selection of those animals best able to respond to the improved conditions. There are biological limits to this process and at some stage there is a slowing down in the rate of progress with an increase in cost per unit of improvement as these limits are approached. It has generally been accepted that wild animals are unlikely to have a part to play in highly intensive management systems. However, this opinion will have to be revised if the introduction of improved technologies provides opportunities for the intensive production of animal products for which wild animals have a higher biological potential than domesticates. Tribe and Pratt (1975) suggested that, although historical evidence indicates that those animals which best satisfy the present needs of man have already been domesticated, there remains the possibility that other species, susceptible to domestication, might yet be more efficient in certain forms of production. They quoted the suggestion of Hutchinson (1954) that, if eggs and meat are to be produced from animals housed in controlled environments, the basic physiological and behavioural advantages of homeotherms (such as poultry) may be discounted to such an extent that it might be advantageous to replace them by the potentially more productive poikilotherms (such as turtles and terrapins).

In his search for more efficient and rapid forms of fish production Nikolsky (1963), as noted by Wilson (1973), found that the time taken to incubate

TABLE 9.6

Comparison of carcase composition of some mature East African ruminants

Species	Sex	n	Carcase hind-quarter wt. as % of carcase wt.		Carcase edible products wt. as % of carcase wt.		Carcase lean wt. as % of carcase wt.		Carcase fat wt. as % of carcase wt.	
			%	±S.D.	%	±S.D.	%	±S.D.	%	±S.D.
Hippopotamus	M	4	45.6	1.7	82.4	2.2	75.0	2.9	7.0	2.7
	F	4	47.7	1.4	81.8	1.5	70.5	0.6	10.9	2.0
Buffalo	M	8	41.5	1.4	80.4	1.2	74.4	1.2	5.6	1.0
Eland	M	5	44.2	3.4	83.6	3.3	79.0	2.3	4.2	2.4
Zebu bulls	M	10	46.6	1.0	85.4	2.9	68.7	2.1	13.7	3.3
Zebu steers	M	70	52.7	1.5	86.0	1.5	54.8	4.0	28.6	4.6
Zebu fat cows	F	4	52.5	1.4	88.4	0.3	53.6	3.1	32.9	3.3
Zebu thin cows	F	9	52.2	1.0	79.7	2.7	64.7	1.5	13.4	5.9
Wildebeest (N)	M	10	46.4	1.3	85.8	1.0	78.6	2.0	6.8	2.1
	F	10	50.5	0.8	84.7	0.9	77.0	3.1	7.3	3.5
Waterbuck	M	10	48.6	1.5	84.0	1.5	82.6	1.4	1.0	0.4
	F	10	56.4	1.3	83.0	1.9	78.9	2.7	4.0	1.9
Wildebeest (S)	M	10	46.4	1.2	81.4	2.1	78.3	1.9	2.7	3.0
	F	10	51.4	2.2	82.0	1.6	75.1	2.5	6.4	3.0
Oryx	M	10	50.2	1.5	84.2	1.9	80.4	2.7	2.9	1.2
	F	10	52.7	1.0	84.7	1.1	77.3	2.8	7.1	3.5
Kongoni (Hartebeest)	M	5	47.3	1.7	83.1	1.1	80.6	1.1	2.2	0.3
	F	5	50.0	0.2	83.2	0.02	79.0	1.4	3.9	1.6
Topi	M	10	46.1	0.8	84.4	0.7	81.7	0.6	2.3	0.8
	F	10	49.1	0.9	83.9	0.9	81.6	1.0	1.9	0.8
Kob	M	10	52.5	1.5	85.7	1.1	82.8	0.7	2.6	1.1
	F	10	60.7	0.6	85.2	1.5	80.8	1.9	4.0	1.5
Lesser kudu	M	10	50.9	2.0	84.4	2.6	80.5	2.6	3.3	1.5
Warthog	M	10	43.5	1.1	85.3	1.9	82.9	2.2	1.8	0.2
	F	10	47.6	1.2	86.2	1.3	83.9	1.2	1.8	0.7
Grant's gazelle	M	6	50.0	1.6	83.0	1.1	79.6	1.0	2.8	0.6
	F	5	58.0	0.8	83.2	1.2	77.4	3.7	5.1	4.2
Impala	M	10	52.2	1.4	83.8	1.9	81.4	2.2	1.9	0.5
	F	10	57.1	0.6	83.3	0.7	80.8	1.0	2.0	1.2
Gerenuk	M	5	56.2	5.9	83.1	2.4	80.6	2.4	2.0	0.1
Thomson's gazelle (N)	M	10	52.7	0.7	84.7	0.6	82.0	1.7	2.0	1.3
	F	10	59.4	0.7	82.6	1.2	79.7	1.3	2.2	5.6
Thomson's gazelle (S)	M	10	54.6	1.5	82.7	1.6	80.0	1.4	2.0	0.8
	F	10	59.8	1.0	82.0	2.3	76.1	2.1	5.1	2.2

From Ledger (1968). N = Nairobi Game Park; S = Serengetti Game Park.

trout eggs decreased in proportion to increases in the water temperature. It is possible that the efficiency of reproduction and growth of other species of fish and amphibia might be substantially improved under controlled systems of more intensive management. Similarly, it remains possible that the application of modern techniques of animal husbandry in systems of wildlife management might raise yields and efficiencies of wildlife production to levels which would be competitive with farm animals. For example, the introduction of artificial insemination practices might help to overcome

Chapter 9 references, p. 188

184

breeding difficulties experienced with newly captive animals, whilst hormone treatments might help induce multiple births and increase the rate of production of lean meat.

11. CHANGING IDEAS OF WILD ANIMAL MANAGEMENT

In recent years there has been a notable change in the approach taken by scientists in their quest to utilize wild animals for the benefit of man. It has long been recognized that wild ungulates are likely to be used to their best ecological advantage if they are free to range at will within their own environments. There is evidence to show that under these circumstances some wild species, at least, are more productive per unit area of land than are introduced species such as cattle and sheep.

However, experience has shown that the costs of harvesting, transporting and marketing meats from free-ranging wild animals often negate any genetic and territorial advantages they may possess. In addition, as markets become progressively more sophisticated, it becomes increasingly difficult to satisfy statutory standards of meat inspection and of hygiene with animals that cannot be transported to and processed through a licensed abattoir.

For these reasons it is becoming increasingly accepted that only those species which can be periodically herded and handled can be seriously considered potentially suitable as meat producers. Game-cropping projects, where animals are shot in the field, are likely to continue as a form of game culling and population control and, in these circumstances, local markets will absorb these limited and spasmodic supplies.

'Wild' animals are not all equally intractable or fractious. Numerous instances have been recorded of 'tameness' in most species, particularly where very young animals have been captured and hand-reared. The outstanding example among the 'wild' ungulates is the eland (*Taurotragus oryx*) which has been successfully tamed and managed for the production of both meat and milk (Posselt, 1963; Treus and Kravchenko, 1968). Other attempts to domesticate wild herbivores have been listed by Talbot et al. (1965). Recently oryx (*Oryx beisa*) have been captured and, in as short a time as 4 months from the date of capture, they have been domesticated to the extent of being herded by a single herdsman in open country (King et al., 1977).

The most successful of recent commercial attempts to introduce a species of wild animal to farming has been reported from New Zealand (Drew and McDonald, 1976) where deer have been used to meet the new demand for leaner meat. A New Zealand Deer Farmers' Association was formed in 1975 with a membership of 246 farmers licensed 'to farm, display or capture deer'. At first the objective was to export venison obtained from wild deer slaughtered in their natural habitat. However, although legislative changes were introduced to ease meat inspection regulations for deer carcases, these modifications proved to be insufficient for the growing and selective export industry. It was found in practice that the expansion of venison into lucrative export markets depended upon the registration of licensed premises for the slaughter of farmed deer and this meant that live deer had to be delivered, undamaged, to an appointed abattoir. Under these circumstances the main outlet for captured wild deer is now for breeding to increase the number of 'farmed' herds managed in fenced paddocks.

12. INEDIBLE ANIMAL PRODUCTS

More recently it has been found that deer by-products also have a considerable export value. For example, the antlers in high velvet are highly prized among Chinese communities for inclusion in medicines and health foods. In Taiwan and Korea, where deer have long been farmed intensively, deer velvet is harvested annually. Antlers, tails, sinews and testicles are also used widely in the preparation of medicines and soups.

The identification of these markets for deer by-products has added a new dimension to the developing industry in both New Zealand and Australia, and stags are now considered almost as valuable for their by-products as for their meat. An example of the relative values of hinds and stags, which illustrates the importance of by-products, is given by Wallis and Faulks (1977) (Table 9.7). The case of the deer illustrates the more general principle that the value of skins, horns, trophies and other inedible products is often particularly high in the case of wild animals.

A good example of a wild animal which was domesticated in order to farm it for the yield of an inedible product is the ostrich.

At the beginning of the 20th century ostrich farming was popular in South Africa and the fashion vogue of the time stimulated a large demand in Europe and North America for ostrich feathers. With changes in fashion the industry has declined considerably and now it exists mainly for the production of skins.

The world market for hides, skins and trophies continues to be supplied by hunters rather than farmers, but sporting products are often of indifferent quality and in spasmodic supply. This is due partly to the seasonal nature of hunting and its increasing costs, and partly to the progressive reduction in the variety and numbers of wild animals that can be hunted. Satisfactory markets can only be developed and maintained if a reliable supply of a high quality product is available.

TABLE 9.7

Relative production value of hinds and stags (N.Z. $)

Hinds: say 45 kg carcase less skin at $2.42/kg			$110.00
	Skin	$5.00	
	Tail	6.00	
	Sinews	1.00	
	Tusks	3.50	
		15.50	15.50
			Total return $125.50
Carcase	88%		
By-products	12%		
Stags: say 82 kg carcase less skin at $2.42/kg			$198.00
	Skin	$5.00	
	Tail	6.00	
	Sinews	1.00	
	Tusks	7.00	
	Pizzle	9.00	
	Velvet		
	2.3 kg at $66	150.00	
		178.00	178.00
			Total return $376.00
Carcase	53%		
By-products	46%		

From Wallis and Faulks (1977).

Chapter 9 references, p. 188

Interest is being shown in Australia in the domestication of the emu for its eggs and skin, and the farming of crocodiles for high quality skins. The emu egg is of particular value because its thick shell can be carved in the style of a cameo, its gradations in colour in shell thickness from turquoise to white being well suited for this purpose.

13. EDUCATIONAL AND AESTHETIC USE OF WILD ANIMALS

Although zoos have a part to play in the preservation of endangered species, this is probably a minor role in comparison with their educational value. Situated, as many are, near or in large cities, they provide the only direct experience for many people of nature in the wild. This personal contact invites curiosity and helps to produce a better appreciation of the complexity, wealth and problems of our environment. Zoos have also provided valuable opportunities for research into the nutrition, reproduction and health of wild species, the results of which provide a better understanding of the requirements of animals in their natural habitats. The management of wildlife in parks has done much to lessen charges of cruelty which have been levelled at the management of caged animals. Those who criticize zoo systems of animal management tend to forget the suffering imposed by nature on animals ranging freely in their natural habitats.

Television and cinemas have done much to increase public awareness of wildlife, and the popularity of nature films and wildlife documentaries is evidence of man's interest in the conservation and management of wild animals. Such programmes have contributed to the expansion of the tourist industries in those parts of the world where a variety of animals can still be viewed in their natural surroundings. Similarly, improvements in photographic equipment and techniques used by the amateur have provided an alternative to hunting and increased the range of tourist participation in wild animal appreciation. This trend has been particularly welcome because surviving animal populations are progressively unable to withstand the increased mobility and sophisticated weaponry of the modern hunter. Nevertheless, in many parts of the world, notably the U.S.A., wildlife preservation owes much to the efforts of organized hunting associations and communities who for many years have been at pains to ensure the well-being and multiplication of their eventual quarry.

In several locations in the world, where land is unsuited to agriculture, the viewing of wild animals by tourists has become an economically viable form of land use. Marginal lands of this type extend over large areas and are to be found in arid or semi-arid zones or in environments which are cold, wet and often of high altitude.

Thornthwaite (1933) estimated that some 31% of the world's land surface is accounted for by arid and semi-arid areas, the division between the two categories being 15.3 and 15.2%, respectively. Most semi-arid areas occur in the tropical latitudes, with smaller regions being found within the temperate zones and at higher altitudes (Hare, 1961; Stamp, 1961; Payne, 1965). It is in this type of country that the greatest range of potentially interesting wild animals is to be found, particularly on the African and Australian continents.

Such country is too deficient or erratic in its rainfall for sustained crop production or, alternatively, the evaporation rate may be too high or the terrain too rugged or inaccessible. Drinking-water for livestock is often absent, sparsely distributed or only intermittently available, and some form of climatic stress occurs, usually in terms of high daily heat loads or wide diurnal variations of temperature. Additionally, some areas with a higher production potential may be incapable of supporting domesticated stock because of the presence of predators, pests or diseases, e.g., trypanosomiasis.

The value of the tourist industry to the economic development of relatively infertile rangelands, as well as more fertile forest lands, is evidenced by the world-wide expansion of national game parks, nature reserves and recreation areas.

Tourism is now a significant factor in the foreign exchange earnings of many developing countries. For example, the demands of 'animal viewing' tourism in Kenya have resulted in an expansion of its game parks, the provision of good hotel accommodation and the improvement of its park road systems, and has yielded foreign exchange earnings comparable to those of its major agricultural exports.

However, the rapid expansion of tourism and recreational activities has its own array of problems. The enclosure of animals in ever-decreasing areas of land combined with the control of ever-increasing numbers of tourists, vehicles and hotels, involves difficult and controversial issues of both animal and people management. As standards of living increase and more people have resources to spend on recreation, the problems of 'over-stocking' with tourists will become even more acute and visitor quotas may have to be introduced into game parks. Already, in Europe and the U.S.A. new criteria are being developed to describe the capability of land (or water) to support tourist and recreational activities. For example, Olschowy (1969) has proposed the following 'stocking rates' for recreational activities:

Purpose	Stocking rate
(a) A quiet rest in the woods	25 persons/ha
(b) Optimal use of woods	100 persons/ha
(c) Sail boats	1 boat/ha of water surface
(d) Canoes	10 boats/ha of water surface
(e) Tent sites, varying according to vegetation	70—100 persons/ha
(f) Camping areas	60—75 persons/ha

Such a response is not really so surprising when reference to Wallace (1956) shows that, in a completely different context, interest in wild animals more than 20 years ago was responsible for sportsmen in Washington State spending more on hunting, shooting and fishing than did the whole state population on the sales of packaged spirits, beer and wine (U.S.$71 million) (Tribe and Pratt, 1975).

14. CONCLUSIONS

If any single conclusion can be drawn from this consideration of the rational use of wild animals it must be that any management decision has to be made in the light of a range of interrelated socio-economic and ecological factors. As with domesticated animals, the possible uses of wildlife are many and variable. However, before the potential of any wildlife resource can be realized on a sustained and ecologically sound basis, an understanding of the biological attributes and limitations of the species is essential. For example, we need to appreciate the factors which affect an animal's ability to breed, grow, utilize food, withstand disease and adapt to or tolerate environmental variations. This knowledge cannot then be used effectively without an adequate technology to harvest and handle the animals and to convert their products into marketable goods. In this context 'marketable

Chapter 9 references, p. 188

goods' include the presentation of animals for viewing by tourists, either in the wild or in captivity.

The extent to which any programme of wildlife use can be developed depends also upon the levels of education, interest and management expertise among local populations. Frequently, social factors such as the system of land tenure or the availability of financial credit limit the nature and extent of an enterprise. This is especially true in countries where both expertise and finance are in short supply, where land is held on a communal basis, and where the financial returns from wildlife industries are not sufficiently enjoyed by the local communities whose land and resources are essential to the enterprise. The economic feasibility of wild animal projects of course involves a consideration of existing or promoted market demands and the costs of production and marketing.

Unfortunately these basic requirements for the efficient and long-term use of wild animal resources cannot yet be satisfied. Although much useful research in this field has been reported, particularly since 1965, there is still insufficient information on the ecological, technological and socio-economic problems of wildlife management, harvesting and marketing to support adequately the planning and operation of a profitable utilization project on a sustained-yield basis. Virtually all attempts in practice have failed to achieve commercial success and most have been discontinued. For example, one of the most ambitious and costly wildlife research projects, the Kenya Wildlife Management Project, was wound up after almost 7 years, having failed to convince commercial interests that wildlife could be profitably used for the production of meat, skins and trophies.

Where a wildlife resource has been used successfully, either for tourism as in East Africa or for meat and by-products as in New Zealand, projects have involved high levels of management, well-defined objectives, the support of government infrastructures, and considerable capital investment.

15. REFERENCES

Berg, R.T. and Butterfield, R.M., 1976. New Concepts of Cattle Growth. Sydney University Press, Sydney, or International Book Distributors, Hemel Hempstead, England.

Berry, R.J., 1969. The genetical implications of domestication in animals. In: P.J. Ucko and G.W. Dimbleby (Editors), The Domestication and Exploitation of Plants and Animals. Gerald Duckworth and Co., London.

Boice, R., 1973. Domestication. Psychol. Bull., 80: 215—230.

Bowman, J.C., 1977. Animals for Man. Edward Arnold, London.

Callow, E.H., 1944. The food value of beef from steers and heifers and its relation to the dressing-out percentage. J. Agric. Sci., 34: 177—189.

Callow, E.H., 1947. Comparative studies of meat. I. The chemical composition of fatty and muscular tissue in relation to growth and fattening. J. Agric. Sci., 37: 113.

Coop, I.E. and Lamming, R., 1976. Observations from the Lincoln College Deer Farm. In: K.R. Drew and M.F. McDonald (Editors), Deer Farming in New Zealand, Progress and Prospects. Editorial Services Ltd., Wellington, New Zealand.

Crawford, M.A. (Editor), 1968. Comparative Nutrition of Wild Animals, Symposia of the Zoological Society of London, No. 21. Academic Press, New York, NY.

Darwin, C., 1868. The Variations of Animals and Plants under Domestication. John Murray, London.

Drew, K.R. and McDonald, M.F. (Editors), 1976. Deer Farming in New Zealand, Progress and Prospects. Editorial Services Ltd., Wellington, New Zealand.

Eibl-Eibesfeldt, I., 1970. Ethology. Holt Rinehart Winston, New York, NY.

Forss, D.A., 1976. The chemical composition of meat from wild and domesticated animals. In: K.R. Drew and M.F. McDonald (Editors), Deer Farming in New Zealand, Progress and Prospects. Editorial Services Ltd., Wellington, New Zealand.

Glover, J., 1963. The elephant problem in Tsavo. East Afr. Wildl. J., 1: 30.

Hale, E.B., 1969. Domestication of the evolution of behaviour. In: E.S.E. Hafez (Editor), The Behaviour of Domestic Animals. Bailliere, Tindall and Cox, London.

Hare, F.K., 1961. A history of land use in arid regions. Arid Zone Res., 17: 25. UNESCO, Paris.

H.R.H. The Prince Philip, Duke of Edinburgh and Fisher, J., 1970. Wildlife Crisis. Hamish Hamilton Ltd., London.

Huntington, E., 1925. The distribution of domestic animals. Econ. Geog., 1:143—172.

Hutchinson, J.C.D., 1954. Heat regulation in birds. In: J. Hammond (Editor), Progress in the Physiology of Farm Animals. Butterworths, London.

Kannel, W.B., McGee, D. and Gordon, T., 1976. A general cardiovascular risk profile. The Framingham Study. Am. J. Cardiogr., 38: 46.

King, H.D. and Donaldson, H.H., 1929. Life processes and size of body and organs of the Grey Norway Rat during the generations in captivity. Am. Anat. Mem. No. 14.

King H.D. and Donaldson, H.H., 1939. Life processes in Grey Norway rats during fourteen years in captivity. Am. Anat. Mem. No. 17.

King, J.M., Heath, B.R. and Hill, R.F., 1977. Game domestication for animal production in Kenya. Theory and practice. J. Agric. Sci. Camb., 89: 445—457.

Laflamme, L.F. and Burgess, T.D., 1973. Effect of castration, ration and hormone implants on the performance of finishing cattle. J. Anim. Sci., 36: 762.

Ledger, H.P., 1968. Body composition as a basis for a comparative study of some East African Mammals. In: M.A. Crawford (Editor), Comparative Nutrition of Wild Animals. Academic Press, London, New York, NY.

Ledger, H.P., Sachs, R. and Smith, N.S., 1967. Wildlife and food production. World Rev. Anim. Prod. 3: 13—37.

Moore, G.H. and Brown, C.G., 1977. Growth performance in farmed red deer. N.Z. J. Agric. Sci., 11: 175—181.

Morris, D., 1965. The Mammals. Harper and Row, New York, NY.

New Elizabethan Dictionary, 1960. George Newnes Ltd., London.

Nikolsky, G.V., 1963. The Ecology of Fishes. Academic Press, New York, NY.

Olschowy, G., 1969. The capacity of the biosphere and its limits. In: Problems of the Rational Use and Conservation of the Resources of the Biosphere. Deutsche UNESCO Kommission, Cologne.

Payne, W.J.A., 1965. Specific problems of semi-arid environments. Qual. Plant. Mater. Veg., 12: 269—294.

Posselt, J., 1963. The domestication of the eland. Rhod. J. Agric. Res., 1: 81—87.

Pratt, D.J., 1968. Rangeland development in Kenya. Ann. Arid Zone, 7: 177—208.

Ratner, C.R. and Boice, R., 1975. Effects of domestication on behaviour. In: E.S.E. Hafez (Editor), The Behaviour of Domestic Animals, 3rd edn. Balliere Tindall, London.

Regan, T. and Singer, P., 1976. Animal Rights and Human Obligations. Prentice-Hall Inc., Englewood Cliffs, NJ.

Ritzman, E.G. and Colvos, N.F., 1943. Physiological requirements and utilization of protein and energy by growing dairy cattle. Tech. Bull. New Hamps. Agric. Exp. Stn., No. 80.

Rogerson, A., 1968. Energy utilisation by the Eland and Wildebeest. Symp. Zool. Soc. London, 21: 153—161.

Simon, N. and Géroudet, P., 1970. Last Survivors. The World Publishing Company, New York, NY and Cleveland.

Singer, P., 1975. Animal Liberation. A New York Review Book, New York, NY.

Smellie, W., 1838. The Philosophy of Natural History. Hilliard, Gray, Boston.

Stamp, L.D., 1961. A History of Land Use in Arid Regions. Arid Zone Res., 17: 17. UNESCO, Paris.

Talbot, L.M., Payne, W.J.A., Ledger, H.P., Verdcourt, L.D. and Talbot, M.H., 1965. In: The Meat Production Potential of Wild Animals in Africa. Tech. Comm. No. 16, Commonwealth Bureau of Animal Breeding and Genetics, Commonwealth Agricultural Bureaux, Farnham Royal, Bucks., England.

Taylor, G.R., 1970. The Doomsday Book. The Camelot Press Ltd., London.

Thornthwaite, C.W., 1933. The climate of the earth. Geogr. Rev., 23: 433—440.

Thornton, D.D., 1971. The effect of complete removal of hippopotamus on grassland in the Queen Elizabeth National Park, Uganda. E. Afr. Wildl., 9: 47.

Treus, V. and Kravchenko, D., 1968. Methods of rearing and economic utilization of Eland in the Askaniya-Nova Zoological Park. Symp. Zool. Soc. London, 21: 395—411.

Tribe, D.E. and Peel, L., 1963. Body composition of the Kangaroo (Macropus sp.). Aust. J. Zool., 11: 273—289.

Tribe, D.E. and Pratt, D.J., 1975. Animal production in relation to conservation and recreation. In: Proc. III World Conf. Anim. Prod. Sydney University Press, Sydney.

Ucko, P.J. and Dimbleby, G.W. (Editors), 1969. The Domestication and Exploitation of Plants and Animals. Gerald Duckworth and Co. Ltd., London.

Wallace, P.F., 1956. An evaluation of wildlife resources in the State of Washington. Economic and Business Studies Bull. 28. State College of Washington, U.S.A.

Wallis, I. and Faulks, J., 1977. Production and marketing of deer by-products. N.Z.J. Agric. Sci., 11: 195—201.

Wilson, P.N., 1973. Livestock physiology and nutrition. Philos. Trans. R. Soc. London, Ser. B, 267: 101—112.

World Bank Operations, 1972. Sectorial Programs and Policies. John Hopkins University Press, Baltimore, MD and London, pp. 407—513.

Wright, N.C., 1954. The ecology of domesticated animals. In: J. Hammond (Editor), Progress in the Physiology of Farm Animals. Butterworths, London.

Zeuner, F.E., 1963. A History of Domesticated Animals. Harper and Row, New York, NY and Evanston.

Chapter 10

Modifying Growth: an Example of Possibilities and Limitations

P.L.M. BERENDE and E.J. RUITENBERG

1. INTRODUCTION

Anabolic agents with hormonal action are used on a large scale to improve intensive meat production. Anabolism is the constructive metabolism of body tissues and is the reverse of catabolism which is defined as their destructive metabolism. Among the most effective anabolic agents are hormones, secretions of the endocrine glands. Hormones are transported to their sites of action by blood and other body fluids. They affect morphological characteristics and a variety of metabolic processes and functions.

Tissue synthesis is a complex affair and this chapter is concerned with only one aspect of it: the increase in protein deposition in farm animals as a result of increased nitrogen retention. This can be achieved in two ways: firstly, the extent to which tissues are synthesized may be increased, and secondly, the extent to which they are broken down may be decreased. Anabolic agents, by their influence on the metabolism of ingested protein, cause the former effect to exceed the latter. Their action can be measured directly by nitrogen balance determinations and total carcase analysis, though only a few experiments of this type have been carried out with large animals. Generally, the anabolic effect on protein deposition has been determined indirectly by comparative growth studies using liveweight gain and feed conversion efficiency as criteria. Slaughter quality of the carcases has also been used as an index. In some experiments the effects of certain substances on changes in body composition (protein, fat) and on changes in blood characteristics (urea, protein, amino acids, hormones) have been studied.

Anabolic agents can be produced synthetically or extracted from organs — in the case of growth hormone, from the pituitary. In the following, the expression 'anabolic agent' will generally be used instead of 'hormone'. The action of anabolic agents is complex. Various specific anabolic agents influence protein or fat deposition, and these anabolic agents may influence each other (Table 10.1).

Anabolic agents which will be discussed here are shown in Fig. 10.1, and fall into two groups:

   (i) natural steroidal anabolic agents;
   (ii) xenobiotic anabolic agents.

Of particular importance are the following three xenobiotic agents:

   (a) trenbolone acetate — a treanic steroid with androgenic activity;
   (b) zeranol — a non-steroidal chemical derivative of zearalenon, a compound derived from various fungi; zeranol has weak estrogenic activity;
   (c) synthetic stilbenes — including diethylstilbestrol, hexestrol and dienestrol — non-steroidal compounds that have powerful estrogenic activity.

Chapter 10 references, p. 225

TABLE 10.1

Action of 'anabolic agents' on protein and fat deposition (Karg, 1966)

Gland	Anabolic agent	Protein deposition[a]	Fat deposition[a]
Adenohypophysis	Growth hormone	+	—
Pancreas	Insulin	+	+
Adrenal medulla	Adrenalin	(+)	—
Adrenal cortex	Glucocorticoids (a low dose)	(—)	(+)
Adrenal cortex	Glucocorticoids (a large dose)	—	—
Thyroid gland	Thyroxine (a low dose)	+	(—)
Thyroid gland	Thyroxine (a large dose)	—	—
Gonads	Androgens	+	(—)
Gonads	Estrogens (a low dose)	+	(—)
Gonads	Estrogens (a large dose)	—	+

[a]+, Anabolic effect; —, catabolic effect; ( ), uncertain.

Each group is represented by one or more substances which are used on a larger scale, e.g., diethylstilbestrol, hexestrol and estradiol-17β. Some products are applied unconjugated, such as estradiol-17β and testosterone, although conjugated forms are also used (estradiol-17β benzoate and testosterone propionate).

The forms of anabolic agents (or combinations of them) which are (were) used on a more or less large scale for stimulating protein deposition in practice are given in Table 10.2. In Section 2, first the effects of the individual anabolic agents on zootechnically important criteria will be discussed. Thereafter, the effects of combinations of anabolic agents — in most cases an estrogen plus an androgen or a gestagen — are described. In Section 3 the factors which may influence the effects of the anabolics are dealt with, in Section 4 some thoughts on the mode of action are given, in Section 5 some remarks on the effects of antihormones are made, and in Section 6 some toxicological aspects with subsequent risk evaluation for the consumer together with the legal consequences are discussed. In Section 7 conclusions are drawn regarding the use of anabolic agents.

In this chapter the following (sometimes unofficial) abbreviations are used:

GH	: growth hormone	MGA	: melengestrol acetate	Prog.	: progesterone
DES	: diethylstilbestrol	Trenb.	: trenbolone acetate	Eth.est.	: ethinyl-estradiol
Hex.	: hexestrol	Est.	: estradiol-17β	Zer.	: zeranol
Dien.	: dienestrol	Test.	: testosterone		

2. ANABOLIC AGENTS

The effects of anabolic agents on young calves, bulls, steers, heifers, sheep, pigs and poultry will be discussed. The results of a number of experiments carried out at ILOB (the Institute for Animal Nutrition Research, Wageningen), are described, and are presented in the Tables and/or Figures. The Figures are all drawn in the same way: on the horizontal axis the age of the animal is given and on the vertical axis the average differences in protein deposition (= N retention $\times$ 6.25) or weight gain between the treated group(s) and the untreated control group from the start of the experimental period onwards, i.e. the cumulative effect(s) of the treatment(s). The figures for the protein deposition or the weight gain of the untreated control animals

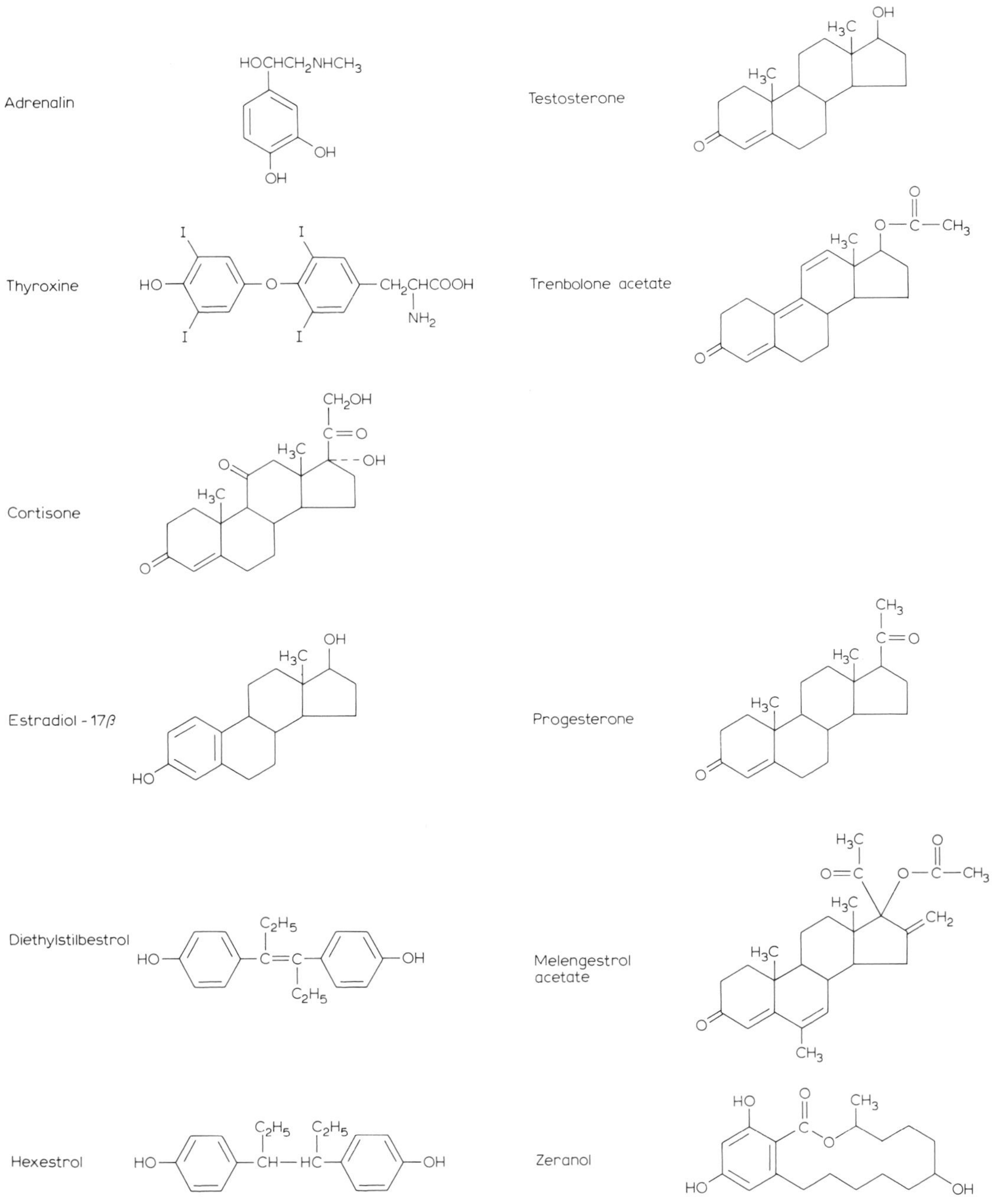

Fig. 10.1. Anabolic agents.

are represented by the horizontal zero line. The cumulative effect on protein deposition is expressed in g/animal, the cumulative effect on liveweight gain in kg. The arrows indicate the time of treatment (implantation, injection or start of oral application) with the anabolic agents.

2.1. Growth hormone

Growth hormone (GH) is essential for protein synthesis and for the formation of bone tissues (NAS, 1959); its anabolic effect is an increase in nitrogen retention. GH increases the protein and ash content of tissues and decreases the fat content. Hence, GH is responsible for a body composition which is

TABLE 10.2

Anabolic agents with approximate doses and applications

Chemical name	Trade name or abbreviation	Dose (mg)	Application	Species
Diethylstilbestrol	DES	12, 24, and 36	Implant	Steers
Diethylstilbestrol	DES	$10\ \text{animal}^{-1}\ \text{day}^{-1}$	Oral	Steers
Diethylstilbestrol	DES	2 to c. 24	Implant	Lambs
Diethylstilbestrol	DES	0.5 to c. 4.7	Oral	Lambs
Dienestrol diacetate	Lipamone	23/kg feed	Oral	Poultry
Hexestrol	Hex.	1000—2000[a]	Implant	Steers
Hexestrol	Hex.	15—25[a]	Oral	Steers
Hexestrol	Hex.	15—30[a]	Implant	Poultry
Trenbolone acetate	Finaplix	300	Implant	Heifers
Melengestrol acetate	MGA	$0.4\ \text{animal}^{-1}\ \text{day}^{-1}$	Oral	Heifers
Estradiol-17β benzoate/testosterone propionate	Synovex-H	20 and 200	Implant	Heifers
Estradiol-17β/testosterone	Implix-BF	20 and 200	Implant	Female cattle
Estradiol-17β/trenbolone acetate	Revalor	20 and 140	Implant	Veal calves
Estradiol-17β benzoate/progesterone	Synovex-S	20 and 200	Implant	Steers
Estradiol-17β/progesterone	Implix-BM	20 and 200	Implant	Steers, male (veal) calves
Estradiol-17β benzoate/progesterone	Synovex-L	2.5 and 25	Implant	Lambs
Diethylstilbestrol + methyltestosterone (DES + MT)	Maximim	2.2/kg feed	Oral	Pigs
Diethylstilbestrol/testosterone	Rapigain	24 and 120	Implant	Steers
Zeranol	Ralgro	12	Implant	Lambs
Zeranol	Ralgro	36	Implant	Steers, veal calves

[a] According to Merck Index 1968.

characteristic of the young animal (Karg, 1966). Injection of a large dose of GH will result in general in hyperglycaemia and glucosuria (Van Oordt, 1958). The action of the hormone is not the same for all species. For example, dogs and cats react differently from rats (Van Oordt, 1958). GH results in a very clear increase in carbohydrate and fat metabolism in dogs and cats, but a decrease in rats. The effect of GH depends also on the origin: each species has its specific GH which reacts only on the pituitary of the same or closely related 'higher' species (Tausk and Boerman, 1961). GH cannot be given orally but must be injected daily, which restricts its practical use. Furthermore, it cannot be synthesized.

2.2. Insulin and adrenalin

Insulin has a positive effect on protein synthesis and increases the appetite (Tausk and Boerman, 1961). GH and insulin act synergistically in protein synthesis, although in fat synthesis their action is antagonistic (Karg, 1966). Insulin is essential for the metabolism of glycogen to fat and promotes fat synthesis in the peripheral tissues, the extent of which can be influenced by adrenalin, an antagonist of insulin. Injection of adrenalin causes a decrease of fat and glycogen content in the animal.

Dzalalov (1968) performed experiments with bulls and steers implanted with 36 and 48 mg insulin at 13—14 months of age. Average weight gain, measured over a period of 107 days, increased by 25 g/day and feed conversion was 0.4 feed units lower for the treated animals than for the controls. The effect with bulls was somewhat greater than with steers. There was no difference in effect between 36 and 48 mg insulin.

If adrenalin is used for promoting protein synthesis and lowering fat deposition, it should be injected daily; therefore an alternative could be a substance that, given orally, would stimulate the release of adrenalin. Such a substance is nicotine. Cunningham and Friend (cited by Karg, 1966) fed pigs

5 mg nicotine sulphate/kg feed. Weight gain and feed conversion were not influenced significantly, but the carcase contained 10% less fat and 5.3% more protein. Karg (1966) also mentioned experiments which did not confirm these results.

2.3. Thyroxine

Thyroxine intensifies intermediary metabolism. To increase the efficiency of feed utilization and to accelerate growth, thyroxine can be injected, but thyroproteins such as iodinated casein may be fed (Winchester, 1953; NAS, 1959). The dosage used is very important. Positive effects of small amounts of thyroxine or thyroprotein on weight gain in calves, pigs and poultry have been reported by NAS (1953, 1959), Grashuis (1962) and Karg (1966). However, the results were not uniform. The effect depends on the dosage and also on the age of the animal, the physiological endocrine status of the animal, the composition of the ration, and the temperature of the environment; at lower temperatures the effects are greater. The effect of thyroxine and iodinated protein on ruminants was less pronounced than in swine (Winchester, 1953; Grashuis, 1962). Susceptibility to these products varies from animal to animal, as the margin between an effective dose of the hormone and an overdose is narrow. An overdose of thyroxine causes a breakdown of the fat depots and results in hypertrophy of heart, kidneys, adrenals and liver.

2.4. Glucocorticoids

Glucocorticoids affect a wide variety of metabolic events, including the regulation of carbohydrate, protein, nucleic acid and lipid metabolism. The effects are related to the organs in which glucocorticoids are active (Schulster et al., 1976). Some metabolic effects of glucocorticoids are:

in the liver — increased glycogen deposition, increased glucose production, increased amino acid uptake and increased protein synthesis;
in adipose tissue — decreased glucose utilization;
in muscle — decreased glucose utilization, decreased protein synthesis and/or increased protein breakdown;
in lymphatic tissue — decreased glucose utilization, decreased protein synthesis and/or increased protein breakdown.

A large amount of glucocorticoids given to animals results in a greater breakdown of protein (catabolism) and an increase in glyconeogenesis. This causes a larger excretion of nitrogen via the urine and hyperglycaemia (steroid-diabetes) and glucosuria (Meites and Nelson, 1960). This can result in a decrease in feed utilization. The effects mentioned are not uniform for all animal species; a high dose of glucocorticoids in guinea pigs does not cause a decrease in body weight and an increase in nitrogen excretion in the urine.

Only a few experiments have been reported with cortisone. Karg (1966) cited an experiment with steers injected three times a week with 1 g cortisone acetate, and lambs injected three times a week with 25 and 300 mg over 103 days. Feed conversion was less efficient, fat deposition was greater, and protein content of the carcase was lower. These results correspond very well with those of Ellington et al. (1964) who gave wethers an injection of 100 mg cortisone acetate three times a week for 42 days. Weight gain was higher for the treated animals than for the controls and more fat was formed, as was indicated by the thicker backfat and the lower specific gravity of the body.

Chapter 10 references, p. 225

2.5. Estrogens

2.5.1. Endogenous estrogens and derivatives of endogenous estrogens

Endogenous estrogens and their derivatives are substances of steroid structure produced in the animal body, and substances of similar structure produced synthetically, such as estradiol-17β benzoate. Of these estrogens only estradiol-17β and its derivatives are used in practice. However, this hormone is seldom applied as a single substance, but nearly always in combination with androgens or gestagens (Section 2.8).

Estrogens can have a strong anabolic effect but this depends on the dosage (Karg, 1966). At a relatively low level they act anabolically, but at high levels they decrease weight gain and are lipogenetic. The effect also depends on the species, as was shown clearly by Meites and Nelson (1960) and Poppe (1965). In cattle and sheep they increase N retention and weight gain and improve feed conversion efficiency. In poultry, however, an increase of fat deposition was observed with little or no influence on weight gain. In rats estrogens seem to decrease the feed intake, resulting in a lower weight gain.

In experiments at ILOB with male veal calves of 90—100 kg the implantation of Est. effected a considerable increase in N retention and weight gain, and a lower, more favourable feed conversion (Figs. 10.2 and 10.3). The effects were, however, less than with the application of combinations of Est. and androgens (Section 2.8). These findings were confirmed in comparative growth experiments by Grandadam et al. (1975a) and Gropp et al. (1976b) in which the implantation doses were 20 and 500 mg Est., respectively.

Hale and Ray (1973) performed a series of experiments with steers of initial weight 200—300 kg. Each group consisted of 16—56 animals. In these experiments oral dosages of 10, 15, 20, 40 and 80 mg Est. were tested. The results of these trials indicated that Est. was an effective growth promotor for both growing and fattening steers. An oral dose of 40 mg Est. resulted in improvement in weight gain and feed conversion comparable to those obtained with 10 mg oral DES. These results do not agree with the assumption that the oral activity of steroidal estrogens is much less than that of stilbenes (NAS, 1966).

Only a few experiments have been done with oral application of ethinylestradiol to pigs. In our experiments with castrated male pigs the effects on N retention, weight gain and feed conversion were small (E.J. Van Weerden and J. Huisman, unpublished data, 1978). Fowler et al. (1978) concluded that the results for the calculated amount of lean tissues and rate of N retention showed that the anabolic properties of Eth. est. were confined mainly to the castrated male pigs. At physiologically active levels, estrogens markedly stimulate food intake and sometimes weight gain in chickens (Nesheim, 1976), resulting in an increase in fat deposition (Megally et al., 1969; York and Mitchell, 1969).

2.5.2. Synthetic exogenous estrogens

'Synthetic exogenous estrogens' are compounds with potent estrogenic activity having no steroid structure. The most important synthetic estrogens are (NAS, 1953, 1959, 1966):

substances derived from stilbestrol, like diethylstilbestrol (DES), stilbestrol dipropionate, dioxydiethylstilbene propionate;
hexestrol, diallylhexestrol;
dienestrol, dienestrol diacetate;
dimethylethers of substances mentioned above, like dianisylhexene, dianisylhexane, dianisylhexadiene.

Large quantities of DES have been used in practice. DES has two isomers, *cis*-DES and *trans*-DES, with different biological activity (Preston et al., 1970, 1971; Bradley et al., 1972). Because in most publications the type of isomer is not specified, only the term DES is used here.

Positive effects of the implantation of stilbestrol in poultry have been known since the early nineteen-forties. Later, an enormous number of publications became available, especially with regard to the effect of implantation and oral application in steers. The effects of DES, hexestrol and dienestrol will be discussed for calves, bulls, steers, heifers, sheep, pigs and poultry.

In experiments with male veal calves from 8 to 18 weeks of age the maximum improvement of liveweight gain was from 4 to 10 kg (Van der Wal et al., 1971; Berende et al., 1973). Thomas et al. (1970) implanted male calves of 74 kg liveweight with 12 mg DES and found an improvement in weight gain of 3 kg. Patton and Ralston (1968) implanted male calves at birth with 12 mg DES; after 3 months the weight gain of the treated animals was significantly higher than that of the controls. Martin and Stob (1978) also implanted bull calves at birth and again at 3, 6, 8 and 10 months of age. Weight gain of the animals treated in this way was improved by an average of 0.11 kg/day until the age of 383 days. Hendrickson et al. (1957) could detect an improvement in weight gain of 12% in beef calves at an age of 6.5 months fed 5 mg DES/day. With young bulls Aguirre et al. (1968) found an increase in weight gain of 9.1 kg over a period of 112 days after implantation with 24 mg DES.

More experiments have been carried out with older bulls. Andrews et al. (1956b) fed bulls 10 mg DES for 179 days and found an improvement in weight gain of 0.18 kg/day. Approximately the same improvement was found by Dvoracek and Markovic (1965) in bulls fed 5—10 mg DES/day for 360 days. Klosterman et al. (1955) could improve weight gain by 0.12 kg per animal per day by implantation with 84 and 132 mg DES. Also Hunsley et al. (1967) found a significant increase in weight gain after oral application of DES. Baker and Arthaud (1972) collected results of 34 experiments in which DES was implanted. The effect of the DES treatment by comparison with the controls varied from —0.03 to +0.16 kg weight gain/day. In five experiments in which DES was given orally, the differences between treated animals and their controls varied from —0.06 to +1.40 kg weight gain/day. These results correspond well with those of the literature, as reviewed by Preston and Willis (1970).

Many experiments have been performed with steers (Andrews et al., 1954; Clegg and Cole, 1954; Burroughs et al., 1955; Perry et al., 1955, 1958; Beeson et al., 1956; Kolari et al., 1960; Ogilvie et al., 1960; Nesler et al., 1966; Goodrich et al., 1967; Stob et al., 1968; Ralston et al., 1969; Ray et al., 1969; Kliewer, 1970; Preston et al., 1971; Oltjen et al., 1973; Preston, 1975). Preston and Willis (1970) and Szumowski and Grandadam (1976) have collected literature data about the effects of DES in steers. Preston and Willis mention increases in weight gain of 12—18% under drylot circumstances. Szumowski and Grandadam (1976) calculate increases in weight gain from 15.6 to 27.0% for implanted steers and from 11.4 to 18.1% for steers fed orally with 2.75—30 mg DES/day.

With heifers the results of DES treatments are more irregular, as shown in the experiments of Hawkins et al. (1967), Ralston et al. (1968), Goodrich and Meiske (1969), and Utley and McCormick (1974). In their study of the literature, Preston and Willis (1970) found that in one experiment weight gain of the treated animals remained 6% below that of the control group and in 12 experiments DES improved weight gain from 6 to 50%.

Chapter 10 references, p. 225

Vanschoubroek (1965) and NAS (1966) mentioned the so-called side-effect in heifers, e.g., depression of the loin, elevation of the tailhead, stimulation of the mammae, riding, stimulation of the vulva, and in some cases vaginal eversion. According to Wickersham and Schultz (1964) and Preston and Willis (1970), implantation or oral application of DES did not have any influence on reproduction.

In the period 1950—1970 many experiments were done on sheep (lambs) (Whitehair et al., 1953; Clegg et al., 1955; Hale et al., 1955; Andrews et al., 1956a; Preston and Burroughs, 1958; McLaren et al., 1960; Oxley et al., 1960a, 1960b; Ellington et al., 1964; Garret, 1964; Preston et al., 1965; Ogandzanov and Paduceva 1969; Vetter, 1969; Grebing et al., 1970; Huber, 1970; Simms et al., 1970; Davis and Garrigus, 1971; Hafs et al., 1971; Hutcheson and Preston, 1971). The effects of DES in sheep are similar to those in cattle (Vanschoubroek, 1968): with implantations of 2—36 mg DES and with oral application of 0.5—47 mg DES, clear positive effects on weight gain were observed (NAS, 1966).

With pigs, fewer experiments have been done with DES alone. The available data show that the application, especially parenterally, of DES to pigs gives very variable results (NAS, 1953, 1959, 1966; Teague et al., 1964; Vanschoubroek, 1968). Part of this variation may be attributed to the different experimental conditions such as differences in age, sex (boars, gilts, castrated males) and dosage (1.5—96 mg implanted and 1.0—5.0 mg fed orally). Sometimes the objective of using DES in boars was only to suppress the sex odour of the carcases. Plimpton et al. (1967) and Plimpton and Teague (1972) found an increase in weight gain with boars implanted with DES and, in addition, the amount of lean was increased. Day et al. (1960), Hale et al. (1960) and Hale and Johnson (1970) observed hardly any positive effect of an implantation of 6 mg DES or of an oral application of 2 mg DES/day in castrated males, but the slaughter quality of the animals was better, as indicated by a lower backfat thickness and a higher percentage of lean cuts. These results were confirmed in an N balance study by Melliere et al. (1970), who found an increase in N retention after an oral application of 4.4 ppm DES.

The levels of DES applied to poultry were implantations of 12 and 15 mg. In this species estrogens influence the fat metabolism in particular. Brahmakshatriya et al. (1969) saw an increase in plasma glycerides and cholesterol in adult hens after oral administration of DES. DES had no influence on egg production, but improved feed conversion. Krista et al. (1969) found a higher fat content in carcases of turkeys after injections of DES.

Hexestrol has been used as an implant and via oral application in cattle, sheep, pigs and poultry. In an experiment with 2-year-old steers, Perry et al. (1955) found clear increases in weight gain after oral application of 10 mg Hex. The same effect was seen after oral application of 19 and 30 mg diallyl-Hex. fed to steers by Dyer et al. (1960). Also a very clear effect of the implantation of 30 and 60 mg Hex. in steers of 230 kg was recorded by MacDearmid and Preston (1969). NAS (1966) concluded, on the basis of a literature study, that Hex. was not as effective as DES.

Preston et al. (1965) mentioned positive effects of Hex. in sheep. NAS (1966) reported five experiments in which Hex. was tested. It seems that Hex. can increase weight gain in sheep (lambs) to a similar extent as DES.

With pigs very few experiments have been performed to study the effect of Hex. (Yoshimoto et al., 1968).

In the years 1950—1960 Hex. was used in poultry for increasing the slaughter quality of broilers (Vanschoubroek, 1965). Ryley et al. (1970) confirmed this effect and found a significant increase in feed consumption and weight gain in broilers implanted with 15 mg Hex.

In an experiment of Andrews et al. (1954) with steers of 300—400 kg, dienestrol improved weight gain as much as DES. Both agents were implanted as tablets of 60 and 80 mg, respectively, 140 days before slaughter. This result does not correspond with those quoted in the literature study of NAS (1966), in which it has been concluded that Dien. is less effective than stilbestrol.

Couch (1969) and Herrick et al. (1970) mentioned results of experiments with broilers fed rations with Dien.-acetate. Dien. improved weight gain and feed conversion of birds on a ration with a low fat content. Male and female broilers reacted differently to Dien. in a ration with a high fat (5%) content. Males gained faster and females more slowly and gave unfavourable conversion figures when Dien. was fed.

2.6. Endogenous androgens, derivatives of endogenous androgens and synthetic exogenous androgens

Endogenous androgens, their derivatives, and synthetic exogenous androgens are androgenic substances produced by the animal itself, and synthetic compounds with an androgenic activity, such as methyltestosterone, trenbolone acetate, etc.

In practice the application of single androgens is very limited; androgens, especially testosterone (Test.), testosterone propionate and methyltestosterone, are used on a large scale in combination with estrogens (Section 2.8). For experimental purposes nortestosterone and androstenedion have been used. Two commercial androgens have been produced for broilers, namely Bolfortan (testosterone nicotine acid ester) and Dianabol (1-dihydro-17α-methyltestosterone) (Schneeberger and Schürch, 1961; Leibetseder, 1966). During recent years experiments have been performed with the synthetic androgen trenbolone acetate (Trenb.), especially in female cattle.

Gropp et al. (1976a) observed a 9% improvement of weight gain in young female calves after an implantation of 140 mg Trenb. In male veal calves, however, hardly any effect of implantation with Trenb. on N retention was found in an ILOB experiment (E.J. Van Weerden and J. Huisman, unpublished data, 1979).

Samberev et al. (1967) tested Dianabol in an experiment with steers and found an improvement in weight gain. Heitzman et al. (1977) found a significant improvement in weight gain and feed conversion from an implantation with Trenb. in the case of steers. In three tests with 1190 steers in total, Roche and Davis (1978a,b) obtained an improvement in weight gain of 15—24% with an implantation of 300 mg Trenb. Bastiman and Scott (1977) performed eight field trials with steers implanted with Trenb. In two trials there was no response, but in the other six the daily gain in body weight was increased by 6—27%.

In female cattle especially, androgens have a positive effect on protein deposition (NAS, 1959). Rako and Mikulec (1968) confirmed this and mentioned that the carcases of heifers treated with Test. contained more meat than did the controls. Heitzman and Chan (1974) implanted heifers with 300 mg Trenb.; this implantation increased weight gain by 70%. Szumowski and Grandadam (1976) calculated an average increase in weight gain of 24.8% in heifers implanted with 200/250 mg Trenb.; an implantation with 300 mg Trenb. increased average weight gain by 27.0%. The effects found by Galbraith and Miller (1977) were similar. In an experiment with cull cows Béranger and Malterre (1968) found that daily gain was improved significantly after implantation with Trenb. and the amount of adipose deposit in the carcase was noticeably less. In an experiment with heifers

Chapter 10 references, p. 225

(Berende et al., 1977) oral doses of 4 and 8 mg/day of coated Trenb. improved weight gain by 13 and 17%, respectively.

Szumowski and Grandadam (1976) calculated an average growth improvement in castrated male lambs of 18.3 and 15% from an implantation of 40 and 60 mg Trenb. Implantation of 40 and 60 mg Trenb. in female lambs increased weight gain on average by 6.2 and 17.3%, respectively.

Schneeberger and Schürch (1961) and Leibetseder (1966) did not observe a clear improvement of Bolfortan and Dianabol on weight gain in broilers.

2.7. Endogenous gestagens, derivatives of endogenous gestagens and synthetic exogenous gestagens

Endogenous gestagens, mainly progesterone, are widely used in combination with estrogens. The synthetic progestational product, melengestrol acetate (MGA), has been used on a larger scale than any single progestative agent. MGA is fed orally in a dose of 0.4 mg animal^{-1} day^{-1}.

In experiments at ILOB with male veal calves (Berende, unpublished data, 1977) some derivatives of progesterone (Prog.) were tested; none of them improved weight gain.

With steers few experiments were done with MGA. Klinskij and Pucnin (1969) saw a very sharp increase in weight gain after MGA application. These results were not in agreement with those of Hawkins et al. (1967, 1972), who did not find a positive effect on weight gain.

Several authors found a significant improvement in weight gain and feed conversion by adding MGA to the feed of heifers (Hawkins et al., 1967, 1972; Davis and Truesdale, 1968; Ralston et al., 1968; Ray et al., 1968; Klosterman et al., 1970; Purchas et al., 1971a,b; Utley et al., 1972). Weight gain improvements varied from 6 to 10%; O'Brien et al. (1968b) even found an effect of 21%. In contrast to these results, Young et al. (1967), studying different levels of MGA, found hardly any effect on weight gain.

O'Brien and Miller (1967, 1970) and O'Brien et al. (1968a) fed MGA to female lambs and did not find any effect on weight gain.

2.8. Combinations of estrogens with androgens or gestagens

The literature describes a great number of experiments in which the effect of combinations of estrogens with androgens or gestagens was studied. In this section, first the results with the combination of estradiol and testosterone will be discussed and, next, the combinations of estradiol and trenbolone, estradiol and progesterone and DES and testosterone. The combinations estradiol-17β (benzoate) with testosterone (propionate), estradiol-17β (benzoate) with progesterone, and DES with testosterone are being used on a large scale.

In ILOB experiments (Berende et al., 1973; Van Weerden et al., 1973; Van der Wal, 1976) the combination Est./Test. proved to be effective in male veal calves as well as in steers. In Figs. 10.2 and 10.3 the results of an N balance and a comparative growth trial with male veal calves are summarized; in these experiments the effects of different combinations were compared with that of the application of Est. alone.

Gropp et al. (1976a,b) and Boehncke and Gropp (1976), in experiments with a small number of male veal calves, found an improvement in weight gain of 12% and in N retention of 24%, with an implantation of Est./Test. Weight gain improvement in female veal calves was increased to a similar extent. Schulz et al. (1976) performed an experiment with 1236 male and female veal calves under practical field conditions. Weight gain improvement

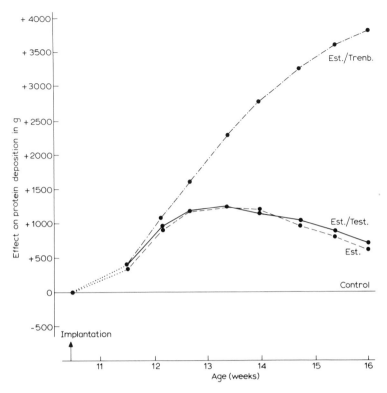

Fig. 10.2. Cumulative effect of different anabolic agents on protein deposition in male veal calves.
Liveweight controls: At age of 10.5 weeks: c. 97 kg
                      At age of 16 weeks: c. 145 kg
No. of animals per group: Control and Est. group: 6
                               Est./Test. and Est./Trenb. group: 3

was 11% in female calves and 1% in male calves in the liveweight range from about 70 to about 165 kg.

In older bulls only a few experimental data are available, in which Est./Test. have been tested. Preston (cited by Baker and Arthaud, 1972) found an improvement in weight gain of only 40 g/day.

In heifers of 307 kg liveweight Est./Test. improved weight gain significantly, according to trials of Ralston et al. (1968) and Ray et al. (1969).

Oxley et al. (1960a) observed an improvement in weight gain of about 50% in castrated lambs after an implantation of 5 mg Est. benzoate/50 mg Test. propionate.

In recent years a large number of experiments with the combination Est. and the synthetic androgen Trenb. have been performed. In ILOB experiments (Van der Wal et al., 1975a,b; Van der Wal, 1976; Berende, 1978) this combination proved to be very effective in male veal calves, bulls and steers (Figs. 10.2, 10.3 and 10.4, Table 10.3).

Grandadam et al. (1975a,b) and Szumowski and Grandadam (1976) found, in male veal calves implanted with Est./Trenb., a growth increase varying from 1.5 to 11%. These results correspond well with those of Gropp et al. (1976a,b) and Boehncke and Gropp (1976). In field trials with large numbers of veal calves Schulz et al. (1976) found an average improvement in weight gain of 6% in male and 16% in female calves.

In three trials with older bulls, Grandadam et al. (1975b) observed an average improvement in weight gain of 146.4% during a period varying from 50 to 81 days after implantation with Est./Trenb. But in two of the three trials weight gain of the control animals was very low (735 and

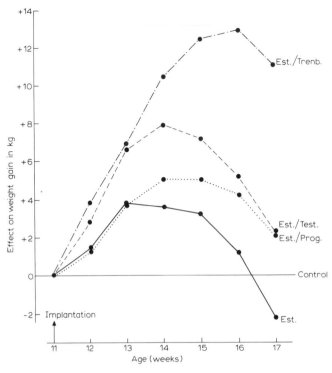

Fig. 10.3. Cumulative effect of different anabolic agents on weight gain in male veal calves.
Liveweight controls: At age of 11 weeks: 97.8 kg
                        At age of 17 weeks: 157.7 kg
No. of animals per group: 23

750 g/day). Szumowski and Grandadam (1976) mention in their literature survey the results of four and six experiments of two authors who found improvements in weight gain averaging 15.4 and 16.5%, respectively.

Heitzman (1976) and Heitzman et al. (1977) obtained a significant improvement in weight gain in steers implanted with Est./Trenb.

In female and male lambs, Szumowski and Grandadam (1976) concluded from literature data there would be an improvement in weight gain of 4.5—13% caused by an implantation with Est./Trenb. These authors and Grandadam et al. (1975b) found in castrated male lambs a growth stimulation of 10—27%.

The results of an implantation with Est. and Est./Trenb. in castrated male pigs are given in Fig. 10.5 (Van Weerden and Grandadam, 1976). In this experiment the combination Est./Trenb. was very effective. In later experiments the results were irregular and, on average, lower.

The combination Est. (benzoate) with Prog. is used on a large scale. In ILOB experiments with male veal calves a clear improvement in weight gain was detected (Berende et al., 1973; Van der Wal et al., 1975a). In an experiment with male veal calves Gropp et al. (1976a) found an improvement in weight gain of 12%. Grandchamp (1968) studied the implantation of Est./Prog. in bulls of different age and liveweight. The implantation improved weight gain, but the weight gain of the treated groups was more variable than that of the controls. Preston (cited by Baker and Arthaud, 1972) observed a weight gain improvement of 50 g/day.

Deans et al. (1956) implanted steers of about 450 kg with an unusual combination of Est. and Prog., viz. 50 mg and 1500 mg, respectively. Over a period of 140 days, average weight gain was improved by approximately 30%. Weight gain of the implanted animals was even better than that of a group

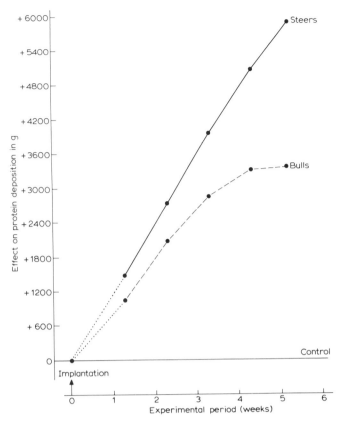

Fig. 10.4. Cumulative effect of estradiol/trenbolone acetate on protein deposition in bulls and steers.

Liveweight control bulls:
At age of c. 52 weeks: 320 kg
At age of c. 57 weeks: c. 360 kg
No. of animals per group: 6

Liveweight control steers:
At age of c. 104 weeks: 454 kg
At age of c. 109 weeks: 489 kg
No. of animals per group: 6

TABLE 10.3

Average values for protein deposition in control and anabolic-treated steers (period 3—36 days after implantation)

	No. of animals	Protein deposition	
		(g/day)	(%)
Control	7	200.6	100
Est./Trenb.	6	382.5	191
Est.	6	320.0	160
Trenb.	6	255.0	127
Est./Test.	6	329.4	164

fed 10 mg DES/day. This corresponds very well with the results of Ray and Child (1965).

Oxley et al. (1960a,b) implanted castrated male lambs with 2.5 mg Est. benzoate/25 mg Prog. and observed an improvement in weight gain of approximately 10% in their first experiment and 40% in the second. Andrews et al. (1956a) found an improvement of 46% in young lambs implanted with 10 mg Est. benzoate/250 mg Prog.

In an experiment with castrated male pigs of approximately 50 kg, an implantation with 3.3 mg Est. benzoate/166.7 mg Prog. had hardly any effect on weight gain and feed conversion (Day et al., 1960), but backfat thickness

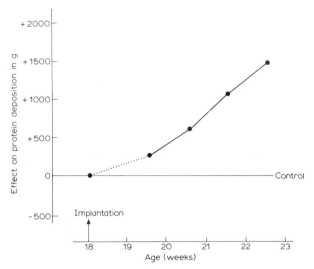

Fig. 10.5. Cumulative effect of estradiol/trenbolone acetate on protein deposition in castrated male pigs.
Liveweight controls: At age of 18 weeks: c. 55 kg
                              At age of 22 weeks: c. 80 kg
No. of animals per group: 6

was decreased significantly. There was a tendency for the treatment to decrease the amount of body fat and increase the amount of meat.

Combinations of estrogens with other hormones widely used in practice are preparations containing 24 mg DES and 120 mg Test. as implants in steers, and the combination of DES with methyltestosterone (MT), given orally in dosages of 2 or 2.2 g of each substance per ton (finisher) feed for pigs.

The combination of DES with Test. was studied in steers by Ray et al. (1969) and Brethour (1970). The results of these experiments were not conclusive; the effects were small and depended on the season (winter or summer). Beeson et al. (1956) studied the combination of 5 mg DES + 50 mg MT in steers of approximately 350 kg. Weight gain improvement with the oral application of these anabolic agents during a period of 179 days was the same as obtained with an oral application of 10 mg DES.

Glimp and Cundiff (1971) studied the implantation of 24 mg DES/120 mg Test. in a trial with 547 heifers at 1 and 2 years of age. In this experiment no real control group without treatment was available. Their conclusion was that this combination was not as effective as the progestative substance MGA. In a similar experiment Goodrich and Meiske (1969) came to the same conclusion.

In a number of experiments with castrated male pigs DES + MT decreased weight gain, gave lower backfat thickness and a more favourable feed conversion (Jordan et al., 1965; Trasher et al., 1967; Heldt and Lucas, 1970; Wahlstrom, 1970; Lucas et al., 1971, 1973; Bidner et al., 1972; Plimpton and Teague, 1972). Baker et al. (1967), Lucas et al. (1967) and Walker (1972) also found a lower weight gain of castrated males fed DES + MT, while feed conversion was hardly affected and backfat thickness was lower.

These effects can be explained by an increase in N retention and a decrease in fat deposition. This phenomenon was confirmed by the results of N balance studies of Melliere et al. (1970), who found a higher N retention when DES + MT was fed. The effects on N retention of DES and of the combination DES + MT were almost identical. In gilts the same tendency in the effects on feed conversion and slaughter quality (backfat thickness) of DES + MT was seen as with castrated males, though the differences were less pronounced in all cases (Jordan et al., 1965; Baker et al., 1967; Thrasher et al., 1967; Plimpton et al., 1971; Bidner et al., 1972).

2.9. Other anabolic agents

In this section only the results with zearalanol or zeranol (Zer.) will be discussed. The dosage used was in most cases 36 mg as an implant for cattle and 12 mg as an implant for lambs.

Some publications describe experiments in which Zer. was tested in veal calves. Zucker et al. (1972) noticed a distinct improvement in weight gain during a period of 5 weeks after implantation with 24 mg Zer. in male and female veal calves at a liveweight of 63 kg. Ralston (1978) found no effect in male calves implanted at birth with 36 mg Zer. In our experiments a small effect on N retention and weight gain was seen when male veal calves were implanted with 36 and 72 mg Zer. at the age of 5 and 11 weeks. Thomas et al. (1970) found with calves of this age a growth promotion of 2—5% in three trials, in which an implantation of 36 mg Zer. was performed. In a trial with 437 veal calves Brown (1970) found an average improvement in weight gain of 9.0%.

Most experiments in which Zer. was tested have been carried out with steers. The following positive effects of an implantation with 36 mg Zer. are mentioned: Thomas and Armitage (1970) 6.9 and 4.6%; Perry et al. (1970) 8.9%; Thomas (1970) 19%; Woods (1970) 14%; Roche and Davis (1978a,b) 29.3, 21.3 and 14.0%; and Shorrock et al. (1978) 11.9, 24.5 and 12.3%, and under feedlot conditions 17.9%. Sharp and Dyer (1971) and Bastiman and Scott (1977) performed a series of experiments with implantations of 36 mg Zer.; the effects varied between 4.9 and 25.9% and from 0 to 36%, respectively. Brown (1970) mentioned a study with 5000 beef cattle and 1380 (veal) calves. In these trials an average growth promotion of 10% was found. The following authors mention effects of Zer. similar to an implantation with DES: Dunbar (1969); Perry et al. (1970) and Sharp and Dyer (1971). The differences in effect between 36 and 72 mg Zer. seem to be small (Thomas and Armitage, 1970; Perry et al., 1970).

Sewell (cited by Bridson, 1972) established that the effects of Zer. were similar in heifers and in steers. This was confirmed by the experiments of Sharp and Dyer (1971). Algeo et al. (1970) found, with heifers of about 238 kg, small effects varying from 0 to 12%, and Utley et al. (1976) noticed a growth promotion of only 2.6%, but in their trial the combination Est./Test. had no effect at all; this is in contrast with the results of several authors who found positive effects of this combination.

Brown (1970) calculated from a number of experiments with 2280 suckling lambs and 6000 feedlot lambs in total (males or castrated males?) an average growth promotion of 11.4 and 15.4% after an implantation of 12 mg Zer. The same effects were found by Wilson et al. (1972b) in castrated male and female lambs. Jordan and Hanke (1970), Wilson et al. (1972a,b), Vipond and Galbraith (1978) and Wiggins et al. (1979) established smaller effects, namely 5, 7, 0 and 5%, respectively. A small, non-significant improvement in weight gain with an implantation of 12 mg Zer. was indicated by Wilson et al. (1972a) in an experiment in which rations of 9.5, 11.0 and 12.5% crude protein were fed.

Chapter 10 references, p. 225

3. FACTORS INFLUENCING ANABOLIC EFFECTIVENESS

The following factors can influence the magnitude of the effect of an anabolic agent:
(1) kind of anabolic agent;
(2) dose level;
(3) species, age, breed and sex;
(4) route of administration and carrier used;
(5) reimplantation or reinjection; and
(6) diet

3.1. Type of anabolic agent

In the foregoing section the effects obtained with several anabolic agents or combinations of anabolic agents have been described. In veal calves the effects of DES, Est./Test., Est./Prog. and Est./Trenb. are very pronounced (Figs. 10.2, 10.3, 10.6, 10.7, 10.8, 10.9 and 10.10). In steers DES and Est./Prog., in heifers MGA, and in pigs the combination DES + MT are the effective compounds used in practice.

3.2. Dose level

In three experiments with a total of 180 male veal calves the effects of injection with 10 and 25 mg DES were compared (Berende et al., 1973). The difference in reaction of the animals was small and not significant, though there was a tendency for the 25 mg DES treatment to have a somewhat greater effect.

In Fig. 10.6 the results are given of an experiment (Berende et al., 1973; Van der Wal, 1976) with male veal calves in which 25 or 100 mg DES was administered subcutaneously in the neck. Two groups were treated once and two groups were treated twice with the doses mentioned. From this experiment it may be concluded that an increase in the total dose of DES from 25 to 2×100 mg resulted in only a relatively small increase in effect.

In another experiment with male veal calves (Berende et al., 1973) two oral doses of DES and two oral doses of Dien. were tested (Fig. 10.7). DES and Dien. were fed during the period 10—17 weeks of age. The growth responses to these treatments were compared with that to an injection of 25 mg DES subcutaneously at 10 weeks of age. The reaction of the group of calves injected with 25 mg DES was not identical to that of the calves in other experiments after the same treatment. In this experiment the maximum effect was lower and at the end of the experiment no positive influence could be observed.

The administration of 5 mg DES with the milk replacer during 7 weeks had a slight, non-significant, positive effect on weight gain, whereas with 10 mg DES weight gain was improved more ($P \leqslant 0.20$). The addition of 1 and 5 mg Dien. improved weight gain similarly to 5 mg DES. These results correspond well with those collected by Preston and Willis (1970), who observed that increasing the oral dose of DES from 10 to 24 mg and the implantation dose from 24 to 60 mg scarcely influenced the magnitude of the effect.

3.3. Species, age, breed and sex

Anabolic agents prove to be very effective in cattle (male and female calves, heifers, bulls and steers) and effective in sheep. The effects in castrated male pigs are very irregular and, on average, moderate, and in poultry it

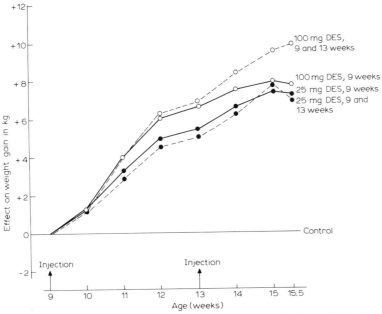

Fig. 10.6. Cumulative effect of different dosages of DES on weight gain in male veal calves.
Liveweight controls: At age of 11   weeks: 79.8 kg
At age of 15.5 weeks: 141.6 kg
No. of animals per group: 10

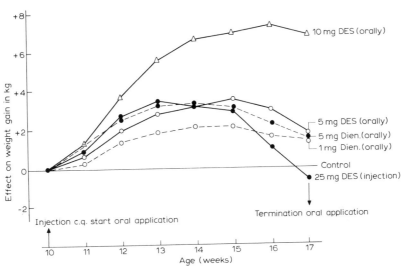

Fig. 10.7. Cumulative effect on weight gain of male veal calves of two dosages of DES and of two dosages of dienestrol.
Liveweight controls: At age of 10 weeks: 92.8 kg
At age of 17 weeks: 157.6 kg
No. of animals per group: 10

seems that there is a marked increase in fat deposition and generally a decrease in N retention (Nesheim, 1976). In Table 10.4 the results of experiments performed at ILOB are compiled in which Est./Trenb. was applied to different species and categories of farm animals.

Figs. 10.8, 10.9 and 10.10 show the effect of treatments with anabolic agents at different ages (Berende et al., 1973). From Fig. 10.8 it can be concluded that, during the period studied, the maximum response to DES

TABLE 10.4

Average values for weight gain in control and estradiol/trenbolone acetate treated farm animals (test period 0 until 7—8 weeks after implantation)

	Cattle					Sheep			Pigs		
	Female calves	Male calves	Heifers[a]	Bulls	Steers	Male lambs[b]	Ewes	Wethers	Females	Males[c]	Castrated males
No. of animals/group	8	57	16	18	18	30	15	15	32	5	32
Liveweight at the start of the experiment (kg)	75.6	97.3	366.6	384.2	424.9	26.9	53.1	58.2	52.6	57 5	54.6
Wt. gain of the controls (g animal⁻¹ day⁻¹)	1163	1388	932	1275	1007	222	63	86	727	908	696
Wt. gain of the treated animals (as % of the controls)	122.8	107.0	120.9	118.5	147.5	110.9	87.0	164.2	102.2	97	104.1

[a] 8 mg Trenb. (oral) animal^{-1} day^{-1}.
[b] Extrapolated from 4 to 7 weeks.
[c] N Balance trial; N retention of the treated animals 102.3% of the controls.

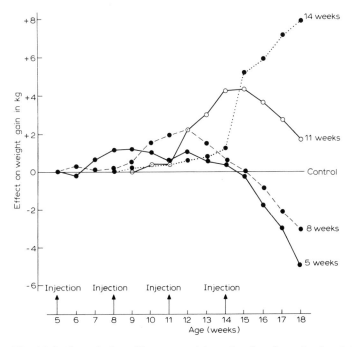

Fig. 10.8. Cumulative effect on weight gain of male veal calves injected with 10 mg DES at different ages.

Liveweight controls: At age of   5 weeks: 52.5 kg
                       At age of 11 weeks: 96.0 kg
                       At age of 18 weeks: 168.9 kg
No. of animals per group: Groups treated at   5 and   8 weeks: 20
                            Groups treated at 11 and 14 weeks: 10

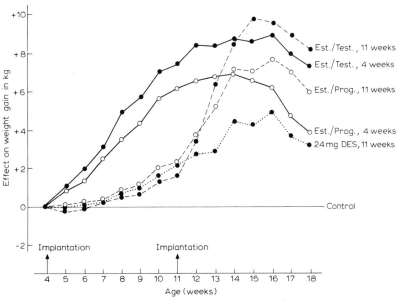

Fig. 10.9. Cumulative effect on weight gain of male veal calves of combinations of anabolic agents implanted at different ages.

Liveweight controls: At age of   4   weeks: 49.6 kg
                       At age of 17.5 weeks: 163.2 kg
No. of animals per group: 10

Chapter 10 references, p. 225

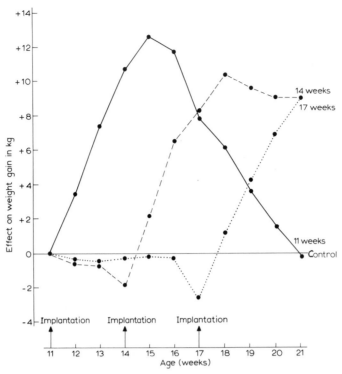

Fig. 10.10. Cumulative effect of estradiol/trenbolone acetate on weight gain in male calves implanted at different ages.
Liveweight controls: At age of 11 weeks: 96.9 kg
                                         At age of 21 weeks: 197.9 kg
No. of animals per group: 12

was higher and was obtained more quickly in older calves. The results given in Fig. 10.9 show that, taking into account the approximately 2 kg lead in weight gain of the groups implanted at 11 weeks over the control group just before implantation, the maximum response was not higher for the calves implanted at 11 weeks, but the effect was again obtained in a shorter period of time. Fig. 10.10 shows that after an implantation with Est./Trenb. at 11, 14 or 17 weeks of age the effects were almost identical and the maximum was reached in approximately the same period of time.

With castrated male pigs in the liveweight range 50—75 kg, no clear influence of liveweight or age could be detected (Berende, unpublished data, 1977; E.J. Van Weerden and J. Huisman, unpublished data, 1978).

From the experiments of C. Van Eenaeme (personal communication, 1980) it was very clear that the effect of an anabolic agent was more pronounced in dual-purpose-type animals than in 'double-muscling'-type animals.

The influence of sex on the effect was studied in two comparative growth experiments and two N balance trials with bulls and steers and in one N balance trial with male and castrated male pigs (Van Weerden and Grandadam, 1976; Berende, 1978; E.J. Van Weerden and J. Huisman, unpublished data, 1979).

In Fig. 10.11 the effects of the anabolic agents on weight gain are given for the two comparative growth experiments with bulls and steers. The bulls were implanted with 40 mg Est./200 mg Trenb. and the steers with 20 mg Est./140 mg Trenb. Weight gain of the control steers stayed far behind that of the control bulls, being 994 and 1145 g/day, respectively, in the experimental period. The effect of the anabolic agent was much higher for the steers than for the bulls, and, as a result, weight gain of the treated steers and the treated bulls was nearly identical during the whole experimental period.

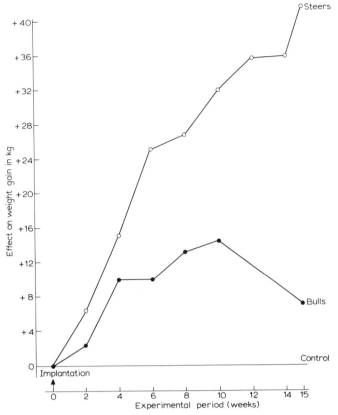

Fig. 10.11. Cumulative effect of estradiol/trenbolone acetate on weight gains in bulls and steers.

Liveweight control bulls:
At age of c. 52 weeks: 384.2 kg
At age of c. 67 weeks: 528 kg*
No. of animals per group: 18
*estimated.

Liveweight control steers:
At age of c. 70 weeks: 424.9 kg
At age of c. 85 weeks: 529.3 kg
No. of animals per group: 18

In the N balance trials (Fig. 10.4) the bulls and steers were implanted with 40 mg Est./200 mg Trenb. Protein deposition for the control bulls was 7912 g and for the control steers 4259 g over a period of 6 weeks. The effect of the anabolic treatment on protein deposition was much higher for the steers than for the bulls. Total protein deposition in this 6-week period was 11 300 g for the treated bulls, and 10 200 g for the treated steers.

The results of an experiment with male and castrated male pigs are shown in Fig. 10.12 (E.J. Van Weerden and J. Huisman, unpublished data, 1978). In this experiment the pigs were fed a ration to which Eth.est. and Trenb. were added at a level of 2 ppm of each. The anabolic agents were fed for a period of 5 weeks. As could be expected, protein deposition was higher for the control male pigs than for the control castrated male pigs. The effect of the anabolic agents was almost negligible in the intact pigs and small for the castrated animals.

3.4. Route of administration and carrier used

The initial method of estrogen administration was by subcutaneous implantation. This route is still used extensively and provides for an efficient use of the hormones under a wide variety of conditions (NAS, 1966). This subcutaneous implantation can be done at several sites in the body, for example the neck, the dewlap, the earshell, the base of the ear, the scrotum,

Chapter 10 references, p. 225

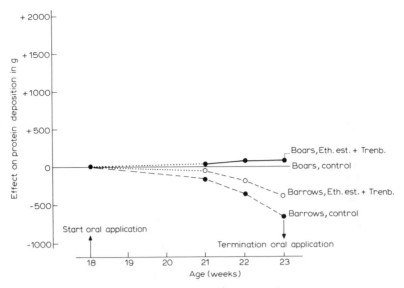

Fig. 10.12. Cumulative effect on protein deposition of male and castrated male pigs of the oral application of ethinyl-estradiol + trenbolone acetate.

Liveweight control boars:	*Liveweight control barrows*:
At age of c. 18 weeks: 57 kg	At age of c. 18 weeks: 57 kg
At age of c. 23 weeks: 91 kg	At age of c. 23 weeks: 85 kg
No. of animals per group: 5	*No. of animals per group*: 5

and the eyelid. Other methods of administration practised are injection, subcutaneous as well as intramuscular, and, for special purposes, intra-peritoneal and oral application, and in females intravaginal devices (sponges).

From the results of our experiments we have a strong impression that implantation in the earshell, in the base of the ear and in the dewlap give almost identical results. Similarly, Rumsey et al. (1974, 1975) did not find differences in effect when the substances were implanted at different sites. However, the site of an injection plays a more important role because in general a substance is injected in such a form that it will act more quickly than an implantation. Therefore, it is conceivable that an injection applied intramuscularly acts more quickly than when given subcutaneously.

The type of carrier used depends on the period of time during which the anabolic agent must be released from the depot. When it has to be released relatively slowly, e.g., in steers treated 4—5 months before slaughter, another carrier must be used than in veal calves in which the anabolics must be released quickly. The maximum time between treatment and slaughter in veal calves is 3 months. The choice of the carrier depends not only on the time between application and slaughter, but also on the species. Pellet composition and selection of a suitable vehicle is of more importance in the anabolic treatment of sheep than of cattle. Sheep are more sensitive to anabolics than cattle with respect to the occurrence of rectal prolapse and increased incidence of urinary calculi (Bell et al. and Jordan, cited by NAS, 1966). In a hard pellet the anabolic(s) will be released more slowly than in a soft pellet.

3.5. Reimplantation or reinjection

In an experiment with male veal calves, 25 and 100 mg DES were injected subcutaneously once and twice (Berende et al., 1973). With the 25 mg dose no extra weight gain was obtained after a second injection (Fig. 10.6). A second injection with 100 mg DES had a minor positive influence on weight gain in comparison with one injection of 100 mg DES. During the first 5 weeks of this experiment in which DES was injected, the results are in agreement with those of an experiment in which male veal calves of about the same age were implanted once or twice with the combination Est./Trenb. (Fig. 10.13; P.L.M. Berende, unpublished data, 1979).

In this latter experiment, treatment with the anabolic agents resulted in a growth depression compared with untreated calves, from 7 weeks after implantation onwards. Reimplantation was successful in this experiment, but the time between reimplantation and slaughter was too long to give any benefit in liveweight at slaughter. At slaughter (23 weeks of age) weight gain of the double-treated animals was also lower than that of the controls. The same was found by O'Mary et al. (1956) in an experiment with 50 steers. Reimplantation with 36 mg DES 42 days after the first implantation increased weight gain during the following weeks, but the effect of reimplantation over the whole fattening period was small.

3.6. Diet

In the literature no consistent effect of treatment with anabolic agents on appetite is suggested, although Clegg and Cole (1954) could detect an increased feed consumption after administration of DES. In ILOB experiments the interaction of anabolic effect and protein supply was studied. In Fig. 10.14 the results of an experiment with male veal calves given diets with different protein levels are shown (P.L.M. Berende, unpublished data, 1979). At normal crude protein levels of 18 and 20.5% and at a

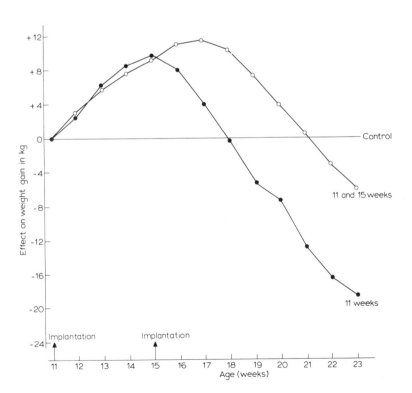

Fig. 10.13. Cumulative effect of estradiol/trenbolone acetate on weight gain in male veal calves implanted once or twice.
Liveweight controls: At age of 11 weeks: 94.5 kg
                At age of 23 weeks: 207.1 kg
No. of animals per group: 8

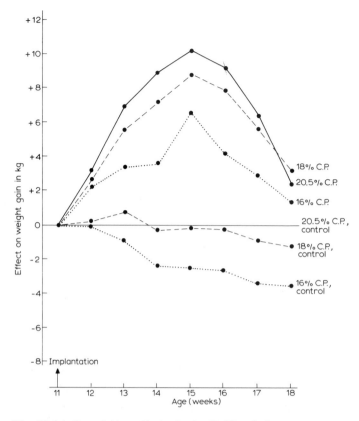

Fig. 10.14. Cumulative effect of estradiol/trenbolone acetate on weight gain in male veal calves fed rations with 16, 18 and 20.5% crude protein (C.P.).
Liveweight controls: At age of 11 weeks: 96.1 kg
                                          At age of 18 weeks: 161.6 kg
No. of animals per group: 18

somewhat lower level of 16% crude protein, the magnitude of the effect of the anabolic agent was about the same.

In an N balance study (Fig. 10.15) the difference in protein content of both rations was more extreme (E.J. Van Weerden and J. Huisman, unpublished data, 1979); 12% crude protein is far below the protein requirement of veal calves of this age. In this experiment the effect of the anabolic agent at the low protein level was much lower, in absolute as well as in relative terms, than at the 21% protein level. The same conclusion can be drawn from an experiment with pigs (Berende, unpublished data, 1977) in which the animals were fed rations with 13 and 16% crude protein (Fig. 10.16).

From experiments of E.J. Van Weerden (unpublished data, 1980) it can be concluded that the amino acid requirements of veal calves are not influenced by treatment with an anabolic agent. These results are in good agreement with those of Klosterman et al. (1959) who found that the protein requirement of steers was not significantly affected by stilbestrol treatment. The experiment of Klosterman et al. also showed that maximum response to DES was dependent on the presence of adequate crude protein in the diet; the feeding of additional protein over the normal requirement was not beneficial.

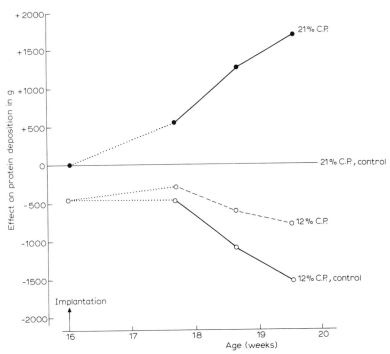

Fig. 10.15. Cumulative effect of estradiol/trenbolone acetate on protein deposition in male veal calves fed rations with 12 and 21% crude protein (C.P.).

Liveweight controls (21% C.P.):   *Liveweight controls* (12% C.P.):
At age of 16 weeks: 149 kg   At age of 16 weeks: 143 kg
At age of 20 weeks: 192 kg   At age of 20 weeks: 178 kg
No. of animals per group: 5   *No. of animals per group*: 5

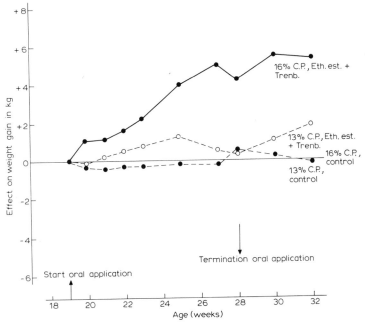

Fig. 10.16. Cumulative effect of 2 ppm ethinyl-estradiol + 2 ppm trenbolone acetate on weight gain in castrated male pigs fed rations with 13 and 16% crude protein (C.P.).

Liveweight controls: At age of c. 19 weeks: 56.0 kg
                 At age of c. 32 weeks: 119.6 kg

No. of animals per group: 24

Chapter 10 references, p. 225

4. MODE OF ACTION

In the foregoing pages it has been shown that the anabolic agents under discussion in these sections are capable of increasing protein deposition in cattle, sheep and pigs. The mechanism by which these agents act is illustrated in Figs. 10.17 and 10.18 and Table 10.5.

In Fig. 10.17 (Van Weerden, 1972) the results of 192 N balance studies with male veal calves of c. 40—c. 140 kg are summarized. It is evident that with increasing intake of digestible N — that is, with increasing age — the N retention calculated as % of the intake of digestible N gradually decreases. The efficiency with which the digestible N is converted to body N decreases from approximately 70% in calves of 40 kg to approximately 35% in calves of 130—140 kg. In pigs the phenomenon of decreasing efficiency of protein deposition with increasing age is indicated in studies of Oslage (1965) and Nielsen (1970).

There is general agreement in the literature that anabolic agents do not affect the digestibility of the diet. In none of our N balance studies has any effect of anabolic treatment on N digestibility been found. As indicated in Fig. 10.18 (Van Weerden et al., 1973), anabolic agents act by influencing the intermediary N metabolism in such a way that the share of the digested feed-protein converted into body-protein is raised. In the veal calf experiment mentioned in Fig. 10.18, the efficiency of the conversion of digested feed-protein into body-protein was temporarily increased from approximately 40% to 60% after the calves were implanted with Est. and with Est./Test.

The increase of N retention after application of anabolic agents was not accompanied by an equivalent increase of energy retention, as is shown in N balance—energy balance experiments in Table 10.5 (pigs: Van Weerden and Grandadam, 1976; veal calves and steers: ILOB experiments, unpublished results).

Table 10.5 shows that while, in general, energy deposition per 1000 kcal metabolizable energy intake is not greatly influenced by the application of

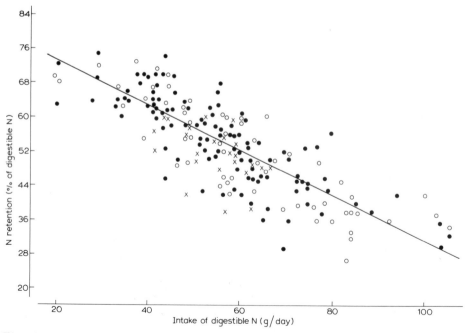

Fig. 10.17. Relation between intake of digestible N and N retention in veal calves.

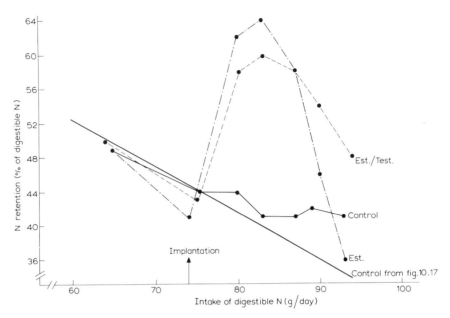

Fig. 10.18. Relation between intake of digestible N and N retention in control and in anabolic-treated veal calves.

anabolic agents, protein deposition is distinctly increased and, consequently, fat deposition remains essentially unaffected or is decreased. The conclusion might be that in cattle and pigs the application of anabolic agents does not result in a distinct increase in the retention of energy, but causes a shift to a higher protein and lower fat deposition.

This anabolic action which, from a zootechnical point of view, is very desirable, is also reflected in the results of the determinations of slaughter quality, a summary of which is given in Tables 10.6, 10.7 and 10.8. In steers the visual appraisal for meatiness of the carcases of the treated animals was more favourable than for the untreated controls. In veal calves also, dressing percentage was higher for the treated animals, and in pigs the measurement of backfat thickness showed lower values for the anabolic-treated animals.

The exact mechanism by which anabolic agents affect the observed shift from fat to protein deposition is not known. A number of workers have tried to find indications of the mode of action of anabolic agents by analyzing blood (plasma) for: estradiol (several metabolites), testosterone, growth hormone, thyroxine (PBI), insulin, glucose and urea (Borger et al., 1973; Hoffmann et al., 1976; Trenkle, 1969; Trenkle and Burroughs, 1978). In several studies carried out to explain the mode of action of hormones, the weight of several organs was studied: thymus and testicles (P.L.M. Berende, unpublished data, 1979), adrenals (Clegg and Cole, 1954; Cahill et al., 1956), anterior pituitary (Struempler and Burroughs, 1959), thyroid gland (Trenkle, 1969), seminal vesicles and bulbo-urethrales (Preston and Burroughs, 1960). No logical explanation of the mode of action of hormones resulted from these studies. Consequently, the way in which the growth-promoting substances bring about their effects remains uncertain. The fact that they produce similar results does not of course necessarily mean that all act in the same way. Androgens and estrogens affect protein metabolism in very different ways (Scott, 1978).

Estrogens, while inducing the production of specific proteins in tissues such as the uterus, liver and chick oviduct, do not appear to induce muscle protein deposition directly, but presumably act by altering the pattern of endogenous anabolic hormones (Buttery et al., 1978). Trenkle (1969),

Chapter 10 references, p. 225

TABLE 10.5

Average values for energy, protein and fat deposition in control and anabolic-treated male veal calves, steers and castrated male pigs (in g and kcal per 1000 kcal metabolizable energy intake)

Kind of animal	No. of animals	Treatment	Liveweight at start of experiment (kg)	Energy deposition kcal	%	Protein deposition g	kcal	%	Fat deposition g	kcal	%
Male calves	6	Control	101	381	100	22.5	128	100	26.6	253	100
	6	120 mg DES (p.i.)	104	386	101	31.8	181	141	21.6	205	81
Steers	2	Control	462	203	100	9.4	54	100	15.7	149	100
	2	20 mg Est.	466	177	87	14.4	82	152	10.0	95	64
	2	20 mg Est./200 mg Test.	466	227	112	15.6	89	165	14.5	138	93
	2	20 mg Est./140 mg Trenb.	462	192	95	18.6	106	196	9.1	86	58
Castrated male pigs	6	Control	69	486	100	14.5	83	100	42.5	403	100
	6	20 mg Est./140 mg Trenb.	72	458	94	22.1	126	152	35.0	332	82

TABLE 10.6

Average values for slaughter quality criteria in control and anabolic-treated male veal calves

Treatment:	DES		Zer.		Est.		Est./Prog.		Est./Test.		Est./Trenb.	
Experimental groups:	Control	Test	Control	Test	Control	Test	Control	Test	Control	Test	Control	Test
No. of animals	70	220	25	50	23	23	33	43	58	68	120	177
No. of experimental groups	5	22	2	4	2	2	3	4	5	6	11	16
Liveweight at slaughter (kg)	158.4	160.4	151.9	153.6	156.0	153.1	156.0	158.9	154.4	158.1	171.5	175.5
Warm carcase wt. (kg)	105.2	107.7	101.2	103.0	104.2	103.1	104.0	107.0	102.9	105.9	113.9	116.6
Dressing percentage	66.4	67.2	66.6	67.0	66.8	67.3	66.7	67.3	66.6	67.0	66.4	66.5
Meatiness[a]	7.6	8.0	6.8	7.2	7.0	7.1	7.0	7.6	6.9	7.3	6.7	7.2

[a] Appraisal 1—10; the higher the figure the better the appraisal.

TABLE 10.7

Average values for slaughter quality criteria in control and anabolic-treated bulls and steers

	Bulls		Steers	
	Control	40 mg Est./200 mg Trenb.	Control	20 mg Est./140 mg Trenb.
No. of animals	18	17	18	18
Warm carcase wt. (kg)	295.3	297.7	297.7	319.1
Wt. of kidney and pelvic fat (g)	3864	3479	8784	7564
Meatiness[a]	3.9	3.9	3.5	3.8
External fatness[b]	2.6	2.5	3.3	3.1
Internal fatness[b]	2.4	2.3	3.3	3.0

[a]Rating from 0.67 to 6.33; the higher the value the better the appraisal.
[b]Rating from 0.67 to 6.33; optimum between values of 2.0—3.5.

TABLE 10.8

Average values for slaughter quality criteria in control and anabolic-treated castrated male pigs

Treatment:	Eth.est.[a]		Est./Trenb.[b]		Est./Trenb. and Eth.est. + Trenb.		
Experimental groups:	Control	Test	Control	Test	Control	Test[b]	Test[c]
No. of animals	24	23	34	36	29	31	31
Liveweight at slaughterhouse (kg)	118.9	122.1	104.3	106.1	118.0	117.9	118.7
Warm carcase wt. (kg)	97.3	99.3	84.6	87.1	97.0	96.6	96.7
Dressing percentage	81.8	81.6	81.1	82.1	82.2	81.9	81.5
Ave. backfat thickness (mm)[d]	32.9	30.7	29.6	29.2	38.3	36.8	36.6
Wt. eye muscle:							
in kg			15.44	16.15			
as % of liveweight			14.8	15.2			
Wt. ham:							
in kg			20.46	21.06			
as % of liveweight			19.6	19.8			

[a]Eth.est. : 2 ppm ethinyl estradiol in the ration (oral application).
[b]Est./Trenb. : 20 mg estradiol-17β/140 mg trenbolone acetate (implantation).
[c]Eth.est. + Trenb. : 2 ppm ethinyl estradiol + 2 ppm trenbolone acetate in the ration (oral application).
[d]measured at four sites along the back of the animal.

Trenkle and Burroughs (1978) and Buttery et al. (1978) explained the growth-promoting actions of estrogens by increased thyroid activity, increased growth hormone secretion, increased production of androgens from the adrenal cortex, and a direct effect of estrogens at the tissue level. There is some experimental evidence to support each of these proposals, but it is indirect and quite variable (Trenkle and Burroughs, 1978). Most of the experimental data support the hypothesis that increased secretion of growth hormone in response to estrogens results in faster growth of animals.

The most direct evidence is that estrogen administration is followed by larger pituitary glands containing more growth hormone, and higher concentrations of growth hormone in the plasma. The androgens induce the synthesis of protein by regulation of the ribonucleic acid and the protein biosynthesis system at the microsomal level (Kochakian, 1966; Michel and

Baulieu, 1976). This theory has been further elaborated by Schulster et al. (1976). A possible second mode of action is based on the rate of protein degradation. Androgens are known to displace corticosteroids. Corticosteroids are potent catabolic agents and may serve a regulatory and suppressive role in normal growth. It is not inconceivable that androgens may limit this role in animals by replacing corticosteroids at the receptor site (R.J. Heitzman, unpublished data, 1977).

According to Krüsskemper (1965) steroid hormones are active as co-enzymes in enzyme systems.

At present there is no completely satisfactory explanation for the manner in which anabolic agents influence protein synthesis, although it does appear that androgens may act in the muscle itself, whereas estrogens act indirectly by stimulating growth hormone production. It is not known why certain estrogen/androgen combinations are more effective and act over a longer period than the individual hormones. The combinations used in practice have been developed empirically.

5. THE EFFECT OF ANTIHORMONES

Antihormones are organic substances which may decrease the production, release or action of hormones. The action of antihormones was described by Van Oordt as follows:

> 'When an animal is treated for a long time with, for example, proteo-hormones coming from other species, it can happen that suddenly the animal will not react any more to the treatment. So, antihormones are produced after injection of hormones coming from the adenohypophysis of other species (thyroid stimulating hormone and gonadotropic hormones); these hormones are glycoproteins in nature. At first the hypothesis was that certain organic substances, antihormones, are secreted. These substances were supposed to be antagonistic to the hormones the animals were treated with. Gradually the idea developed that these substances arose in the body as a reaction to a foreign protein. Injection of hormones of simple chemical composition like adrenalin (a phenol) or of sex hormones of the steroid type, did not cause the production of the so called antihormones. Though insulin is a protein, the formation of an antihormone very rarely takes place' (Van Oordt, 1958, translated from Dutch by Berende).

In this survey special attention will be paid to those antihormones which may give an improvement in weight gain. In practice the only agents used are goitrogens, their function being to depress thyroid activity. The following two goitrogens are most frequently used: 1-methyl-2-mercaptoimidazole (Methimazole, Thiamazole or Tapazole) and 2,3-dihydro-6-methyl-2-thioxo-4(1H)-pyrimidinone [methylthiouracil (MTU), methiacil or thyreostat I]. Besides these commercial products there are goitrogenic substances present in such plants as the *Brassica* species. Goitrogens or thyrostats suppress the function of the thyroid gland (Raun et al., 1960); thus the production of thyroxin is lowered. A reduced level of thyroxin in the blood decreases the basal metabolism and thus less feed will be used for maintenance. The opposite effect, stimulation of the thyroid gland, can be achieved by using products such as iodized casein and thyroprotein (Whiteker et al., 1959; Peo and Hudman, 1960; Karg, 1966).

The results of the effect of antithyroid drugs, according to the literature, are very contradictory. Frens (1949, 1958) fed MTU to adult cows; his conclusion was that MTU caused an extra liveweight gain, 50—70% of which could be accounted for by an increase in the contents of the rumen and, in one experiment, by an increase in the intestinal fill also. Wöhlbier and

Schneider (1966) did not agree. Their conclusion, based on a balance and respiration trial, was that water retention was higher (myxoedema) after treatment with MTU. Fields et al. (1971) found an increase in water content of the eye muscle by feeding 800 mg Tapazole/day in steers implanted with DES. Average weight gain and feed conversion were more favourable for the treated animals. This increase in weight gain was not caused by increasing contents of stomach and intestine, because the dressing percentages were equal for both groups. Fox et al. (1973) found that some muscle characteristics and the fatty acid composition of the adipose tissues were changed. The literature surveys of Winchester (1953) and NAS (1959) mention very contradictory results with goitrogens in swine and ruminants.

6. TOXICOLOGICAL ASPECTS, BENEFIT—RISK ASSESSMENT AND LEGAL CONSEQUENCES OF THE USE OF ANABOLIC AGENTS

The use of anabolic agents to increase meat production in farm animals is very controversial and a number of international meetings have been devoted to this subject; in 1975, the FAO/WHO Symposium on Anabolic Agents in Animal Production in Rome (1976), in 1980 the International Symposium on Steroids in Animal Production in Warsaw (1981), in 1981 the EEC Workshop on Anabolic Agents in Beef and Veal Production in Brussels (1981), and the meeting organized by the World Health Organization Regional Office for Europe of a Working Group on Health Aspects of Residues of Anabolics in Meat in Bilthoven.

There is understandable concern that any substance used in animal production shall involve any hazard to the user, the stock or the consumer of the eventual product (Scott, 1978). Estrogens, especially DES, are potential carcinogenic agents (NAS, 1966; IARC, 1974) and for that reason the use of DES is forbidden in most Western European countries and since 1 November 1979 also in the U.S.A. In several countries in Western Europe the use of other anabolic agents is also forbidden; in other countries the use is legalized. In general, the situation is confusing. In some countries 'all' products can be used, in other countries 'all' anabolic agents are forbidden. The future of the use of anabolic agents as described by Scott could be as follows:

> 'the growth promoters at present available for implantation are effective and are regarded in many countries as being acceptably safe in use. Understandably, there are those who have reservations, and it may be that the stimulus they provide to legislation may overtake the situation. However, new substances may be found or synthesized which, although growth promoting, can be clearly shown to present no risks to stock or consumer. In the meantime, the variation in the method of administering these substances by oral feeding, ear clips or nose clips or vaginal inserts which can be removed at a stated time before slaughter, may provide an interim answer' (Scott, 1978).

When using anabolic agents it is necessary to have sensitive methods to demonstrate residues of the agents or their metabolites, especially in the consumable products. When using natural hormones like estradiol-17β, testosterone and progesterone, it is necessary to know the normal levels of these hormones in blood, urine and, especially, in the consumable product. This is essential in determining whether or not the level in the tissues is increased as a result of using them. Analytical methods for the determination of naturally occurring hormones are now being developed at various centres. Interest is focused on excreta, such as urine and faeces, and organs, such as liver and kidney, as well as other edible tissues.

Chapter 10 references, p. 225

In this section, first, some toxicological aspects will be described, next a benefit—risk assessment for the consumer together with the possible legal consequences will be discussed.

6.1. Toxicological aspects

As mentioned earlier in this chapter, the sources and kind of residues of anabolic agents in meat fall into three groups:
  (a) residues of estrogens, progestagens and androgens derived from the animals' own endocrine glands (endogenous natural steroids);
  (b) residues derived from natural steroidal anabolic agents administered to animals during life (exogenous natural steroids); and
  (c) residues derived from xenobiotic agents administered during life (exogenous xenobiotic agents).

From these residues the natural steroids are readily metabolized by the liver, and for this reason have little or no activity when given or taken up by the oral route. From a public health point of view, three different xenobiotic agents have to be considered:
  (i) zeranol which is readily metabolized by the liver and only weakly active by mouth;
  (ii) trenbolone acetate, which is well metabolized by the liver and therefore only weakly active after oral administration; and
  (iii) synthetic stilbenes, including DES, Hex. and Dien. These are active by the oral route, being little metabolized by the liver. Features of these compounds not shared by trenbolone or zeranol are their high bioavailability and low biodegradability.

6.1.1. Naturally-occurring anabolic agents

The levels of endogenous natural steroids in meat from untreated animals vary widely, depending on the sex and maturity of animals and on whether female animals are pregnant (Hoffmann, 1981). The highest levels of androgens are found in the mature bull and the highest levels of estrogens in cows or ewes during the later stages of pregnancy. Under the conditions of their recommended use, the treatment of animals with exogenous natural steroids results in residues in meat and other edible tissues derived from treated animals that are orders of magnitude lower than those which can occur naturally (e.g., in bulls and pregnant animals). Furthermore, residues of natural steroidal hormones, whether derived from untreated or treated animals, are of negligible toxicological potential because they are readily degraded in the liver of the meat consumer and therefore do not reach relevant hormone receptor sites in the human body.

The only cause for concern associated with the treatment of animals with natural steroids is a possible depot at the site of administration. The risk of consumption by humans can be avoided easily by clearly defining and controlling its use. Implantation of a pellet into the base of the ear and removal of it before or at the time of slaughter would constitute a conceivable and safe practice.

6.1.2. Synthetic anabolic agents

Residues of xenobiotic anabolic agents pose questions of safety, firstly because the hormonal activity for which they are used is excessive, and secondly because they themselves or their metabolites may exhibit some toxic effect essentially unrelated to their hormonal activity. The safety of their use, therefore, needs to be assessed in both these respects and if not enough data are as yet available, additional laboratory tests need to be carried out.

6.1.3. Safety assessment from a hormonal and general viewpoint

Some review papers are devoted to this subject (Metzler, 1982; Ross, 1981; and Taylor, 1981). Metzler (1982) has discussed the various adverse effects and potential risks for the consumer of anabolic-containing meat using DES as an example. Ross (1981) has stressed the relevance of metabolic studies in the assessment of the toxicity to man of a test compound. Based on experimental evidence for trenbolone acetate he concluded that various short-term toxicological studies with trenbolone acetate have been completed and that so far no adverse effects, other than those that could be associated with its hormonal activity, have been observed. Furthermore, a feeding study involving feeding tissues from treated veal calves to rats and involving a safety factor of at least 100 gave no indication of any adverse, teratological reproductive or toxicological effect. Taylor (1981) tried to answer the questions (a) whether anabolic steroids are primary carcinogens, and (b) whether if it is shown that some steroids are carcinogenic in laboratory animals, it follows that these steroids are carcinogenic in man. It is recognized that natural steroidal hormones in excessive doses may increase the risks of particular types of cancer in humans and laboratory animals. Thus, where natural steroidal hormones influence cancer risks, they only do so under conditions of dose and duration of exposure resulting in a manifest change in hormonal status. Therefore, it seems reasonable to assume that when there is exposure to doses of only a fraction of the normal dose required for the manifestation of change in hormonal status, there will be no change in cancer risk.

Xenobiotic anabolic agents, which may imitate the hormonal activity of androgens, estrogens and progestogens, can only be regarded as safe from a hormonal viewpoint if (a) by the time of slaughter the hormonal status of the treated animals lies within the wide range encountered in nature (Hoffmann, 1981), and (b) the levels of hormonal activity of the residues are well below the lowest level expected to cause manifest hormonal changes in the meat consumer.

It is important to stress that if the residues of anabolic agents are safe from the viewpoint of hormonal activity, the other aspects of their safety can be judged on the basis of the same procedures as for food additives and food contaminants such as pesticide residues (WHO Regional Office Europe Meeting, 1982). Such testing procedures include firstly, consideration of metabolic pathways in treated animals and the chemical nature and bioavailability of residues in meat, and secondly, data from appropriate laboratory tests for acute and chronic toxicity, mutagenicity, carcinogenicity and reproduction of teratogenicity. A good example is given by Ross (1981) in studying the safety of trenbolone acetate (see above). On the basis of the results of this type of test it should be possible to determine whether an acceptable daily intake (ADI) can be set for xenobiotic residues. However, it remains essential that the possible establishment of an ADI for residues of xenobiotic anabolic agents must be linked to clearly formulated definitions of how they may be used, to exclude any possibility that sites of administration including depots of the agent may be ingested by man or other animals.

With regard to the three groups of xenobiotic agents of most interest, i.e. the stilbenes, trenbolone acetate and zeranol, the following can be stated. In the case of DES and the other stilbenes, banning the use of these compounds is justified on the grounds that, unlike natural steroids, they are not readily destroyed in the liver, so that residues in meat products are mainly bioavailable to humans who ingest them. Furthermore, these compounds are not readily biodegradable so that their use may result in contamination of the food chain and other undesired environmental effects.

Chapter 10 references, p. 225

Finally, the results of mutagenicity and animal studies on these compounds which are structurally related to carcinogens have given rise to suspicion of carcinogenicity unrelated to their hormonal activity (WHO Regional Office Europe Meeting, 1982).

In the case of trenbolone acetate and zeranol, both in use in a number of countries, the bioavailability and toxicological data should be further analyzed to see whether an ADI can be set for residues in meat products of these agents, on which basis their safe use can be recommended (WHO Regional Office Europe Meeting, 1982).

6.2. Benefit—risk assessment and legal consequences

It has been clearly shown in this chapter that under certain production conditions (especially in steers and veal calves) the use of anabolics can yield advantages in terms of increase in weight gain and feed conversion rate. Consequently there is a distinct benefit to the producer (farmer). In many countries they are aware of the possibilities of profitable application of anabolics as growth promoters. As stated in Section 6.1, application of these compounds is only justified when it does not constitute a hazard for the user, the animal and the consumer. So far, in most countries, a total ban on anabolics exists. If, however, a distinction could be made between safe and toxic products, then the necessary regulations can be suggested. In formulating legislation on anabolic agents, the health and protection of the consumer is of paramount importance. Legislation should be based on thorough scientific evaluation of the different compounds, be enforceable, and be reviewed periodically to take into account new scientific information and discoveries.

At the time of writing, it seems clear that stilbene estrogens should be banned in animal production. Natural hormones under supervision and proper control could be allowed, whereas trenbolone acetate and zeranol should be further assessed regarding safety. Legislation should be enforced by proper programmes to control and monitor the legal and illegal use of anabolic agents. In this framework the further development of sufficiently sensitive assay methods to allow detection of banned compounds in excreta or edible parts is essential. Legal control of possibly licensed products should include:
  (a) control of manufacture and distribution of anabolic agents;
  (b) the method of application should ensure that the site of application is discarded and does not enter the human food chain;
  (c) enforcement of a withdrawal period, which can be done by certification of data of implantation and monitoring of carcases for residue levels above the tolerance levels for synthetic compounds.

In conclusion, legislation is essential and should be based on proper scientific evaluation of the safety of growth-promoting anabolics.

7. CONCLUDING REMARKS

In this chapter the possibilities and limitations of the use of anabolics as growth promotors have been discussed. It became clear that anabolic agents with hormonal action may provide a marked benefit in meat production. Thorough zootechnical knowledge for the correct and most profitable use, in terms of increase in weight gain and feed conversion rate, is, however, essential. From the various groups of compounds, the natural and synthetic estrogens alone or in combination with trenbolone acetate, when used appropriately in steers and veal calves, afford a marked promotion of growth. However, their use is restricted by the possible toxicological risks to the consumer posed by the anabolic residues. It was suggested that naturally-occurring anabolic agents can be regarded as safe to the consumer, provided their use is controlled. The control of these agents should include route of administration, dosage and withdrawal period.

The safety of a synthetic anabolic agent needs to be established, firstly in terms of its measurable hormonal activity, and secondly in terms of potential unwanted side-effects. For this latter purpose information on metabolism, bioavailability and general toxicological safety of the compound is required, similar to the data necessary for assessing the safety of food additives and contaminants. Stilbenes, including diethylstilbestrol, because they have carcinogenic potential, certainly will not pass safety assessment; however, other xenobiotic agents, like trenbolone acetate and zeranol, may be judged safe.

On the basis of such assessments the use of certain anabolic agents for growth promotion may be legalized. Enforcement of legislation should be based on monitoring slaughtered animals for residues of anabolic agents. By this means either the legal or illegal use of anabolic agents can be effectively controlled. Only under these conditions can the health and protection of the consumer be guaranteed.

8. ACKNOWLEDGEMENT

The authors gratefully acknowledge the helpful critical remarks made by Dr. L.G. Huis in 't Veld, former Head of the Laboratory of Endocrinology, RIV, Utrecht, The Netherlands, during the preparation of the manuscript.

9. REFERENCES

Aguirre, V.E.L., Zamora, P.J.E. and Carrera, M.C., 1968. [Effect of implantation of diethylstilboestrol in beef calves before weaning on summer grazing]. Rev. Mex. Prod. Anim., 1: 9—16 [Nutr. Abstr. Rev., 40 (1970) 2, abstr. 3882].

Algeo, J.W., Hibbits, A.G., Bris, E.J. and Wooden, G.R., 1970. Effect of two levels of implanted resorcylic acid lactone and two levels of oral zinc bacitracin on the performance of finishing heifers. J. Anim. Sci., 31: 234—235 (abstr.).

Andrews, F.N., Beeson, W.M. and Johnson, F.D., 1954. The effects of stilbestrol, dienestrol, testosterone and progesterone on the growth and fattening of beef steers. J. Anim. Sci., 13: 99—107.

Andrews, F.N., Stob, M., Perry, T.W. and Beeson, W.M., 1956a. The effect of oral subcutaneous estrogen and androgen administration on growth and carcass quality of lambs. J. Anim. Sci., 15: 575—588.

Andrews, F.N., Stob, M., Perry, T.W. and Beeson, W.M., 1956b. The oral administration of diethylstilbestrol, dienestrol and hexestrol for fattening calves. J. Anim. Sci., 15: 685—688.

Baker, F.H. and Arthaud, V.H., 1972. Use of hormone active agents in production of slaughter bulls. J. Anim. Sci., 35: 752—754.

Baker, D.H., Jordan, C.E., Waitt, W.P. and Gouwens, D.W., 1967. Effect of a combination of diethylstilbestrol and methyltestosterone, sex and dietary protein level on performance and carcass characteristics of finishing swine. J. Anim. Sci., 26: 1059—1066.

Bastíman, B. and Scott, B.M., 1977. Growth promoting implants for beef cattle. Anim. Prod., 24: 131—132.

Beeson, W.M., Andrews, F.N., Stob, M. and Perry, T.W., 1956. The effect of oral estrogens and androgens singly and in combination on yearling steers. J. Anim. Sci., 15: 679—684.

Béranger, C. and Malterre, C., 1968. Influence d'un stéroide triènique à activité anabolisante sur l'engraissement des vaches taries. C.R. Soc. Biol., 162: 1157—1164.

Berende, P.L.M., 1978. Effect of estradiol/trenbolone acetate on performance of bulls and steers. In: 3rd World Congr. Anim. Feed. VIII: Symp. 23—27 October 1978. Madrid, p. 71.

Berende, P.L.M., Van der Wal, P. and Sprietsma, J.E., 1973. De invloed van anabool werkende stoffen op groei, voederconversie en slachtkwaliteit van mestkalveren. Landbouwkd. Tijdschr., 85: 395—408.

Berende, P.L.M., Scheid, J.P. and Van der Wal, P., 1977. Effect of the oral administration of trenbolone acetate on performance of heifers. In: Abstr. 69th Annu. Meet. 23—27 July 1977. University of Wisconsin, Madison, pp. 135—136.

Bidner, T.D., Merkel, R.A., Miller, E.R., Ullrey, D.E. and Hoefer, J.A., 1972. Effects of diethylstilbestrol plus methyltestosterone and dietary protein level on swine performance and composition. J. Anim. Sci., 34: 397—402.

Boehncke, E. and Gropp, J., 1976. Der Einfluss von Anabolika auf die N-Retention von Mastkälbern. In: Fortschritte in der Tierphysiologie und Tierernährung. Beih. Z. Tierphysiol. Tierernähr. Futtermittelkd., 6: 18—25.

Borger, M.L., Wilson, L.L., Sink, J.D., Ziegler, J.H. and Davis, S.L., 1973. Zeranol and dietary protein level effects on live performance, carcass merit, certain endocrine factors and blood metabolite levels of steers. J. Anim. Sci., 36: 706—711.

Bradley, N.W., Boling, J.A. and Ludwick, R.L., 1972. Comparative efficacy of the high *cis* and high *trans* isomers of diethylstilbestrol for steers. J. Anim. Sci., 34: 497—500.

Brahmakshatriya, R.D., Snetsinger, D.C. and Waibel, P.E., 1969. Effects of exogenous estrogen and/or androgen on performance, egg shell characteristics and blood plasma changes in laying hens. Poult. Sci., 48: 444—451.

Brethour, J.R., 1970. Rapigain I (testosterone—stilbestrol paste implants). Feedstuffs, 42 (24): 41.

Bridson, R., 1972. Zeranol similar to DES in boosting calf gains. Feedstuffs, 44 (9): 25, 28.

Brown, R.G., 1970. An anabolic agent for ruminants. J. Am. Vet. Med. Assoc., 157: 1537—1539.

Burroughs, W., Culbertson, C.C., Cheng, E., Hale, W.H. and Homeyer, P., 1955. The influence of oral administration of diethylstilbestrol to beef cattle. J. Anim. Sci., 14: 1015—1024.

Buttery, P.J., Vernon, B.G. and Pearson, J.T., 1978. Anabolic agents — some thoughts on their mode of action. Proc. Nutr. Soc., 37: 311—315.

Cahill, V.R., Kunkle, L.E., Klosterman, E.W., Deatherage, F.E. and Wierbicki, E., 1956. Effect of diethylstilbestrol implantation on carcass composition and the weight of certain endocrine glands of steers and bulls. J. Anim. Sci., 15: 701—709.

Clegg, M.T. and Cole, H.H., 1954. The action of stilbestrol on the growth response in ruminants. J. Anim. Sci., 13: 108—130.

Clegg, M.T., Albaugh, R., Lucas, J. and Weir, W.C., 1955. A comparison of the effect of stilbestrol on the growth response of lambs of different age and sex. J. Anim. Sci., 14: 178—185.

Couch, J.R., 1969. Summary and report on recent developments in poultry nutrition. Feedstuffs, 41 (37): 21—24.

Davis, J.K. and Truesdale, R.W., 1968. Effect of melengestrol acetate on feedlot performance and carcass characteristics of heifers. J. Anim. Sci., 27: 1162—1163 (abstr.).

Davis, S.L. and Garrigus, U.S., 1971. Intermittent versus continuous diethylstilbestrol in lambs: effects on plasma urea nitrogen and nitrogen retention. J. Anim. Sci., 32: 107—110.

Day, B.N., Zobrisky, S.E., Tribble, L.F. and Lasley, J.E., 1960. Effects of stilbestrol and a combination of progesterone and estradiol on growing-finishing swine. J. Anim. Sci., 19: 898—901.

Deans, R.J., Van Arsdell, W.J., Reineke, E.P. and Bratzler, L.J., 1956. The effect of progesterone—estradiol implants and stilbestrol feeding on feedlot performance and carcass characteristics of steers. J. Anim. Sci., 15: 1020—1028.

Dunbar, J., 1969. R.A.L. Feedstuffs, 41 (38): 28.

Dvoracek, M. and Markovic, P., 1965. [Different amounts of diethylstilboestrol by mouth for young fattening bulls]. Zivocisna Vyroba, 10: 677—688. [Nutr. Abstr. Rev., 36 (1966) 4, abstr. 6987].

Dyer, I.A., Rupnow, E.H. and Ham, W.E., 1960. Effects of level of 3, 3¹ diallylhexestrol on steer performance and carcass characteristics. J. Anim. Sci., 19: 1009—1012.

Dzalalov, Ja.D., 1968. Primenenie insulina pri oskorme byckor [Insulin for fattening bulls]. Zivotruwodstvo, 6: 84—85 [Nutr. Abstr. Rev., 39 (1969) 1, abstr. 1546].

EEC Workshop, Brussels, 1981. Anabolic agents in Beef and Veal Production, Proceedings of a Workshop held at Brussels, 5 and 6 March 1981, p. 156.

Ellington, E.F., Fox, C.W. and Kennick, W.H., 1964. Effects of cortisone and diethyl-stilbestrol singly and in combination on weight gain and carcass characteristics of wether lambs. J. Anim. Sci., 23: 905 (abstr.).

FAO/WHO Symposium in Rome, 1976. Anabolic Agents in Animal Production. F.C. Lu and J. Rendel (Guest Editors). In: F. Coulston and F. Korte (Editors), Environmental Quality and Safety, Supplement Vol. 5. Georg Thieme Publishers, Stuttgart, 277 pp.

Fields, C.L., Mitchell Jr., G.E., Boling, J.A., Tucker, R.E., Ludwick, R.L. and Bradley, N.W., 1971. Tapazole in steer finishing rations: nitrogen metabolism. J. Anim. Sci., 33: 1375—1380.

Fowler, V.R., Stockdale, C.L., Smart, R.I. and Crofts, R.M.J., 1978. Effects of two androgens combined with oestrogen on the growth and efficiency of pigs. Anim. Prod., 26: 358—359.

Fox, J.D., Moody, W.G., Boling, J.A., Bradley, N.W. and Kemp, J.D., 1973. Effect of 1-methyl-2-mercaptoimidazole (Tapazole[R]) feeding on muscle characteristics, fiber type and fatty acid composition of Charolais—Hereford steers. J. Anim. Sci., 37: 438—442.

Frens, A.M., 1949. Proeven over het practisch effect van de toediening van methyl-thiouracil aan mestkoeien. Landbouwk. Tijdschr., 61: 916—929.

Frens, A.M., 1958. Proef met mestkoeien over het effect van het mestpreparaat 'Vevoron'. Landbouwk. Tijdschr., 70: 138—144.

Galbraith, H. and Miller, T.B., 1977. Effect of trienbolone acetate on the performance, blood metabolites and hormones and nitrogen metabolism of beef heifers. Anim. Prod., 24: 133—134.

Garret, W.N., 1964. Low level diethylstilbestrol implantation for lambs grazing alfalfa. J. Anim. Sci., 23: 430—435.

Glimp, H.A. and Cundiff, L.V., 1971. Effects of oral melengestrol acetate and a testo-sterone—diethylstilbestrol implant, breed and age on growth and carcass traits of beef heifers. J. Anim. Sci., 32: 957—961.

Goodrich, R.D. and Meiske, J.C., 1969. Minnesota reports on nitrates, hormones, bull feeding work. MGA, stilbestrol, rapigain and stilbestrol plus rapigain for feedlot heifers. Feedstuffs, 41 (50): 14.

Goodrich, R.D., Meiske, J.C., Kolari, O.E., Harvey, A.L., Aunan, W.J. and Hanson, L.E., 1967. Stilbestrol studies with beef cattle. Univ. Minn. Agric. Exp. Stn. Bull., No. 486.

Grandadam, J.A., Scheid, J.P., Dreux, H. and Deroy, R., 1975a. Influence de différentes préparations anabolisantes sur la qualité de la viande de veau. Rec. Méd. Vét., 151: 355—362.

Grandadam, J.A., Scheid, J.P., Jobard, A., Dreux, H. and Boisson, J.M., 1975b. Results obtained with trenbolone acetate[R] in conjunction with estradiol 17β in veal calves, feedlot bulls, lambs and pigs. J. Anim. Sci., 41: 969—977.

Grandchamp, G., 1968. Essais d'implants à d'oestrogènes naturels chez le taureau. Schweiz. Arch. Tierheilkd., 110: 439—454.

Grashuis, J., 1962. De toepassing van hormonen bij het vee. A.W. Bruna en Zoon, Utrecht, The Netherlands, 6 pp.

Grebing, S.E., Hutcheson, D.P. and Preston, R.L., 1970. Early reduction in urinary-N by diethylstilbestrol in lambs. J. Anim. Sci., 31: 763—766.

Gropp, J., Matzke, P., Schulz, V., Ferstl, R. and Peschke, W., 1976a. Mast- und Schlachtleistung von Kälbern unter dem Einfluss von Anabolika (Pilotversuch). In: Fortschritte in der Tierphysiologie und Tierernährung. Beih. Z. Tierphysiol. Tierer-nähr. Futtermittelkd. 6: 10—17.

Gropp, J., Boehncke, E., Schulz, V., Sandersleben, J. von, Giesel, O. and Hänichen, T., 1976b. Die Wirkung von 17β-Östradiol und Trenbolonacetat in unterschiedlicher Dosierung auf verschiedene physiologische und morphologische Parameter (Provokationsversuch). In: Fortschritte in der Tierphysiologie und Tierernährung. Beih. Z. Tierphysiol. Tierernähr. Futtermittelkd., 6: 33—52.

Hafs, H.D., Purchas, R.W. and Pearson, A.M., 1971. A review: Relationships of some hormones to growth and carcass quality of ruminants. J. Anim. Sci., 33: 64—71.

Hale, O.M. and Johnson, Jr., J.C., 1970. Effects of hormones and diets on performance and carcass characteristics of pigs during summer and winter. Anim. Prod., 12: 47—54.

Hale, O.M., McCormick, W.C. and Beardsley, D.W., 1960. Response of pigs to diethyl-stilbestrol and testosterone fed in diets high and low in energy and protein. J. Anim. Sci., 19: 646—647 (abstr.).

Hale, W.H. and Ray, D.E., 1973. Efficacy of oral estradiol 17β for growing and fattening steers. J. Anim. Sci., 37: 1246—1250.

Hale, W.H., Homeyer, P.G., Culbertson, C.C. and Burroughs, W., 1955. Response of lambs fed varied levels of diethylstilbestrol. J. Anim. Sci., 14: 909—918.

Hawkins, D.R., Henderson, H.E. and Geasler, M.R., 1967. Melengestrol and stilbestrol for finishing steer and heifer calves. J. Anim. Sci., 26: 1480 (abstr.).

Hawkins, D.R., Henderson, H.E. and Newland, H.W., 1972. Effects of melengestrol acetate on the performance of feedlot cattle receiving corn silage rations. J. Anim. Sci., 35: 1257—1262.

Heitzman, R.J., 1976. The effectiveness of anabolic agents in increasing rate of growth in farm animals; report on experiments in cattle. In: F.C. Lu and J. Rendel (Guest Editors), Anabolic Agents in Animal Production. In: F. Coulston and F. Korte (Editors), Environmental Quality and Safety, Supplement Vol. 5. Georg Thieme Publishers, Stuttgart, pp. 89—98.

Heitzman, R.J. and Chan, K.H., 1974. Alterations in weight gain and levels of plasma metabolites, proteins, insulin and free fatty acids following implantation of an anabolic steroid in heifers. Br. Vet. J., 130: 532—537.

Heitzman, R.J., Chan, K.H. and Hart, I.C., 1977. Liveweight gains, blood levels of metabolites, proteins and hormones following implantation of anabolic agents in steers. Br. Vet. J., 133: 62—70.

Heldt, J.D. and Lucas, L.E., 1970. Effect of hormone combination and protein level on performance and carcass traits in two genetic lines of swine. J. Anim. Sci., 31: 203 (abstr.).

Hendrickson, R.F., Pope, L.S., Nelson, A.B., Stephens, D.F. and Acker, D.C., 1957. Effect of feeding or implanting stilbestrol on the performance of suckling beef calves. J. Anim. Sci., 16: 1079 (abstr.).

Herrick, G.M., Fry, J.L., Damron, B.L. and Harms, R.H., 1970. Evaluation of dienestrol diacetate (Lipamone) supplementation of broiler finisher feeds on pigmentation, growth characteristics and market quality. Poult. Sci., 49: 222—225.

Hoffmann, B., 1981. Levels of androgenous anabolic sex hormones in farm animals. In: Anabolic Agents in Beef and Veal Production, Proceedings of EEC Workshop held at Brussels, 5—6 March 1981, pp. 96—112.

Hoffmann, B., Heinritzi, K.H., Kyrein, H.J., Oehrle, K.L., Oettel, G., Rattenberger, E., Vogt, K. and Karg, H., 1976. Untersuchungen über Hormonkonzentrationen in Geweben, Plasma and Urin von Mastkälbern nach Behandlung mit hormonwirksamen Anabolika. In: Fortschritte in der Tierphysiologie und Tierernährung. Beih. Z. Tierphysiol. Tierernähr. Futtermittelkd., 6: 80—90.

Huber, T.L., 1970. The effect of diethylstilbestrol on nitrogen excretion in sheep. Can. J. Physiol. Pharmacol., 48: 573—574.

Hunsley, R.E., Vetter, R.L., Kline, E.A. and Burroughs, W., 1967. Effects of age, sex and diethylstilbestrol on feedlot performance, carcass characteristics and muscle tenderness of male beef cattle. J. Anim. Sci., 26: 1469—1470 (abstr.).

Hutcheson, D.P. and Preston, R.L., 1971. Stability of diethylstilbestrol and its effects on performance in lambs. J. Anim. Sci., 32: 146—151.

IARC (International Agency for Research on Cancer), 1974. IARC monographs on the Evaluation of Carcinogenic Risk of Chemicals to Man. Sex Hormones, Vol. 6. IARC, Lyon, 243 pp.

International Symposium in Warsaw, 1981. Steroids in Animal Production. H. Jasiorowski (Editor). Roussel-Uclaf/Warsaw Agricultural University, Warsaw, 261 pp.

Jordan, C.E., Waitt, W.P. and Scholz, N.E., 1965. Effects of orally-administered diethylstilbestrol and methyltestosterone on finishing barrows and gilts. J. Anim. Sci., 24: 890 (abstr.).

Jordan, R.M. and Hanke, H.E., 1970. Antibiotics, ensiled corn, Ralgro, sunflower hulls and self-feeding all increase lamb gains. Feedstuffs, 42 (17): 44—45.

Karg, H., 1966. Hormonale Wirkstoffe in der Mast. Z. Tierz. Züchtungsbiol., 82: 154—168.

Kliewer, R.H., 1970. Influence of varying protein and diethylstilbestrol on live performance and carcass characteristics of Holstein steers. J. Dairy Sci., 53: 1766—1770.

Klinskij, Ju.D. and Pucnin, A.M., 1969. [Effect of melengestrol acetate on weight gain of bullocks]. Zivotnovodstvo, 9: 32 [Nutr. Abstr. Rev., 40 (1970) 2, abstr. 3866].

Klosterman, E.W., Cahill, V.R., Kunkle, L.E. and Moxon, A.L., 1955. The subcutaneous implantation of stilbestrol in fattening bulls and steers. J. Anim. Sci., 14: 1050—1058.

Klosterman, E.W., Moxon, A.L. and Cahill, V.R., 1959. Effect of stilbestrol and amount of corn silage in the ration upon the protein requirement of fattening steer calves. J. Anim. Sci., 18: 1243—1249.

Klosterman, E.W., Cahill, V.R. and McClure, K.E., 1970. Additive effect of hormones for finishing heifers. J. Anim. Sci., 31: 246—247 (abstr.).

Kochakian, C.D., 1966. Regulation of muscle growth by androgens. In: E.J. Briskey, R.G. Cassens and J.C. Trautman (Editors), The Physiology and Biochemistry of Muscle as Food. Chapter 7, pp. 81—114.

Kolari, O.E., Harvey, A.L., Meiske, J.C., Aunan, W.J. and Hanson, L.E., 1960. Diethylstilbestrol, oxytetracycline, linseed oil meal, soyabean oil meal and levels of corn silage in cattle fattening rations. J. Anim. Sci., 19: 1041—1048.

Krista, L.M., Sautter, J.H. and Waibel, P.E., 1969. Influence of diethylstilbestrol on the turkey with special reference to histological changes in the aorta. Poult. Sci., 48: 1961—1968.

Krüsskemper, H.L., 1965. Anabole Steroide. 2. überarbeitete Auflage. Georg Thieme Verlag, Stuttgart, 184 pp.

Leibetseder, J., 1966. Über die Wirkung eines anabolen Steroids in der Kükenmast bei oraler Verabreichung. Z. Tierphysiol., Tierernähr. Futtermittelkd., 21: 131—136.

Lucas, E.W., Wallace, H.D., Palmer, A.Z. and Combs, G.E., 1971. Influence of hormone supplementation, dietary protein level and sex on the performance and carcass quality of swine. J. Anim. Sci., 33: 780—790.

Lucas, L.E., Svajgr, A.J. and Peo, Jr., E.R., 1967. Effect of hormone combination and protein level on performance and carcass traits in market pigs. J. Anim. Sci., 26: 908—909 (abstr.).

Lucas, L.E., Peo, Jr., E.R. and Svajgr, A.J., 1973. Effect of a combination of diethylstilbestrol and methyltestosterone and protein level in the diet on performance and carcass traits in an unselected line of swine. J. Anim. Sci., 36: 1094—1098.

MacDearmid, A. and Preston, T.R., 1969. A note on the implantation of intensively-fed beef cattle with hexestrol. Anim. Prod., 11: 419—422.

Martin, T.G. and Stob, M., 1978. Growth and carcass traits of Holstein steers, bulls and bulls implanted with diethylstilbestrol. J. Dairy Sci., 61: 132—134.

McLaren, G.A., Anderson, G.C., Welch, J.A., Campbell, C.D. and Smith, G.S., 1960. Diethylstilbestrol and length of preliminary period in the utilization of crude biuret and urea by lambs. II. Various aspects of nitrogen metabolism. J. Anim. Sci., 19: 44—53.

Megally, M.A., Harrington, R.B. and Stadelman, W.J., 1969. The effect of estradiol-17β-monopalmitate on yields and quality of chicken roasters. Poult. Sci., 48: 130—136.

Meites, J. and Nelson, M.J., 1960. Effects of hormonal imbalances on dietary requirements. Vitam. Horm. (NY), 18: 205—235.

Melliere, A.L., Waitt, W.P. and Jordan, C.E., 1970. Factorial evaluation of DES and MT on nitrogen retention of swine. J. Anim. Sci., 31: 1025 (abstr.).

Metzler, M., 1982. Residues of anabolics in meat: special risk for vulnerable consumer groups. In: Report of WHO Regional Office for Europe Working Group on Health Aspects of Residues of Anabolics in Meat, Bilthoven, 10—13 November 1981. Euro Reports and Studies, 59: 28—35.

Michel, G. and Baulieu, E.E., 1976. An approach to the anabolic action of androgens by an experimental system. In: F.C. Lu and J. Rendel (Guest Editors), Anabolic Agents in Animal Production. In: F. Coulston and F. Korte (Editors). Environmental Quality and Safety, Supplement Vol. 5. Georg Thieme Publishers, Stuttgart, pp. 54—59.

NAS, 1953. Hormonal relationships and applications in the production of meats, milk, and eggs. National Academy of Sciences—National Research Council, Washington, DC, Publ. 266, 54 pp.

NAS, 1959. Hormonal relationships and applications in the production of meats, milk, and eggs. National Academy of Sciences—National Research Council, Washington, DC, Publ. 714, 53 pp.

NAS, 1966. Hormonal relationship and applications in the production of meats, milk, and eggs. National Academy of Science—National Research Council, Washington, DC, Publ. 1415, 87 pp.

Nesheim, M.C., 1976. Some observations on the effectiveness of anabolic agents in increasing the growth rate of poultry. In: F.C. Lu and J. Rendel (Guest Editors), Anabolic Agents in Animal Production. In: F. Coulston and F. Korte (Editors). Environmental Quality and Safety, Supplement Vol. 5. Georg Thieme Publishers, Stuttgart, pp. 110—114.

Nesler, R.J., Essig, H.W. and Pund, W.A., 1966. Stilbestrol during suckling, wintering and fattening as related to steer performance, carcass characteristics, organoleptic qualities and chemical analyses. Proc. Assoc. South. Agric. Workers, Inc. 63rd Annual Convention, Mississippi, p. 130.

Nielsen, A.J., 1970. The energy value of balanced feed rations for growing pigs determined by different methods. Frederiksberg Bogtrykkeri, Copenhagen, 212 pp.

O'Brien, C.A. and Miller, C.F., 1967. Effect of melengestrol acetate (MGA) on the reproductive physiology and feedlot performance of ewe lambs. J. Anim. Sci., 26: 949 (abstr.).

O'Brien, C.A. and Miller, C.F., 1970. Response of fattening ewe lambs to oral melengestrol acetate. J. Anim. Sci., 31: 588—592.

O'Brien, C.A., Miller, C.F. and Stovel, W.C., 1968a. Response of ewe lambs to melengestrol acetate following induced puberty. J. Anim. Sci., 27: 302—303.

O'Brien, C.A., Bloss, R.E. and Nicks, E.F., 1968b. Effect of melengestrol acetate on the growth and reproductive physiology of fattening heifers. J. Anim. Sci., 27: 664—667.

Ogandzanov, V.B. and Paduceva, A.L., 1969. [Effect of synthetic hormones implanted twice in animals]. Him. sel sk. Hoz., 7: 541—542 [Nutr. Abstr. Rev., 40 (1970) 2, abstr. 3999].

Ogilvie, M.L., Faltin, E.C., Hauser, E.R., Bray, R.W. and Hoekstra, W.G., 1960. Effects of stilbestrol in altering carcass composition and feedlot performance of beef steers. J. Anim. Sci., 19: 991—1001.

Oltjen, R.R., Swan, H., Rumsey, T.S., Bolt, D.J. and Weinland, B.T., 1973. Feedlot performance and blood plasma amino acid patterns in beef steers fed diethylstilbestrol under ad libitum, restricted, and compensatory conditions. J. Nutr., 103: 1131—1137.

O'Mary, C.C., Warren, E.P., Davis, T.J. and Pierce, Jr., H.H., 1956. Effects of low level implantations of stilbestrol in steers fattened on drylot rations. J. Anim. Sci., 15: 52—58.

Oslage, H.J., 1965. Untersuchungen zum Stoff- und Energieumsatz wachsender Mastschweine. Landbauforsch. Völkenrode, 15: 107—138.

Oxley, J.W., Kercher, C.J., Nicholls, O.L., Wall, M.W., Patterson, L.C., Cox, P.B. and Hiser, R.G., 1960a. Effect of hormone implants on suckling lambs. J. Anim. Sci., 19: 965—966 (abstr.).

Oxley, J.W., Thompson, R.C. and Kercher, C.J., 1960b. Gain, carcass and liver response of feeder lambs to hormone implants. J. Anim. Sci., 19: 1283 (abstr.).

Patton, W.R. and Ralston, A.T., 1968. Early diethylstilbestrol treatment of bull calves. J. Anim. Sci., 27: 1117 (abstr.).

Peo, Jr., E.R. and Hudman, D.B., 1960. Supplementation of pig starters with thyroprotein. J. Anim. Sci., 19: 477—483.

Perry, T.W., Beeson, W.M., Andrews, F.N. and Stob, M., 1955. The effect of oral administration of hormones on growth rate and deposition in the carcass of fattening steers. J. Anim. Sci., 14: 329—335.

Perry, T.W., Beeson, W.M., Andrews, F.N., Stob, M. and Mohler, M.T., 1958. The comparative effectiveness of oral and subcutaneous implantation of diethylstilbestrol in combination with chlortetracycline. J. Anim. Sci., 17: 164—170.

Perry, T.W., Stob, M., Huber, D.A. and Peterson, R.C., 1970. Effect of subcutaneous implantation of resorcylic acid lactone on performance of growing and finishing beef cattle. J. Anim. Sci., 31: 789—793.

Plimpton, Jr., R.F. and Teague, H.S., 1972. Influence of sex and hormone treatment on performance and carcass composition of swine. J. Anim. Sci., 35: 1166—1175.

Plimpton, Jr., R.F., Cahill, V.R., Teague, H.S., Grifo, Jr., A.P. and Kunkle, L.E., 1967. Periodic measurement of growth and carcass development following diethylstilbestrol implantation of boars. J. Anim. Sci., 26: 1319—1324.

Plimpton, Jr., R.F., Ockerman, H.W., Teague, H.S., Grifo, Jr., A.P. and Cahill, V.R., 1971. Influence of the time following diethylstilbestrol implantation on the palatability, composition and quality of boar pork. J. Anim. Sci., 32: 51—56.

Poppe, S., 1965. Über den Einfluss von Diaethylstilboestrol und anderer Wirkstoffe auf den Stoffwechsel männlicher Ratten. 4. Mitteilung. Einfluss von Diaethylstilboestrolgaben in Intervallen und steigenden Gaben. Arch. Tierernähr., 15: 61—80.

Preston, R.L., 1975. Biological responses to estrogen additives in meat producing cattle and lambs. J. Anim. Sci., 41: 1414—1430.

Preston, R.L. and Burroughs, W., 1958. Stilbestrol responses in lambs fed rations differing in calorie to protein rations. J. Anim. Sci., 17: 140—151.

Preston, R.L. and Burroughs, W., 1960. Physiological actions of diethylstilbestrol in lambs fed varying levels of protein and energy. J. Appl. Physiol., 15: 97—100.

Preston, R.L., Martin, J.E., Blakely, J.E. and Pfander, W.H., 1965. Structural requirements for the growth response of certain estrogens in ruminants. J. Anim. Sci., 24: 338—340.

Preston, R.L., Cahill, V.R. and Klosterman, E.W., 1970. Levels and isomers of stilbestrol for cattle. J. Anim. Sci., 31: 1043 (abstr.).

Preston, R.L., Klosterman, E.W. and Cahill, V.R., 1971. Levels and isomers of diethylstilbestrol for finishing steers. J. Anim. Sci., 33: 491—496.

Preston, T.R. and Willis, M.B., 1970. Intensive Beef Production. Pergamon Press, Oxford, Chapter 8, pp. 281—304.

Purchas, R.W., Pearson, A.M., Pritchard, D.E., Hafs, H.D. and Tucker, H.A., 1971a. Some carcass quality and endocrine criteria of Holstein heifers fed melengestrol acetate. J. Anim. Sci., 32: 628—635.

Purchas, R.W., Pearson, A.M., Hafs, H.D. and Tucker, H.A., 1971b. Some endocrine influences on the growth and carcass quality of Holstein heifers. J. Anim. Sci., 33: 836—842.

Rako, R. and Mikulec, K., 1968. Der Einfluss gonadotroper Hormone auf Mastleistung und Schlachtkörperwert von Färsen. Züchtungskunde, 40: 366—374.

Ralston, A.T., 1978. Effect of zearalanol on weaning weight of male calves. J. Anim. Sci., 47: 1203—1206.

Ralston, A.T., Caster, J.E., Kennick, W.H. and Davidson, T., 1968. Response of feedlot heifers to certain exogenous hormones. J. Anim. Sci., 27: 1117 (abstr.).

Ralston, A.T., Taylor, N.O. and Davidson, T.P., 1969. Effect of diethylstilbestrol on growth and carcass quality of beef cattle. Oreg., Agric. Exp. Stn., Tech. Bull. No. 110.

Raun, A.P., Cheng, E.W. and Burroughs, W., 1960. Effects of orally administrated goitrogens upon thyroid activity and metabolic rate in ruminants. J. Anim. Sci., 19: 678—686.

Ray, D.E., Hale, W.H., Marchello, J.A. and Kuhn, J.O., 1968. Influence of season and sex on response of beef cattle to growth stimulants. J. Anim. Sci., 27: 1135 (abstr.).

Ray, D.E., Hale, W.H. and Marchello, J.A., 1969. Influence of season, sex and hormonal growth stimulants on feedlot performance of beef cattle. J. Anim. Sci., 29: 490—495.

Ray, M.L. and Child, R.D., 1965. Synovex-S or stilbestrol for finishing steers. J. Anim. Sci., 24: 901 (abstr.).

Roche, J.F. and Davis, W.D., 1978a. Effect of trenbolone acetate and resorcylic acid lactone alone or combined on daily liveweight and carcass weight in steers. Ir. J. Agric. Res., 17: 7—14.

Roche, J.F. and Davis, W.D., 1978b. Effect of time of insertion of resorcylic acid lactone and trenbolone acetate and type of diet on growth rate in steers. Ir. J. Agric. Res., 17: 249—254.

Ross, D.B., 1981. Toxicology and residues of trenbolone acetate as a model. In: H. Jasiorowski (Editor), Proceedings of an International Symposium on Steroids in Animal Production, Warsaw, 1980. Roussel-Uclaf/Warsaw Agricultural University, pp. 227—234.

Rumsey, T.S., Oltjen, R.R. and Kozak, A.S., 1974. Implant absorption, performance and tissue analysis for beef steers implanted with diethylstilbestrol and fed an all-concentrate diet. J. Anim. Sci., 39: 1193—1199.

Rumsey, T.S., Oltjen, R.R., Kozak, A.S., Daniels, F.L. and Aschbacher, P.W., 1975. Fate of radiocarbon in beef steers implanted with ^{14}C-diethylstilbestrol. J. Anim. Sci., 40: 550—560.

Ryley, J.W., Moir, K.W., Pepper, P.M. and Burton, H.W., 1970. Effect of hexoestrol implantation and body size on the chemical composition and body components of chickens. Br. Poult. Sci., 11: 83—91.

Samberev, Ju.N., Solovea, V.N. and Goremykina, A.P., 1967. Der Einfluss anabolischer Hormone auf die Jungrindermast. Dokl. TSKhA, Mosk. Skh. Akad. im. K.A. Timirjazeva, 130: 53—59 (in Russian).

Schneeberger, J. and Schürch, A., 1961. Über die Wirkung der Verfütterung eines synthetischen Androgens. Dianabol an Mastküken. Z. Tierphysiol., Tierernähr. Futtermittelkd., 16: 67—75.

Schulster, D., Burstein, S. and Cooke, B.A., 1976. Molecular Endocrinology of the Steroid Hormones. John Wiley and Sons, London, Chapter 14, pp. 245—261.

Schulz, V., Donnerbauer, H., Aigner, R. and Gropp, J., 1976. Der Effekt von Anabolika auf die Mastleistung von Kälbern (Feldversuch). In: Fortschritte in der Tierphysiologie und Tierernährung. Beih. Z. Tierphysiol. Tierernähr. Futtermittelkd., 6: 26—32.

Scott, B.M., 1978. The use of growth promoting implants in beef production. J. Agric. Dev. Adv. Serv. Q. Rev., 31: 185—217.

Sharp, G.D. and Dyer, I.A., 1971. Effect of zearalanol on the performance and carcass composition of growing-finishing ruminants. J. Anim. Sci., 33: 865—871.

Shorrock, C., Capper, B.S., Light, D. and Mlambo, M.M.J., 1978. A note on the performance of fattening steers implanted with zeranol under grazing and feedlot conditions in Botswana. Anim. Prod., 26: 221—224.

Simms, R.H., Perry, T.W. and Andrews, F.N., 1970. Influence of delayed castration and of diethylstilbestrol implantation on the performance of suckling lambs. J. Anim. Sci., 30: 970—973.

Stob, M., Beeson, W.M., Perry, T.W. and Mohler, M.T., 1968. Effects of coumestrol in combination with implanted and orally administered diethylstilbestrol on gains and tissue residues in cattle. J. Anim. Sci., 27: 1638—1642.

Struempler, A.W. and Burroughs, W., 1959. Stilbestrol feeding and growth hormone stimulation in immature ruminants. J. Anim. Sci., 18: 427—436.

Szumowski, P. and Grandadam, J.A., 1976. Comparaison des effets du diéthylstilboestrol (DES) et de l'acétate de trenbolone (TBA) seul ou associé à l'oestradiol 17β (TBA-E$_2$) sur la croissance et l'engraissement des ruminants. Rec. Méd. Vét., 152: 311—321.

Tausk, M. and Boerman, A.J., 1961. De Hormonen. Erven J. Bijleveld, Utrecht, The Netherlands, 175 pp.

Taylor, W., 1981. The relationship between steroids and cancer in man, and its relevance to the use of steroidal anabolic agents in farm animals. In: H. Jasiorowski (Editor),

Proceedings of an International Symposium on Steroids in Animal Production, Warsaw, 1980. Roussel-Uclaf/Warsaw Agricultural University, pp. 235—247.

Teague, H.S., Plimpton, Jr., R.F., Cahill, V.R., Grifo, Jr., A.P. and Kunkle, L.E., 1964. Influence of diethylstilbestrol implantation on growth and carcass characteristics of boars. J. Anim. Sci., 23: 332—338.

Thomas, O.O., 1970. Zearalanol for fattening steers. Feedstuffs, 42 (29): 80.

Thomas, O.O. and Armitage, J., 1970. Zearalanol for growing-fattening steers. J. Anim. Sci., 30: 1039 (abstr.).

Thomas, O.O., Armitage, J. and Sherwood, O., 1970. Zearalanol and stilbestrol for suckling calves. J. Anim. Sci., 30: 1039 (abstr.).

Thrasher, D.M., Vincent, C.K., Scott, V.B. and Mullins, A.M., 1967. Effects of a hormone combination on the feedlot, carcass and reproductive performance of pigs. J. Anim. Sci., 26: 911 (abstr.).

Trenkle, A.H., 1969. The mechanisms of action of estrogens in feeds on mammalian and avian growth. In: The Use of Drugs in Animal Feeds. Proceedings of a Symposium, Publ. 1679. NAS—NRC, Washington, DC, Chapter II, pp. 150—164.

Trenkle, A. and Burroughs, W., 1978. Nutrition and Drug Interrelations. Academic Press, New York, NY, Chapter 21, pp. 577—611.

Utley, P.R. and McCormick, W.C., 1974. Effects of feeding melengestrol acetate in combination with diethylstilbestrol and zeranol implants on the feedlot performance of finishing heifers. Can. J. Anim. Sci., 54: 211—216.

Utley, P.R., Chapman, H.D. and McCormick, W.C., 1972. Feedlot performance of heifers fed melengestrol acetate and oxytetracycline separately and in combination. J. Anim. Sci., 34: 339—341.

Utley, P.R., Newton, G.L., Ritter, R.J. and McCormick, W.C., 1976. Effects of feeding monensin in combination with zeranol and testosterone—estradiol implants for growing and finishing heifers. J. Anim. Sci., 42: 754—760.

Van Oordt, G.J., 1958. Endocrinologie van de Gewervelde Dieren. W. de Haan N.V., Zeist, The Netherlands, 200 pp.

Vanschoubroek, F., 1965. Pro en contra het gebruik van oestrogenen en thyreostatica bij het mesten van herkauwers en pluimvee. Vlaams Diergeneeskd. Tijdschr., 34: 341—357.

Vanschoubroek, F., 1968. Pro en contra het gebruik van oestrogenen en thyreostatica bij het mesten van herkauwers en pluimvee. Tijdschr. Diergeneeskd., 93: 679—691.

Vetter, R.L., 1969. Stilboestrol for fattening sheep and cattle. In: Feeding and Nutrition of Animals, Wholesomeness of Food Products for Men. 4. Int. Symp. Anim. Husb., Milan. April, pp. 142—150.

Vipond, J.E. and Galbraith, H., 1978. Effect of zeranol implantation on the growth performance and some blood characteristics of early-weaned lambs. Anim. Prod., 26: 359.

Van der Wal, P., 1976. General aspects of the effectiveness of anabolic agents in increasing protein production in farm animals, in particular in bull calves. In: F.C. Lu and J. Rendel (Guest Editors), Anabolic Agents in Animal Production. In: F. Coulston and F. Korte (Editors), Environmental Quality and Safety, Supplement Vol. 5. Georg Thieme Publishers, Stuttgart, pp. 60—78.

Van der Wal, P., Van Hellemond, K.K. and Berende, P.L.M., 1971. De toepassing van oestrogenen bij de vleesproduktie. Tijdschr. Diergeneeskd., 96: 1173—1190.

Van der Wal, P., Berende, P.L.M. and Sprietsma, J.E., 1975a. Effect of anabolic agents on performance of calves. J. Anim. Sci., 41: 978—985.

Van der Wal, P., Van Weerden, E.J. Sprietsma, J.E. and Huisman, J., 1975b. Effect of anabolic agents on nitrogen-retention of calves. J. Anim. Sci., 41: 986—992.

Van Weerden, E.J., 1972. Nitrogen balance in relation to protein and amino acid requirements in the veal calf. In: Proceedings of 2nd International Milkreplacer Symposium, Zurich, 26—27 May 1971. National Renderes Association, Brussels, pp. 31—44.

Van Weerden, E.J. and Grandadam, J.A., 1976. The effect of an anabolic agent on N deposition, growth, and slaughter quality in growing castrated male pigs. In: F.C. Lu and J. Rendel (Guest Editors), Anabolic Agents in Animal Production. In: F. Coulston and F. Korte (Editors), Environmental Quality and Safety, Supplement Vol. 5. Georg Thieme Publishers, Stuttgart, pp. 115—122.

Van Weerden, E.J., Van der Wal, P., Berende, P.L.M., Huisman, J. and Sprietsma, J.E., 1973. De invloed van anabool werkende stoffen op de eiwitaanzet van mestkalveren. Landbouwkd. Tijdschr., 85: 409—419.

Wahlstrom, R.C., 1970. A note on the effect of tylosin and a combination of diethylstilbestrol and methyltestosterone on performance and carcass characteristics of finishing pigs. Anim. Prod., 12: 181—183.

Walker, N., 1972. A note on the growth and carcass characteristics of castrated male pigs fed on a ration containing diethylstilbestrol and methyltestosterone. Anim. Prod., 14: 255—257.

Whitehair, C.K., Gallop, W.D. and Bell, M.C., 1953. Effect of stilbestrol on ration digestibility and on calcium, phosphorus and nitrogen retention in lambs. J. Anim. Sci., 12: 331—337.

Whiteker, M.D., Brown, H., Barnhart, C.E., Kemp, J.D. and Varney, W.Y., 1959. Effects of methylandrostenediol, methyltestosterone and thyroprotein on growth and carcass characteristics of swine. J. Anim. Sci., 18: 1189—1195.

WHO, Regional Office for Europe Meeting in Bilthoven, 1982. Report of a Working Group on Health Aspects of Residues of Anabolics in Meat, Bilthoven, 10—13 November, 1981. Euro Reports and Studies, 59: 38 pp.

Wickersham, E.W. and Schultz, L.H., 1964. Response of dairy heifers to diethylstilberstrol. J. Anim. Sci., 23: 177—182.

Wiggins, J.P., Rothenbacher, H., Wilson, L.L., Martin, R.J., Wangsness, P.J. and Ziegler, J.H., 1979. Growth and endocrine responses of lambs to zeranol implants: effects of preimplant growth rate and breed of sire. J. Anim. Sci., 49: 291—297.

Wilson, L.L., Borger, M.L., Peterson, A.D., Rugh, M.C. and Orley, C.F., 1972a. Effects of zeranol, dietary protein level and methionine hydroxy analog on growth and carcass characters and certain blood metabolites in lambs. J. Anim. Sci., 35: 128—132.

Wilson, L.L., Varela-Alvarez, H., Rugh, M.C. and Borger, M.L., 1972b. Growth and carcass characters of rams, cryptorchids, wethers and ewes subcutaneously implanted with zeranol. J. Anim. Sci., 34: 336—338.

Winchester, C.F., 1953. Some uses of drugs and hormones in beef cattle, sheep, and swine husbandry. In: Hormonal Relationships and Applications in the Production of Meats, Milk, and Eggs. Publ. 266. NAS—NRC, Washington, DC, pp. 31—54.

Wöhlbier, W. and Schneider, W., 1966. Der Stoffansatz bei Ochsen unter dem Einfluss von Methylthiouracil. Z. Tierphysiol. Tierernähr. Futtermittelkd., 21: 34—40.

Woods, W., 1970. Zearalanol implants. Feedstuffs, 42 (12): 26, 28.

York, L.R. and Mitchell, J.D., 1969. The effect of estradiol-17β-monopalmitate and surgical caponization on production efficiencies, yields and organic characteristics of chicken broilers, Poult. Sci., 48: 1532—1536.

Yoshimoto, T., Mimura, K. and Tamura, T., 1968. [Study on the influences of hexestrol dicaprylate on fattening male pigs and bull calves.] J. Fac. Fish. Anim. Husb. Hiroshima Univ., 7: 313—325 [Nutr. Abstr. Rev. 39 (1969) 3, abstr. 5940].

Young, A.W., Bradley, N.W. and Cundiff, L.V., 1967. Effects of an oral progestogen on feedlot heifers. J. Anim. Sci., 26: 231 (abstr.).

Zucker, H., Leskova, R. and Tschirch, H., 1972. Wachstum und Serumharnstoff beim Kalb. Z. Tierphysiol. Tierernähr. Futtermittelkd., 29: 316—320.

Chapter 11

Animal Production and Energy Resources

JANE BELYEA and D.E. TRIBE

1. INTRODUCTION

In most countries domesticated livestock are still essentially multipurpose; they are kept for a combination of milk and meat production, for their wool, hair or skins, for their manure, as sources of draught power and as means of transportation. Through the photosynthetic activity of the plants on which they feed, herbivorous animals derive their food energy from the sun, accumulating and transforming it in a variety of ways which, throughout the history of mankind, have been turned by man to his own advantage.

Animals have always occupied a special ecological niche in the energy chain. It is only comparatively recently, and only in the more highly developed parts of the world, that a number of technological innovations have markedly eroded the significance of this niche. In particular, inorganic fertilizers have become more important than animal manures as sources of plant nutrients and the place of animal muscle power has largely been replaced by power generated by the steam engine, the internal combustion engine and the electric turbine, driven by fossil fuels.

In those situations where the multiple uses of animals are no longer relevant, management systems have been developed for the specialized production of milk or meat. Frequently these systems involve intensive technologies in which animals are housed and are fed compounded rations based on cereals with supplements of high protein feeds. The products of these intensive systems are consumed in societies where the longevity of people is commonly limited by the effects of over-eating, including an unnecessarily and, many medical specialists would say, undesirably high consumption of feedlotted beef, factory-farm produced eggs, and pig and poultry meat. Further, in these affluent societies there has been a marked increase in the uses of animals for recreation or companionship which now command considerable financial, material and human resources through many large and expensive supporting services, including feed and pharmaceutical firms, veterinarians, companies which manufacture miscellaneous equipment and supplies, show societies and sporting clubs.

The activities of affluent societies must be compared with those in the rest of the world where countless men, women and children remain hungry and are even dying of starvation and malnutrition.

These contrasts in lifestyles are increasingly attracting attention and the manner in which modern livestock industries have developed has been criticized as wasteful in its use of fossil fuel energy, as morally, ethically and nutritionally without justification, and as ecologically aberrant. Suggestions have been made that the further development of livestock industries

Chapter 11 references, p. 252

should be discouraged, at least in those areas where alternative systems of land use, especially cropping, are possible. This chapter assesses world energy resources and evaluates animal production in terms of the varying contributions which livestock can make to mankind's energy needs.

2. WORLD ENERGY RESOURCES

During the last two centuries there have been basic changes in the pattern of energy use throughout the world. First, as industrialization proceeded, coal largely displaced wood and, in the nineteen-fifties, coal provided 56% of the world's energy needs. More recently the use of oil and natural gas has increased and, in 1972, they provided 64% of world energy consumption, while coal had dropped to 29% (Darmstadter and Schurr, 1974). The present energy economy of the developed world relies on essentially non-renewable reserves of fossil fuels and there is now considerable concern as to the extent of these reserves.

Figures for 'ultimate recoverable reserves' — which represent the sum of cumulative production to date, present proven reserves and estimated undiscovered recoverable reserves (Drake, 1974) — are the subject of frequent speculation and discussion. Estimates of fuel reserves are related to the costs of fuel extraction and therefore they change relative to current operating and economic conditions. Further uncertainties in the figures for ultimate reserves arise because methods for estimating 'possible reserves' all have major limitations (Ion, 1974).

Oil is currently the world's major energy source and Table 11.1 shows that estimates of the world's ultimate reserves of oil have increased with time. However, future expansion on the same scale seems unlikely because, since the nineteen-forties, there have been large increases in the amount, the geographical distribution and the effectiveness of oil exploration (Warman, 1972). Recent available estimates indicate that the ultimate reserves of recoverable oil are between 1600 and 1800 $\times$ 10^9 barrels (9.8—11.0×10^{21} J).

TABLE 11.1

Estimates of world ultimate reserves of crude oil from conventional sources

Year	Source	$\times 10^9$ barrels	$\times 10^{21}$ J
1942	Pratt, Weeks and Stebinger	600	3.8
1946	Duce	400	2.4
1946	Pogue	555	3.4
1948	Weeks	610	3.7
1949	Levorsen	1500	9.2
1949	Weeks	1010	6.2
1953	MacNaughton	1000	6.1
1956	Hubbert	1250	7.6
1958	Weeks	1500	9.2
1959	Weeks	2000	12.2
1965	Hendricks (USGS)	2480	15.2
1967	Ryman (ESSO)	2090	12.8
1968	Shell	1800	11.0
1968	Weeks	2200	13.5
1969	Hubbert	1350—2100	8.3—12.9
1970	Moody (Mobil)	1800	11.0
1971	Warman (BP)	1200—2000	7.3—12.2
1971	Weeks	2290	14.0

After Warman (1972).

One of the most useful methods for comparing these reserves estimates with consumption is by using the 'reserves/production (R/P) ratio' which shows the proved recoverable reserves remaining at the end of a year in relation to the annual production for that year. It is generally accepted that an R/P ratio of 10:1 represents the minimum economic level of production for any field, and a ratio of 15:1 is the minimum planning target for oil fields on a world-wide basis. Fig. 11.1, which illustrates the possible future of oil discovery and production in relation to these two R/P ratios, shows that, in the future, to sustain even minimum economic levels of production of oil, 35×10^9 barrels must be discovered annually.

Here, four factors are of major significance (Warman, 1972):

(1) most major discoveries after 1920 have been in the Middle Eastern region, and it is thought that the majority of these sources have now been identified;

(2) exploration technologies are now sufficiently advanced that it is unlikely that major fields (which constitute 75% of reserves) will remain undetected for long;

(3) the world 'average finding rate' for the 10 years 1960 to 1970 was only about 23×10^9 barrels/year; and

(4) remaining oil tends to be in more inaccessible regions, where exploration and production costs are relatively high.

In the light of these considerations, the inevitable conclusion is that oil cannot sustain its role as our major energy source.

Of course, there are other fossil fuel reserves. Coal is undoubtedly the world's most abundant fossil fuel, with total reserves of the order of 9×10^{12} tonnes (267×10^{21} J) (Armstrong, 1974). World resources of natural gas are theoretically more than sufficient to match present demands, although, for economic, geographical, political and other reasons, supply shortages are already apparent in the U.S.A. and are beginning to emerge in Western Europe (Coppack, 1974).

In future, the extensive utilization of coal or natural gas reserves, or of other fossil fuels such as oil from tar sands or shale deposits, is likely to involve large capital expenditure. Table 11.2 shows that significant costs

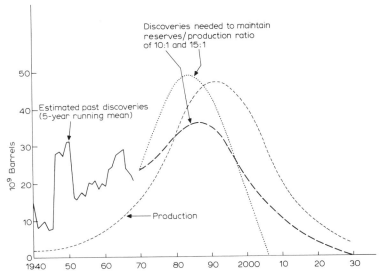

Fig. 11.1. World future annual oil production and discovery requirements. Assumes ultimate world discoveries of 2000×10^9 bbl and production increases of 7.5%/annum to 1981. After Warman (1972).

Chapter 11 references, p. 252

TABLE 11.2

Costs of alternative fossil fuels, equivalent barrel of oil at Persian Gulf Port

Fossil fuel	1973 U.S. $
Conventional crude oil	
Middle East	0.25—1.00
North Sea	4.00—7.00
Synthetic crude oil	
Tar sands	5.00—10.50
Shale	5.50—10.50
Conventional coal	
European	7.50—10.50
U.S.A.	5.50—8.50

From White (1977).

are associated with any of the oil or coal alternatives to Middle Eastern crude oil (White, 1977), and the extent to which natural gas is used is likely to be limited by high transport costs (Coppack, 1974). Increasingly, the possibility of using non-fossil fuel sources of energy is being considered by technologists and economists.

Nuclear sources have been given most attention but there has been a reduction in nuclear power growth because of concern over reactor safety and radio-active waste disposal. However, known uranium reserves have increased and Bowie (1974) concluded: 'The possible utilization of thorium and the introduction of breeder reactors should prevent any shortage developing this century, but only if adequate foresight is exercised throughout the many branches of the uranium industry'. Other potential resources of nuclear energy are shown in Table 11.3 but, although these seem to be large, it must be remembered that the technologies necessary for their utilization have not yet been perfected and to achieve this will involve vast financial expenditures. New methods could be expected to take at least 25 years to reach an economically applicable stage (Australian Academy of Science, 1973). Even for the most technologically advanced resource, nuclear fission, Sir John Hill (1974) concluded that 'it would take quite unexpected shifts in relative economics for nuclear stations completely to supplant conventional systems'. It is also possible that community objections on political and environmental grounds to the development of nuclear fission and fusion technologies may delay such developments further, or even indefinitely.

Turning to other energy sources, there is considerable scope for developing the renewable non-fossil fuel forms of energy. Whilst the adoption of wind,

TABLE 11.3

Potential world energy sources

Resource	10^{21} J	
Fission (fast breeder)	10^2	
Fusion (deuterium—tritium)	10^6	
(deuterium—deuterium)	10^{10}	
Geothermal	10^3	
Solar (per annum) on total surface	10^3	(land)
		(+ sea)

After Australian Academy of Science (1973).

geothermal, hydroelectric or tidal power as an energy source is obviously limited by geographical and geological location, these sources could make significant contributions in some areas. Geothermal energy is presently used in Iceland, Italy, Japan, Mexico, New Zealand and the U.S.A. for space heating and for the generation of electricity. Schemes for using tidal energy are operating in France and are under investigation in Canada and the U.S.S.R. Hydroelectric energy could be utilized much more extensively than at present, with current electricity generation only about 13% of the estimated world potential of 9802 TWh/year (Parker, 1975). There remains a large undeveloped potential for the technological utilization of solar energy. Solar energy is less restricted by locality than other renewable energy sources, and consequently it is under investigation in many countries including the U.S.A., the U.K., France, Australia and the U.S.S.R. It can be used to supply low-grade heat (below 120°C) for water heating, low-pressure steam generation, heating and cooling of buildings, desalination of water, and some industrial drying operations, or can generate electric power through thermal conversion, photovoltaic conversion or thermoelectric generation (Australian Academy of Science, 1973). Here again there are good reasons to have confidence in scientific developments, but large scale practical applications of solar energy technologies are unlikely to reach significance in the near future.

In conclusion, it can be expected that our major energy resource, oil, will become increasingly scarce and expensive in the next two or three decades. Among alternative energy sources available, other fossil fuels or nuclear energy cannot be viewed as certain replacements for oil and undoubtedly will be more costly than present energy sources. In the long term, solar energy could make a significant contribution to the world's energy needs, but its immediate role is limited by:

(1) the large investment in existing energy systems;
(2) the huge capital requirements of large-scale introduction of solar energy technology; and
(3) the technological advances necessary if solar energy is to become competitive economically with the presently important energy sources (Morse and Simmons, 1976).

'Thus, as compared with the past relatively profligate behaviour in regard to energy utilization, it can be expected that with rising energy costs much more attention will be paid to energy conservation and the efficiency of energy utilization' (Newland and Price, 1975).

3. ENERGY USE IN AGRICULTURE

The use of fossil fuels in agriculture has been an integral part of the world-wide increase in their consumption. As late as the nineteen-forties, the farmer, even in the developed world, was still regarded as 'essentially a middleman dealing in energy between the sun and the earth' (Ashby, 1946). Whilst there is no doubt that the sun still provides agriculture with vast amounts of energy — for example, Pimentel and Pimentel (1977) estimated that the solar energy input is nine times greater than the fossil fuel input to U.S. maize production — large inputs of energy from non-renewable sources are now also used in modern agricultural technologies.

The two main purposes of fossil fuel inputs in agriculture are to increase output/ha and output/man (DeWit, 1975). To these ends, labour has been replaced by tractors, animal manure and compost by inorganic fertilizers, roughages by compounded feeds, and so on. The result has been large increases in yield, such as those in the U.K. where relative net output rose

---

Chapter 11 references, p. 252

from a pre-war index of 100, to 128 in 1948 and 198 in 1964 (Hutchinson, 1972). However, with high inputs of support energy there is a strong tendency for diminishing returns (Leach, 1975a). This has been illustrated most dramatically by a study of vegetable production in Hong Kong, where a 585% increase in energy inputs between the late nineteen-fifties and early nineteen-seventies improved yields by only 8% (Newcombe, 1976).

In developed countries inputs of fossil fuels to agriculture amount to about 5—7% of total fossil fuel use, with a further 10% involved in food processing and distribution (Gifford and Millington, 1975; Tribe et al., 1975). This may seem small with respect to total world energy use but Leach (1975b) has estimated that, if the modern farming methods of the Western world are ever adopted world-wide, some 40% of total world energy would then be diverted to agriculture.

In the long term, support energy supplies and costs are likely to be a major determinant of food production capacity and costs (McClymont, 1976). It is possible that higher food costs may be coped with adequately by individuals in developed countries, where expenditures on food range from about 15% (U.S.) to 20—30% (Europe and Oceania) of private consumption expenditures. However, in developing countries, expenditures on food may already represent 50% or more of private consumption expenditures, and for these people a doubling or tripling of the cost of cereals would be a major disaster (USDA, 1974).

Higher production costs are not the only, or even the major problem concerning the continued or expanded use of support energy in agriculture.

'Alternative energy sources are necessary in the long term if present output is to be maintained. In the shorter term, however, energy supplies are not a constraint on production provided it is recognized that the safeguarding of the food supply is a matter of priority and fiscal and other steps are taken to avoid such economy on farms that food production is impaired' (Blaxter, 1977).

In addition to developing alternative energy sources for agricultural use (or alternative means of using traditional energy sources), there is also an increasing need to use support energy inputs with maximum efficiency.

Efficiency is classically regarded as output per unit of input and energetic efficiency can be expressed as an 'energy ratio' (ER) or 'energy quotient' in which the numerator is the metabolizable or digestible energy of the product (meat, milk, egg, cereal or vegetable) and the denominator is the non-solar input energy. This input energy — variously termed the energy subsidy, cultural energy, ancillary energy or support energy — is the sum of the non-solar energy requirements of the whole production process, from the energy content of the fuels and electricity used directly, to the energy involved in the production and distribution of machines, fertilizers, chemicals, and so on. The energy ratios, calculated for various production systems and commodities which are shown in Table 11.4 illustrate the following points.

(1) As a result of their intensive use of labour and their use of few other inputs, primitive agricultural systems are generally more energetically efficient than modern ones. Yield is often correspondingly low, although Leach (1975a) has noted some exceptions.
(2) Crop production systems generally have an ER greater than 1 and the more extensive forms of animal production have efficiencies comparable to those of modern cereal production.
(3) The efficiency of animal production has decreased with intensification, resulting in modern systems of milk, meat and egg production with ER values less than 1, and sometimes less than 0.5.

TABLE 11.4

Range of energy ratios for food products

Commodity			Energy ratio
Tropical crops, pre-industrial			13—38
Tropical crops, semi-industrial			5—10
Wheat, U.K. 1970s			2.2—3.1
Corn, U.S.A. 1970s			2.2
Carrots, U.K. 1970s			1.3
Potatoes, U.K. 1970s			1.1—1.3
Beef (low intensity)		1970s	0.35—5.4
Beef (feedlot intensive)		1970s	0.1
Milk (intensive)	U.K.	1970s	0.33—0.70
Pigs (pork and bacon)	U.K.	1970s	0.63
Poultry (eggs)	U.K.	1970s	0.16—0.26
Poultry (meat)	U.K.	1970s	0.11—0.15

Sources: Leach (1975a, b); White (1975); Pimentel and Pimentel (1977); Slesser (1973).

Modern, intensive systems of animal management are highly productive as means of providing meat, milk or eggs. Their productivity is achieved not only through capital investment in buildings, machines and labour-saving devices but also through a small, highly skilled and well-trained core of management personnel. In terms of their fossil fuel inputs they are clearly costly (some would say extravagant), but economically they are often very profitable. Intensive animal units also have the inherent advantage of high biological efficiency because their housing conditions provide a means of achieving better control of animal nutrition, reproduction and health.

However, in most parts of the world the most plentiful resource available to the livestock industry is an abundant supply of illiterate or semi-illiterate manpower with only traditional skills. Not only is there a shortage of highly skilled personnel, but other scarce resources include capital and credit. Further, in many parts of the world, the aims of livestock production are not only, or even mainly, the economically profitable production of milk, meat and eggs. The non-food contributions of ruminant livestock to mankind's needs have been classified under nine headings (Table 11.5) and only two of these, fibres and skins, are important in international trade, while the remainder serve local or regional needs (McDowell, 1977).

Two of the most vital uses of many domestic animals in the world are as sources of power and manure, and since fuels and fertilizers are the major energy costs of modern agriculture, these uses are highly relevant to our current needs.

4. ANIMALS AS SOURCES OF POWER

Whilst mechanized methods of agriculture in the developed countries are familiar to many, the extent of the use of animals for traction is often ignored. In many parts of the world, animals contribute a significant proportion of the power used in agriculture, and they are likely to be required for power for many years to come (Turner, 1971). Surveys in West Bengal, for example, have shown that 31% of the cattle are adult males, used as working bullocks (Odend'hal, 1972). Predictions of future trends in the use of draught animals are numerous and, whilst some reports project a decline in the use of animal power, others predict an increase in animal power concurrently with increases in tractor power (McDowell, 1977).

Chapter 11 references, p. 252

TABLE 11.5

Classification of non-food contributions of ruminants

Classification	Contribution	Main sources
Fibre	Wool	Sheep, camelids
	Hair	Goats, yak, sheep, camel
Skins	Hides	All ruminants
	Pelts	Sheep, camelids
Inedible products	Inedible fats	Cattle, buffalo, sheep
	Horns, hooves, bones	Cattle, buffalo
	Tankage	Cattle, buffalo, sheep
	Endocrine extracts	Cattle, sheep
Traction	Agriculture	Cattle, buffalo, camel
	Cartage	Cattle, buffalo, yak, camel
	Packing	Camel, yak, buffalo, cattle, reindeer
	Herding	Buffalo, camel
	Irrigation pumping	Buffalo, cattle, camel
	Threshing grains	Cattle, buffalo
	Passenger conveyance	Buffalo, camel, yak, cattle
Waste	Fertilizer	Domestic ruminants
	Fuel (dung)	Cattle, buffalo, yak, camel, sheep
	Methane gas	Cattle, buffalo
	Construction (plaster)	Cattle, buffalo
	Feed (recycled)	Cattle
Storage	Capital	Domestic ruminants
	Grains	Cattle, buffalo, sheep
Conservation	Grazing	All ruminants
	Seed distribution	All ruminants
	Ecological	
	Maintenance	All ruminants
	Restoration	All ruminants
Pest control	Plants in waterways	Buffalo
	Weeds between croppings	Domestic ruminants
	Snails (irrigation canals)	Buffalo
Cultural, including recreation	Exhibitions, including rodeos	Cattle, sheep, goat, buffalo
	Fighting	Cattle, buffalo
	Hunting	Deer, elk, gazelle
	Pet	Goat, sheep, deer
	Racing	Buffalo, cattle
	Riding	Camel, buffalo
	Religious — Instruments — Sacrificial	Buffalo, sheep
	Bride price	Cattle, sheep, goat
	Social status	Cattle, sheep

Species listed in order of importance, if identified. From McDowell (1977).

Certainly, inputs of energy do produce high yields (Leach, 1975a) and the energy inputs to modern agriculture are dominated by the demands of mechanization. For example, in the United Kingdom, farm fuel and electricity represent 41% of the total energy input to agriculture and machines a further 14% (Blaxter, 1975). In the U.S.A. and Australia, respectively, comparable figures are 50 and 67% for fuels and 17 and 11% for machines (Steinhart and Steinhart, 1974; J. Belyea, unpublished data, 1978). There is not, however, general agreement as to whether mechanical cultivation per se directly increases yields. The commonly cited advantages of tractors are those of improved soil and seedbed preparation and more timely operations (e.g., President's Science Advisory Committee, 1967; FAO, 1969a). However, any yield-increasing effects of tractor cultivation are often confounded with those of improved seeds, fertilizers, insecticides, herbicides and irrigation (Clayton, 1973) and it is not clear whether or not hand-operated and animal-drawn implements are equally effective (Marsden, 1973).

Therefore, it is important to identify the objectives of mechanization, and Gordon (1950) stated that the main reason for farm mechanization is to increase production per man-hour. In view of the U.N. predictions that the absolute size of the agricultural populations of developing countries is still increasing, a mechanization strategy which replaces labour by machines would seem to be inappropriate. Nevertheless, the food needs of the peoples of the developing countries are urgent and the FAO (1969b) concluded that

> 'at least 0.50 HP (per hectare) are required to achieve the full potential for high yields ... in the developing regions the present power available from all sources (human, animal and mechanical) is estimated at only 0.05 HP/ha in Africa, 0.19 HP in Asia, and 0.27 HP in Latin America ... the contribution which additional human labour or draught animal power could make to bridging the gap between the present availability of power and future requirements is relatively small ... it is, therefore, of crucial importance to devise a strategy for mechanization which will provide the increased power necessary whilst at the same time reducing labour displacement to the absolute minimum'.

Consequently, developing nations have leant towards policies of mechanization in agriculture, and although some have predicted that scarcity of fossil fuels will slow the rate of increase in tractor numbers, this is not yet apparent (Table 11.6). However, it is almost certainly true that the capital required for mechanization is simply outside the resources of many developing countries. The regional studies of the Indicative World Plan for Agricultural Development (FAO, 1969b) have shown that a formidable investment in equipment and machines would be required to achieve even relatively modest levels of mechanization. For example, in South America, the aggregate investment for the period 1962 to 1985 inclusive would amount to around $15 000 million. This represents approximately 40% of the total identified agricultural investment for the region and is very much greater than any other single investment. In South and Southeast Asia, expenditure on mechanization for the same period would need to be $22 000 million. This represents about 35% of the total aggregate investment, and is second only to the costs of land improvement and development. In addition, rising costs

TABLE 11.6

Annual increases in tractor numbers in selected regions

| Region | % Increase | | | | |
| | Proposed 1965—1985 | Actual | | | |
		1971—1972	1972—1973	1973—1974	1974—1975
Africa[a]	4.8	5.6	3.4	3.9	3.2
Latin America	3.9	2.5	4.3	4.1	4.5
Near East	6.2	11.8	10.9	19.6	16.0
Far East	n.a.	15.9	8.8	10.9	9.8
Asian centrally planned economies	13.0	6.2	6.1	5.6	5.6

[a]Excluding Egypt, Libya, Sudan.
Sources: FAO Production Yearbooks; FAO (1969b).

Chapter 11 references, p. 252

of the necessary petroleum fuels will place mechanized technologies further outside the reach of the resources of the peoples of developing countries.

Therefore, in the foreseeable future, it is likely that animals will be called on to provide much of the power needed to produce food and Rollinson and Nell (1973) have warned that 'if petroleum prices continue to rise, the most serious deterrent to meeting food production goals in S.E. Asia for 1985 will be a shortage of buffalo and cattle for farm power' (cited by McDowell, 1977).

The extent to which animal draught contributes to the power used in agriculture has been estimated by many authors. Phillips (1969) has calculated that animals provided over 95% of the farm power of developing countries in 1965, and estimates for 1969 have indicated that between 66 and 87% of non-human power was of animal origin (FAO, 1969a). However, these statistics appear to include total numbers of animal species, and calculations based on the percentage of animals used for draught purposes suggest that their contribution may actually be in the range of 19% in Africa to 70% in India (President's Science Advisory Committee, 1967). Further, whilst the draught power of buffalo and cattle in the developing regions has increased by 22% and 15%, respectively, since the mid nineteensixties (to 1975), the total number of tractors has increased by 143% (FAO, 1976). Consequently, the proportional contribution of animals to farm power is undoubtedly decreasing. Nevertheless they do make, and will continue to make, a significant contribution to farm power. Cattle are the most widely used source of draught power, followed by horses and buffalo (President's Science Advisory Committee, 1967). In some areas, animals have an inherent advantage over other draught alternatives. For example, the water buffaloes used in the deep muds of Philippine rice cultivation (Marsden, 1973) or the yaks and llamas used as pack animals in the Himalayas and the Andes (Turner, 1971) could only be replaced with great difficulty and expense by tractors or other vehicles. The draught power of various animals is shown in Table 11.7, and comparison with the power requirements of various farming operations shows that a pair of draught animals can provide enough pulling power for most of the tasks performed in small farming conditions. For many operations, even single animals are sufficient, if kept in good condition (Hopfen, 1969). However, the power potential of work animals is seldom fully realized because they are often poorly fed, unhealthy or impeded by a bad harness and crude, inefficient implements and vehicles (Cockrill, 1974; Ensminger, 1977).

There is an urgent need for a far-sighted appraisal of the power requirements of the developing countries and for a fuller utilization of the potential in animal draught power. Current research and extension should be

TABLE 11.7

Normal draught power of various animals

Animal	Ave. wt. (kg)	Approx. draught (kg)	Ave. speed of work (m s^{-1})
Light horses	400—700	60—80	1.0
Bullocks	500—900	60—80	0.6—0.85
Buffaloes	400—900	50—80	0.8—0.9
Cows	400—600	50—60	0.7
Mules	350—500	50—60	0.9—1.0
Donkeys	200—300	30—40	0.7

After Hopfen (1969).

directed towards teaching the proper care, husbandry and feeding of animals, towards the design of better implements and harnesses, and to the breeding of animals for draught purposes.

5. ANIMALS AS A SOURCE OF MANURE

Before the introduction of inorganic fertilizers, animal manures were an important source of nutrients for crops and pastures. During the 17th and 18th centuries in England farmyard manures were the most commonly used fertilizers and on arable farms the main purpose of animals was often to produce manure (Peel, 1976). Throughout the 19th century livestock continued to be kept and especially fed to produce manure, and feeding trials, such as those of Lawes and Gilbert at Rothamsted in 1895, were conducted to determine the manure-producing values of various feedstuffs (Peel, 1976).

However, there was a gradual shift from manure, first to imported organic materials — oilcake, bones and guano — and then to inorganic minerals which were more uniform, readily available and easier to transport and handle than town refuse and other local materials.

Since 1938 there has been a rapid increase in the agricultural use of inorganic fertilizers, particularly of nitrogenous fertilizers, with the 1975/1976 level of nitrogen (N) consumption 17.3 times greater than the 25×10^5 tonnes of N consumed in 1938, and 1.8—2.0 times greater than the 1975/1976 level of phosphate (P_2O_5) or potash (K_2O) consumption (Fig. 11.2). This has resulted in significant increases in the fossil energy used in agriculture since fertilizer manufacture requires energy inputs of 77 GJ/tonne for N, 14 GJ/tonne for P_2O_5 and 8 GJ/tonne for K_2O (White, 1975). Consequently, fertilizer now represents a major energy input to most modern agricultural systems and constitutes 38, 21 and 13% of the total fossil energy input to agriculture in the U.K., U.S.A. and Australia, respectively (Steinhart and Steinhart, 1974; Blaxter, 1975; J. Belyea, unpublished data, 1978).

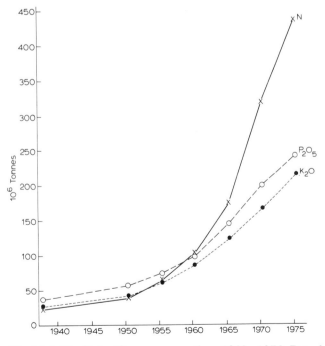

Fig. 11.2. World fertilizer consumption, 1938—1976. Data for years prior to 1957 exclude U.S.S.R. Source: FAO Production Yearbooks; Peel 1976.

Chapter 11 references, p. 252

So far, the proportion of inorganic fertilizers consumed in developing countries is relatively small and in 1975/1976 these regions used only 18% of the N, 15% of the P_2O_5 and 9% of the K_2O consumed worldwide. However, if the 1985 food needs of the peoples of the developing countries are to be met by domestic production, the investment in fertilizers must increase (at constant 1962 prices) by a massive 1080% from the 1962 level of U.S.\$664.2 million to U.S.\$7838.3 million in 1985 (FAO, 1969b). The ability of the developing countries to acquire these quantities of fertilizer may be limited. Although their fertilizer consumption grew almost 14% annually in the late nineteen-sixties, this was slower than had been anticipated. Further, it is still uncertain whether production capacities can meet future demands for fertilizer and there is considerable variation in projections of the future fertilizer demand—supply balance. For example, although a World Bank study suggested a need for about 14 million tonnes of added N and P_2O_5 capacity in developing countries, USDA estimates for the same period indicated a possible surplus of production capacity in the world (USDA, 1974).

The main difference in these alternative projections was that the World Bank study assumed that much of the planned production increase in the developed countries would not materialize because of high energy costs. Certainly, recent high prices for fossil fuel energy have raised fertilizer production costs substantially, especially those of N fertilizer which requires feedstocks of hydrocarbon such as natural gas, naphtha or coal, but the USDA (1974) considered that increased demand and limited capacity would have a more profound influence on production costs than would energy costs. This is probably especially relevant in developed countries such as the U.K. where economic policies and price subsidies for natural gas have so far delayed any reactions in N fertilizer price to actual increases in world fuel prices (Blaxter, 1977). It is anticipated that the effects of possible future limits on supply and capacity will be that developing countries will be unable to bid fertilizer away from the developed countries that produce it (USDA, 1974). Even if the developing countries become self-sufficient in fertilizer supply (and there are important arguments for and against such a situation), the large capital investment and the need to import raw materials would still result in significant financial investments in fertilizer supply (FAO, 1969b).

Consequently, there is considerable scope for the expanded utilization of animal wastes as a source of plant nutrients and already approximately 40% of the farmers of the world depend wholly or in large part on animal wastes to enhance soil fertility (McDowell, 1977).

Both the organic properties and the nutrient contents of manures are useful in improving soil fertility, and manures have been shown to improve soil tilth, increase water-holding capacity, improve aeration, promote soil microbial activity, and act as a source of nutrients (Miner and Smith, 1975). Values for the nutrient concentrations of manures are shown in Table 11.8. These show that manures contain the three major plant nutrients (N, P and K), and they also contain many valuable trace elements (Adriano, 1975). The major problems in the use of these wastes as fertilizers are associated with handling procedures, with nutrient losses during storage, treatment and handling, and with the variable nutrient compositions of manures.

The main problems with handling arise in developed countries, where the introduction of large, intensive beef, poultry and pig units has created the need to dispose of huge amounts of manure annually. For example, in the U.S.A. about 1.7×10^9 tonnes of manure are produced annually, over 50% of it in feedlots and confined rearing situations (Pimentel et al., 1973). Although most manures are generated in a semi-solid condition, labour

TABLE 11.8

Nutrient content of swine and cattle manures

Type of manure	% On dry matter basis		
	N	P	K
Pig faeces[a] — range	1.8—5.0	1.4—4.6	0.6—1.6
— mean	3.0	2.6	1.0
Swine sludge[b] — range	2.3—3.1	0.3—0.6	0.1—0.2
Beef feedlot[c] — range	1.3—3.1	0.3—1.1	0.6—1.7
— mean	2.3	0.7	1.2

[a]Hilliard and Pearce (1978).
[b]Inoescu-Sisesti et al. (1975).
[c]Adriano (1975).

requirements and the lack of suitable solids handling equipment have caused many producers to use slurry or liquid waste systems where wastes can be removed as a liquid and transported by pumps, spreaders or irrigation equipment (Loehr, 1974).

Table 11.9 shows that a considerable loss of nutrients, particularly nitrogen, occurs from manures but the extent of these losses is markedly influenced by the system of storage, treatment and handling. In general, mechanization of manure handling and the associated trend towards liquid systems leads to increased nutrient losses. Losses of N by the volatilization of N_2 and N_2O as products of denitrification and of NH_3 from manure are, in a sense, not recoverable, but losses of P and K are usually caused by sedimentation and could be recovered if the expense of doing so was warranted (Miner and Smith, 1975).

Depending upon the type of feeds used in any system of animal production, the faeces produced can have an extremely variable nutrient content, with ranges having been recorded in values for pig faeces of 140% for N, 183% for P and 275% for K, and similarly large variations in the nutrient

TABLE 11.9

Estimated nutrient losses during storage, treatment and handling for various waste management systems

System	Nutrient loss (%)		
	N[a]	P_2O_5	K_2O
Deep-pit storage, liquid spreading	30—65		
Anaerobic lagoon, irrigation or liquid spreading	60—80	30—50	b
Oxidation ditch, anaerobic lagoon, irrigation or liquid spreading	70—90	b	b
Bedded confinement, solid spreading	30—40		
Open lot, solid spreading, runoff collected and irrigated	50—60	b	b

From Vanderholm (1975)
[a]Assumes N is incorporated in soil within a few hours. Soil injection would reduce losses; delays in incorporation would increase losses.
[b]Losses are likely, but have not been estimated.

Chapter 11 references, p. 252

contents of manure from beef feedlot systems (Table 11.8). These variations cause difficulties in the management of land applications of manure but it has been suggested that these may be overcome by using statistical techniques for estimating manure composition from the composition of the diet (Hilliard and Pearce, 1978). In view of the large potential for using manures as sources of plant nutrients, further research on the problems associated with the use of manures as fertilizer is badly needed.

The usefulness of manures is not limited to their role as fertilizers. They can also be used as a substrate for methane production, as a fuel and as a feed for ruminants. Methane can be generated by the combined action of acid- and methane-forming bacteria on an anaerobically contained organic substrate. Such a process has been used for over 100 years to stabilize sludge from municipal waste treatment systems in Europe and the United States, the methane produced being used to operate lighting and heating systems (Hassan et al., 1975). More recently, attention has been directed towards the treatment of animal wastes, not only with a view to digesting the organic solids in the manure but also as a means of producing significant amounts of energy for domestic and mobile purposes (Neyeloff and Gunkel, 1975; McDowell, 1977). Farm-scale digesters have been operated in Europe, Asia, Africa and India (Kroeker et al., 1975) but are most abundant in the Asian countries, with 29 000 in South Korea and 20 000 in India in 1975 (McDowell, 1977). In the developed regions, work is still largely in the experimental stage, although a small number of digesters are operating commercially, particularly in the United States (e.g. Abeles, 1975; Fischer et al., 1975).

A major drawback in the introduction of methane-producing technologies is the capital cost of digester systems (Morris et al., 1975). Even though several countries have directed research efforts towards reducing component costs, McDowell (1977) has predicted that biogas plants will not be used at anywhere near projected levels unless the costs of fossil fuel alternatives escalate. Nevertheless, anaerobic digestion will become a more significant tool for waste management. In developed countries, efforts to recover fuel from animal manure are likely to be concentrated in the large intensive animal units where the quantities of manure to be processed could reduce costs of methane production to levels comparable to those of other energy sources. For developing countries, biogas—electricity schemes could be introduced at the village level, where farmers would be paid for the organic residues they deliver (Makhijani and Poole, 1975).

Apart from methane production, some of the benefits of anaerobic digestion are that the manure organic matter is stabilized and the odour and pathogen contents are reduced (Morris et al., 1975). Since fertilizer nutrients, especially nitrogen, are conserved, the digested manure is a valuable fertilizer. Alternatively, the effluent can be decomposed in aerated lagoons or used as a substrate for the growth of algae, bacteria or yeasts which could then be used as protein sources in animal feeds (Abeles, 1975; Hassan et al., 1975).

Manure is also used directly in some countries as a fuel for heating, cooking and lighting. It is particularly significant in India where most of the energetic output derived from the cattle population is in the form of manure for cooking fuel (Odend'hal, 1972). Cattle dung represents 25% of the 200 million tonnes of non-commercial fuels used annually in India (Makhijani and Poole, 1975) and, in areas where there is no firewood, the per capita use of manure can be as high as 1 ton/year (McDowell, 1977). With an energy content of 19.2—19.8 MJ/g dry faeces, the amount of fossil fuel needed to replace dung would be prohibitive and McDowell (1977) estimated that India would need to spend over U.S.$3000 million to purchase oil or coal alternatives.

Since animal wastes contain considerable energy and nutritive value, they can also be reused as a feed for animals. Manures from poultry, pigs and cattle have been successfully incorporated into the diets of both ruminants and non-ruminants (e.g., Johnson and Mountney, 1969; Fontenot and Webb, 1974; Bhattacharya and Taylor, 1975) after a variety of physical processes (drying or anaerobic ensiling) or after treatment with chemical compounds such as sodium hydroxide or sodium chlorite (Smith et al., 1969). Practices of refeeding animal wastes are likely to be most important in the large intensive animal systems of developed countries where the use of energy- and nutrient-dense diets results in the production of manures with high nutrient values (McDowell, 1977).

These uses of manures — as fertilizers, substrates for methane generation, fuels for cooking, lighting and heating, feeds for animals, and the other contributions they make as materials for house construction in some countries — illustrate their importance to the peoples of many countries of the world. The production of manures constitutes a significant part of the overall role of animals in the various agricultural systems of the world.

6. ANIMALS AS ACCUMULATORS OF FOOD ENERGY

Of the 2800 million meat-producing ruminants (cattle, buffalo, sheep and goats) in the world, 1808 million are in the Asian, African and Latin American countries (FAO, 1976). Most of these are fed on roughages as illustreated by the following account of animal husbandry in China:

> '(Pigs) are fed on waste materials not suitable for human food: vegetable refuse, ground and fermented rice hulls, corn husks, sweet potato and soybean vines, water hyacinths and so forth. Grains are used only to a limited extent ... Pasture is almost non-existent, and (dairy cattle) are fed vegetable wastes, grass trimmed from roadsides and ditchbanks, grassy weeds pulled from crop fields, plus a small amount of grain and grain by-products' (Sprague, 1975).

In contrast to these practices, some major sectors of the animal industries of developed countries have undergone a transition to extremely intensive systems of husbandry. The reasons for these developments have been largely economic, and reductions in feed and labour requirements in animal production have justified investments in buildings, machines and labour-saving devices (Blaxter, 1973). Further, for many years high prices for livestock products compared with those of grains encouraged the feeding to animals of high-energy diets based on cereals and supplemented with protein.

In 1969—1971, the 372 million tonnes of cereal used annually as animal feed in the developed countries accounted for 57% of total cereal production in those areas (Greenhalgh, 1976). In the United States concentrated feeds (cereals and oilseed residues) provided 46% of the feed energy for farm animals in 1973. Non-ruminants, with 85—100% of their diet from high-protein feeds, were more reliant on concentrates than ruminants, which consumed 63—89% forages (Wedin et al., 1975).

However, in animals there is an enormous, largely undeveloped, scope for utilizing feeds unsuitable for consumption by man. Grazing animals consume both forage and pasture species, and potential for the improvement of existing forage lies in the areas of genetic improvement of forage species and development of cultural practices and management techniques for optimal use of the available forage (Hodgson, 1976; NRC, 1977). The role of pasture also needs reappraisal in the areas of:

(1) the place of legumes in grass utilization systems;
(2) the use of 'buffer feeds' to supplement grass and allow higher stocking rates; and
(3) policies of pasture and crop conservation (Foot, 1977).

Further, there is a rapidly growing awareness of the ability of both ruminants and non-ruminants to convert otherwise useless agricultural and industrial by-products into high-quality animal protein as well as into draught power and manures. The earlier account of pig production in China is not the only record of close integration between crop, animal and industrial systems, and another of the many examples is in animal husbandry practices of West Bengal:

> 'All cattle were stall-fed throughout the year. Grazing areas are extremely limited, being restricted to the banks of canals and along roadsides, paths and railroad tracks.... The major constituent of the diet was rice straw. Other important items included mustard oil cake, wheat bran, rice hulls, and chopped banana tree trunks. A total of 24 different kinds of foods were fed, but many of the items depended on seasonal availability. Fodder crops were virtually non-existent and almost all of the food was by products of crops grown for human consumption.... There is almost no competition between cattle and humans for land or food supply. Basically, the cattle convert items of little direct human value into products of immediate human utility' (Odend'hal, 1972).

Similar feeds are also available in developed countries, where the 'wastes' or 'by-products' of agricultural, industrial and urban areas are often suitable as animal feed. Of crop residues available on farms, cereal straw has received the most attention (e.g., Jackson, 1977), though many other by-products, such as rootcrop tops, legume haulms, the stalks of maize, millet and sorghum, and soya bean vines can be used successfully (Mohamed et al., 1971; Foot, 1977). Industrial wastes, such as bagasse and molasses from the sugar industry, sawdust and wood pulps from the timber industry, and fruit, vegetable and crop wastes from the flour and food processing industries also have an important role (and an even greater potential) in many systems of animal production (e.g., El-Shazly and Naga, 1977).

Whereas Table 11.10 lists these and other feeds according to their protein or energy value, they could be categorized in a number of other ways related to either their value as a feed (fibre or nutrient content, in vivo digestibility, etc.) or to their origin (from plants or animals, from agriculture or industry, from rural or urban areas), and so on. They can be fed untreated or can be processed in one of a number of physical, chemical or microbial ways varying in complexity from simple methods of soaking, pounding or grinding to more sophisticated techniques such as high-energy irradiation or chemical degradation (e.g., Dyer et al., 1975; Jackson, 1977).

There is an enormous potential for using these products as animal feeds and it is essential that their value is fully realized, both in the practical management of plant—animal systems and in research priorities and programmes. Currently, research is too often directed to only one aspect of production. For example, whilst the development of high-yielding grain varieties has been extremely successful in terms of increasing grain yield, it reduced both the quantity and the feeding quality of straw used by peasant farmers in many parts of the world. Similarly, modern industrial processes are rarely geared towards treating residues as potential products for consumption. An appreciation of the extent to which residues are currently used, and could potentially be used, will lead to more integrated practices in both research and management.

TABLE 11.10

Examples of by-product feeds for animal feeding

High protein	High energy	Low energy
Shrimp meal	Broken rice	Tree pods
Feather meal	Rice bran	Tree parts
Poultry by-product meal	Rice polishings	Whole trees
Leather meal	Dried bakery products	Peanut hulls
Blood meal	Distiller's dried grains	Soybean hulls and
Crab meal	with solubles	bran flakes
Fish solubles	Wheat bran	Cottonseed hulls
Hydrolyzed hog hair	Wheat middlings	Sugarcane bagasse
Distillers solubles	Citrus molasses	Sugarcane tips
Corn gluten feed	Cane molasses	Cassava leaves
Citrus seed meal	Beet molasses	Spent coffee grounds
Yeasts	Wood molasses	Banana stems
Single-cell protein feeds	Wet brewer's grains	Pineapple wastes
Animal and human wastes	Dried brewer's grains	Fruit wastes
Cheese whey	Beet pulp	Vegetable wastes
	Citrus pulp	Corn stalk silage
	Homing feed	Wood sawdust
	Cull peas	Newspapers
	Animal fats	Malt sprouts
	Kapok and shea nuts	
	Waste bananas	
	Cassava	

From NRC (1977).

7. CONCLUSION

In some systems of animal production, particularly those practised in developed countries, many of them intensive, economically efficient and technologically advanced, the primary purpose is to produce meat, milk or eggs for sale. However, it remains a fact that most farm animals in the world are managed in far less sophisticated systems, in which the most vital objectives are the production of food for subsistence purposes and the transformation of solar energy, which has been accumulated in crops and pastures, into draught power and manures, which in turn can be utilized in the further production of food and, possibly, cash crops.

With alternative sources of energy for farm power and fertilizer production becoming increasingly scarce and therefore expensive, the traditional and special niche for livestock in the energy chain of agriculture is likely to become increasingly obvious and important. Suggestions that animals are so biologically inefficient that they should be confined to areas that are either 'too hot and dry' or 'too cold and wet' for crop production are misleading. On the contrary, the need in most regions of the world, particularly in developing countries, is for research and education aimed to encourage the growth of more efficient and better integrated mixed farming systems of both crop and animal production. With the growth of human populations, particularly rural populations, and the inevitable decreases of farm size, it follows that crop production will be extended to as much land as possible. Yet, the extent, the efficiency and the intensity of cropping depends in turn upon the availability of fertilizers or manures and mechanization or animal power. On socio-economic and ecological grounds it is likely that in most parts of the world, systems of crop production will continue to involve animal manures and draught power, not as exclusive but certainly as vital inputs.

Chapter 11 references, p. 252

It follows that the animals needed in peasant farming systems will not be able to command large areas of land for pasture or fodder-crop production. The basis of animal husbandry in this context must be the use as feed of crop, animal and urban by-products and residues, combined with such weeds, grasses or occasional crop surpluses as may be periodically available. Animal research throughout the world is giving increasing emphasis to problems related to this general circumstance. The results of these investigations will help determine the extent to which the world food crisis will be satisfactorily solved — and, incidentally, they will also help to determine the future of livestock in world agriculture.

8. REFERENCES

Abeles, T.B., 1975. Energy and economic analysis of anaerobic digesters. In: W.J. Jewell (Editor), Energy, Agriculture and Waste Management. Ann Arbor Science, Michigan, pp. 353—360.

Adriano, D.C., 1975. Chemical characteristics of beef feedlot manures as influenced by housing type. In: Managing Livestock Wastes: Proceedings of the Third International Symposium on Livestock Wastes. American Society of Agricultural Engineers, Michigan, pp. 347—350.

Armstrong, G., 1974. World coal resources and their future potential Philos. Trans. R. Soc. London, Ser. A, 276: 439—452.

Ashby, E., 1946. Nutrition and the World Engine. 4th A.B. Cunning Lecture on Nutrition. Royal Australian College of Physicians, Sydney.

Australian Academy of Science, 1973. Solar Energy Research in Australia, Report No. 17. Griffin Press, South Australia, 63 pp.

Bhattacharya, A.N. and Taylor, J.C., 1975. Recycling animal waste as a feedstuff: a review. J. Anim. Sci., 41: 1438—1457.

Blaxter, K.L., 1973. The nutrition of ruminant animals in relation to intensive methods of agriculture. Proc. R. Soc. London, Ser. B, 183: 321—326.

Blaxter, K.L., 1975. The energetics of British agriculture. J. Sci. Food Agric., 26: 1055—1064.

Blaxter, K.L., 1977. Energy and other inputs as constraints on food production. Proc. Nutr. Soc., 36: 267—273.

Bowie, S.H.U., 1974. Natural sources of nuclear fuel. Philos. Trans. R. Soc. London, Ser. A, 276: 495—505.

Clayton, E.S., 1973. Mechanisation and employment in East African agriculture. In: Mechanisation and Employment in Agriculture. International Labour Office, Geneva, pp. 19—44.

Cockrill, W.R., 1974. The working buffalo. In: W.R. Cockrill (Editor), The Husbandry and Health of the Domestic Buffalo. Food and Agriculture Organisation, Rome, pp. 313—328.

Coppack, C.P., 1974. Natural gas. Philos. Trans. R. Soc. London, Ser. A, 276: 463—483.

Darmstadter, J. and Schurr, S.H., 1974. World energy resources and demand. Philos. Trans. R. Soc. London, Ser. A, 276: 413—430.

DeWit, C.T., 1975. Agriculture's uncertain claim on world energy resources. Span, 18: 2—4.

Drake, Sir E., 1974. Oil reserves and production. Philos. Trans. R. Soc. London, Ser. A, 276: 453—462.

Dyer, I.A., Riquelme, E., Baribo, L. and Couch, B.Y., 1975. Waste cellulose as an energy source for animal protein production. World Anim. Rev., 15: 39—43.

El-Shazly, K. and Naga, M.A., 1977. Animal technology in the world today. In: Proceedings of the World Food Conference 1976. Iowa State University Press, Ames, IA, pp. 485—504.

Ensminger, D., 1977. Constraints to millions of small farmers in developing countries risking changes in farming practices and family living patterns. In: Proceedings of the World Food Conference 1976. Iowa State University Press, Ames, IA, pp. 553—566.

FAO, 1969a. Smaller farmlands can yield more. World Food Problems No. 8 Food and Agriculture Organisation, Rome, 73 pp.

FAO, 1969b. Provisional Indicative World Plan for Agricultural Development, Vols. 1 and 2. Food and Agriculture Organisation, Rome, 672 pp.

FAO, 1976. Production Yearbook Vol. 30. Food and Agriculture Organisation, Rome.

Fischer, J.R., Sievers, D.M. and Fulhage, C.D., 1975. Anaerobic digestion in swine wastes. In: W.J. Jewell (Editor), Energy, Agriculture and Waste Management. Ann Arbor Science, Michigan, pp. 307—316.

Fontenot, J.P. and Webb, K.E., 1974. The value of animal wastes as feeds for ruminants. Feedstuffs, 46 (14): 30—33.

Foot, A.S., 1977. Under-used products from crops and animals. Philos. Trans. R. Soc. London, Series B, 281: 221—230.

Gifford, R.M. and Millington, R.J., 1975. Energetics of agriculture and food production with special emphasis on the Australian situation. Bulletin 288, CSIRO, Melbourne.

Gordon, A.S., 1950. Essential Considerations in the Mechanisation of Farming. FAO Development Paper No. 5. Food and Agriculture Organisation, Washington, DC, 11 pp.

Greenhalgh, J.F.D., 1976. The dilemma of animal feeds and nutrition. Anim. Feed Sci. Technol., 1: 1—7.

Hassan, A.E., Hassan, H.M. and Smith, N., 1975. Energy recovery and feed production from poultry wastes. In: W.J. Jewell (Editor), Energy, Agriculture and Waste Management. Ann Arbor Science, Michigan, pp. 307—316.

Hill, Sir J., 1974. Future trends in nuclear power generation. Philos. Trans. R. Soc. London, Ser. A, 276: 587—601.

Hilliard, E.P. and Pearce, G.R., 1978. Limitations of guidelines governing rates of application of pig manure to land. Agric. Environm., 4: 65—75.

Hodgson, H.J., 1976. Forages, ruminant livestock, and food. BioScience, 26: 625—630.

Hopfen, H.J., 1969. Farm Implements for Arid and Tropical Regions, revised edn. FAO Agricultural Development Paper No. 91. Food and Agriculture Organisation, Rome, 159 pp.

Hutchinson, Sir J., 1972. Farming and Food Supply. University Press, Cambridge, 143 pp.

Ion, D.C., 1974. Conventional primary energy reserves: review and discovery potential (world-wide). Philos. Trans. R. Soc. London, Ser. A, 276: 431—438.

Ionescu-Sisesti, Vl., Jinga, I., Roman, Gh. and Pricop, Gh., 1975. The efficiency of using sludge from pig growing complexes as organic fertilizer. In: Managing Livestock Wastes: Proceedings of the Third International Symposium on Livestock Wastes. American Society of Agricultural Engineers, Michigan, pp. 271—273.

Jackson, M.G., 1977. Review article: the alkali treatment of straws. Anim. Feed Sci. Technol., 2: 105—130.

Johnson, T.H. and Mountney, G.J., 1969. Poultry manure: production, utilisation and disposal. World's Poult. Sci. J., 25: 202—217.

Kroeker, E.J., Lapp, H.M., Schulte, D.D. and Sparling, A.B., 1975. Cold weather energy recovery from anaerobic digestion of swine manure. In: W.J. Jewell (Editor), Energy, Agriculture and Waste Management. Ann Arbor Science, Michigan, pp. 337—352.

Leach, G., 1975a. Energy and food production. Food Policy, 1: 62—73.

Leach, G., 1975b. The energy costs of food production. In: F. Steele and A. Bourne (Editors), The Man—Food Equation. Academic Press, London, pp. 139—163.

Loehr, R.C., 1974. Agricultural Waste Management. Academic Press, New York, NY, 576 pp.

Makhijani, A. and Poole, A., 1975. Energy and Agriculture in the Third World. Ballinger, Cambridge, MA, 168 pp.

Marsden, K., 1973. Technological change in agriculture, employment and over-all development strategy. In: Mechanisation and Employment in Agriculture. International Labour Office, Geneva, pp. 1—18.

McClymont, G.L., 1976. Animal production in a grain hungry world. S. Afr. J. Anim. Sci., 6: 129—137.

McDowell, R.E., 1977. Ruminant Products: More Than Meat and Milk. Winrock International Livestock Research and Training Centre, Arkansas, 28 pp.

Miner, J.R. and Smith, R.J., 1975. Livestock Waste Management with Pollution Control. North Central Regional Research Publication 222. Midwest Plan Service Handbook MWPS-19. 88 pp.

Mohamed, A.A., El-Shazly, K. and Abou Akkada, A.R., 1971. The use of some agricultural by-products in feeding of farm animals. Alexandria J. Agric. Res., 19: 25—32.

Morris, G.R., Jewell, W.J. and Casler, G.L., 1975. Alternative animal waste anaerobic fermentation designs and their costs. In: W.J. Jewell, (Editor), Energy, Agriculture and Waste Management. Ann Arbor Science, Michigan, pp. 317—335.

Morse, F.H. and Simmons, M.K., 1976. Solar Energy. Annu. Rev. Energy, 1: 131—158.

Newcombe, K., 1976. The energetics of vegetable production in Asia, old and new. Search, 7: 423—430.

Newland, E.V. and Price, G.G., 1975. The peaking of the oil age. Span, 18: 4—6.

Neyeloff, S. and Gunkel, W.W., 1975. Methane—carbon dioxide mixtures in an internal combustion engine. In: W.J. Jewell (Editor), Energy, Agriculture and Waste Management. Ann Arbor Science, Michigan, pp. 397—408.

NRC, 1977. World Food and Nutrition Study, Vol. 1: Supporting Papers. National Research Council, Washington, DC.

Odend'hal, S., 1972. Energetics of Indian cattle in their environment. Hum. Ecol., 1: 3—22.

Parker, A., 1975. World energy resources: a survey. Energy Policy, 3: 58—66.

Peel, L.J., 1976. Science, energy and agriculture since 1800. Fourth International Congress of Agricultural Museums, Reading.

Phillips, R.W., 1969. Factors favouring animal production. In: Proceedings of the Second World Conference on Animal Production. American Dairy Science Association, Minnesota, pp. 15—23.

Pimentel, D. and Pimentel, M., 1977. Counting the kilocalories. Ceres, 59: 17—21.

Pimentel, D., Hurd, L.E., Belloti, A.C., Forster, M.J. Oka, I.N., Sholes, O.D. and Whitman, R.J., 1973. Food production and the energy crisis. Science, 182: 443—449.

President's Science Advisory Committee, 1967. The World Food Problem, Vol. 3. U.S. Government Printing Office, Washington, DC, 332 pp.

Slesser, M., 1973. How many can we feed? Ecologist, 3: 216—220.

Smith, L.W. Goering, H.K. and Gordon, C.H., 1969. Influence of chemical treatments upon digestibility of ruminant faeces. In: Animal Waste Management: Cornell University Conference on Agricultural Waste Management, New York, NY, pp. 88—97.

Sprague, G.F., 1975. Agriculture in China. Science, 188: 549—555.

Steinhart, J.S. and Steinhart, C.E., 1974. Energy use in the U.S. food system. Science, 184: 307—316.

Tribe, D.E., Sidey, D.J., McColl, J.C. and Connor, D.J., 1975. Plants and animals: accumulators and transformers of solar energy. Agric. Environm., 2: 85—96.

Turner, H.N., 1971. Conservation of genetic resources in domestic animals. Outlook Agric., 6: 254—260.

USDA, 1974. The World Food Situation and Prospects to 1985. Foreign Agricultural Economic Report No. 98. Economic Research Service, United States Department of Agriculture.

Vanderholm, D.H., 1975. Nutrient losses from livestock wastes during storage, treatment and handling. In: Managing Livestock Wastes: Proceedings of the Third International Symposium on Livestock Wastes. American Society of Agricultural Engineers, Michigan, pp. 282—285.

Warman, H.R., 1972. The future of oil. Geogr. J., 138: 287—297.

Wedin, W.F., Hodgson, H.J. and Jacobson, N.L., 1975. Utilising plant and animal resources in producing human food. J. Anim. Sci., 41: 667—686.

White, D.J., 1975. Energy in agricultural systems. Agric. Eng., 30: 52—58.

White, N.A., 1977. The cost of energy over the next decade. Energy World, 42: 2—9.

Chapter 12

Animal Production and the World Food Situation

J.H.G. HOLMES

1. INTRODUCTION

The problems of inadequate food supply involve issues of hunger, poverty and degrading living conditions; high infant mortality and permanent mental and physical stunting of many of the survivors; uneven distribution of resources and political power; corruption; population dynamics; and religious, political and cultural systems which bear on these issues. These are literally life-and-death matters for hundreds of millions of people. Many authors have found it difficult to maintain dispassionate objectivity when considering such emotive issues. Their desperate urgency clouds perspective and has resulted in some extreme positions being adopted and some simplistic solutions vigorously promoted and defended, often in the popular news media. This chapter considers the major causes of inadequate food availability and attempts to put into perspective the role of animal production in the radical and conventional solutions which have been proposed.

The role of animal production in providing an adequate diet, particularly for the 'poorest of the poor', involves fundamental issues of resource allocation which are receiving a great deal of attention from agricultural scientists. Morley (1969) presents a provocative case for the development of all possible sources of food, animal, plant and microbial, with a plea for animal production specialists to maintain as their objective the satisfaction of the real needs of mankind rather than the promotion of their own particular industries. Unfortunately, such discussions often degenerate into a conflict between plant scientists and animal scientists arguing extreme positions. Plant scientists (few of whom are vegetarians) argue that the undeniably greater yields of energy, protein and other nutrients from cropping will inevitably lead to the elimination of animal production as a luxury an over-crowded and starving world cannot afford. As plant products supply all the essential nutrients in the human diet except for vitamin B12 (which can be produced by fermentation of plants), animal products have no essential characteristics. Since energy, not protein, is the primary limiting nutrient in the diets of most under-fed people (McLaren, 1974) the high quality and content of protein in animal products are of no special benefit. Therefore animals have no special place in agriculture or human nutrition and indeed it is immoral to feed animals on grain while man starves. Animal scientists reply that most of the world's surface is non-arable and is suited best to animal production; more than half of the plant material in most crops is usable only for ruminant, not human food. While plant products can easily supply an adequate diet for healthy adults, the needs of children, especially the unhealthy, are more easily met by highly palatable and nutritious animal products; and in

Chapter 12 references, p. 283

a world where uneven distribution of income means that riches and poverty will always exist (Mark, ch. 4 v. 7), the demands of the rich for preferred products such as meat and the need of the poor to earn cash by meeting this demand will ensure the continuation of animal industries.

However, under real farming conditions, from the rice paddies of Southeast Asia, ploughed by buffalo fed on rice straw, to the wheat farms of Australia where stubbles are removed by grazing sheep and cattle, to the sugarcane fields of Cuba, where molasses forms the basis of the diet of lot-fed cattle, integrated plant and animal production systems flourish. There can be no doubt that the major sources of energy and protein in the diet of most people must be of plant origin. But both plant and animal production are needed, there is no one 'correct' balance between the two; the balance must vary from region to region depending upon political factors as well as production potential and demand. The development of yet more efficient systems of production and of more precise matching of adapted varieties of plants, animals and agricultural systems with local environments will require an increase in diversity of these plants, animals and systems. An increase in integration of plant and animal production appears to be a more likely outcome than the reverse.

The problems of food supply for an ever-increasing population are constantly presented as serious now, and likely to become worse, in daily news media and 'Doomsday' books. In some who work in this area there is an attitude of unvoiced despair which has led to many non-critical analyses of the problems and solutions presented. On the one hand, descriptions of the enhanced agricultural capabilities of the nineteen-eighties and beyond, such as the increased efficiency of management due to the use of systems analysis, computers and earth satellites, are decried by pessimists. On the other hand, simplistic solutions proposed as panaceas are enthusiastically seized upon, leading to the delusion that the food problem can be overcome without serious effort (Boerma, 1969). Such concepts as single cell proteins, fish flour, extracted plant proteins, 'appropriate technology' offer partial solutions in specific situations. If they fail to live up to what in hindsight were often unrealistic and extravagent claims, they may be relegated to an obscurity they do not deserve, thus depriving the world of a component of the solution to food problems. The 'Green Revolution' is a case in point: its success was limited to situations where the necessary resource inputs (fertilizers, pesticides, irrigation, but also continuing research, education and marketing development) were available and it benefited mainly the more affluent of the rural population. This was already understood at its commencement (FAO, 1969). Its apparent 'success', widely publicized in the early nineteen-seventies, has led some to believe that the solution of the food supply problem lies in 'just one more major breakthrough', leading to worldwide nutritional affluence. Despite the considerable increases in grain production which have occurred, subsequent claims of 'failure' of the Green Revolution lead now to a renewed pessimism.

This chapter will demonstrate that the problems of the world food supply are not only extremely grave: they are extremely complex and there is no one solution in any one branch of science or technology. The diversity of agricultural resources (environmental, biological, cultural and economic) requires an equivalent diversity in the solutions developed. Inadequate food production is only part of the problem. It is not enough to simply produce more food, especially if the population is expanding as rapidly as the food supply, although this is a daunting challenge in itself, requiring acceleration of current agricultural development. Each family needs to be able to produce, or earn access to, its own food supply. The world's food problem is, to a great extent, the problem of poverty and uneven distribution of resources. The necessary degree of accelerated agricultural development involves a degree of control and economic participation by the needy which is difficult to achieve. In most countries, 'developed' and 'developing', powerful minorities perceive such necessary changes as being inimical to their own interests. Without these changes agricultural scientists may contribute to an increase in total food production and G.N.P. But their contribution to the solution of food-supply problems of 'the poorest of the poor' will be small. History books and today's newspapers are full of the results of failure to meet the food needs of mankind: food-price riots, coup d'etat, revolutions show that the best interests of the affluent as well as of the poverty stricken seem to be, in the long run, in increased economic and social justice for the impoverished. Agricultural development efforts must be aimed at the enrichment of life of the needy, not at the promotion of specific animal or plant industries. 'Animal production'-oriented scientists must come to understand that circumstances exist where increased animal production is actually contra-indicated and many situations occur where appropriate animal production systems have not yet been developed. They and 'plant production'-oriented scientists must perceive that integrated agricultural systems may often constitute the optima when social and cultural factors and long-term environmental stability are considered along with the more conventional objectives of production and profit.

2. THE CURRENT WORLD FOOD SITUATION

Large-scale protein-calorie malnutrition occurs in three major situations: in multitudes of political refugees, acute famine and in chronic under-production and under-consumption of food by poverty stricken people. The voluntary protein-calorie malnutrition of the obese and the diet-conscious of the developed countries will not be considered here. Nor will this chapter consider deficiency diseases whose aetiology, prevention and cure do not involve animal production, diseases such as cretinism, goitre and vitamin A-deficiency blindness which, unnecessarily, still afflict millions despite the availability of simple, cheap and long-term controls.

The nutritional plight of refugees is a result of the rapid congregation of a population with little in the way of assets, employment, food-supply systems or resources for food production. In the short term they represent a problem of food distribution, not production. The numbers of people involved at the time of writing are estimated at 16 million. Acute famines are caused ostensibly by climatic changes (drought, unusually cold or hot weather, flood), insect plagues (e.g., locusts) or outbreaks of disease in plants or livestock. Famine is basically a failure of a previously adequate food-supply system to provide the required quantity of food. It is an acute manifestation of chronic under-nutrition, and the number of famished is an index of a far larger number of under-fed. It does not occur overnight, without warning. Unpredictable natural disasters will always occur, but the occurrence of famine increasingly results from breakdown of a food supply system overloaded by demands of increasing population. For example, in the Sahelian drought of the early nineteen-seventies the most affected areas were reported to be those which had undergone the greatest recent development and experienced the greatest prior increase in human and livestock population. The numbers involved world-wide may be tens of millions. But the greatest cause of malnutrition, yet the least dramatic and least likely to promote a rapid and adequate response by affluent countries, is the malnutrition or under-nutrition of the chronically poor, those without adequate food production resources or cash income to obtain their needs. Mayer (1976) estimates that 500 million people suffer from malnutrition and another 1000 million would benefit from an improved diet, a total approaching 40% of the human population. The origin of their poverty and lack of farm land is frequently attributed to the rapid population increases of the last several decades. Improved hygiene and sanitation, vaccination and control of insect vectors of disease have reduced the death rate and permitted the population to expand until the next limitation is approached, the supply of food. However, such people are often citizens of recently decolonialized countries, which have several characteristic inadequacies. In some, the population is so small as to preclude efficiencies of scale in administration, education, and production. Agricultural production, research and extension were often well developed for crops for export to the colonial power, but little is known about the subsistence agriculture of the people. Land and labour often are tied up in production of export crops, thus providing for the landless poor a small income dangerously vulnerable to international trade fluctuations. In many ex-colonies, education is often lacking and power has been in the hands of foreigners so that, on removal of the expatriate colonial administration, there are few with knowledge or experience needed for government, education or business development. The often violent process of decolonization has frequently cut political and business ties with industrialized nations so that sources of imports and markets for export of the small number of cash crops may have been lost. These economic and social disruptions are major causes of poverty.

The section of population most vulnerable to malnutrition is the youngest. The baby born of a malnourished mother is more likely to be premature, is born smaller and weaker, and is more prone to suffer death or permanent mental disfunction, as is the child malnourished during infancy. Since brain cell multiplication and interconnection are largely complete by 2 years of age, a consequence of infant malnourishment is an increased incidence of physical and mental retardation, an additional inhibition for national development. The statistics on infant mortality are unreliable in detail but the broad picture is unequivocal (Mayer, 1976). In developed countries, infant mortality to 12 months ranges from 10 to 20 per 1000; in Brazil it is 95 per 1000, in India and Pakistan 140, and Zambia and Bolivia 250. Total child mortality to 5 years may reach 400 per 1000, as in Ethiopia even before the famine (Miller and Holt, 1975). Similar data are available for many other developing countries, and the true picture may be much worse, since deaths of infants and particularly still-births are frequently unreported. The fatalistic attitude of poor villagers may be mistaken for a callous disregard for human life, but it too often is a realistic assessment of the situation, which is that many babies will die and nothing can be done about it.

Malnutrition at weaning is a common problem, due to the change in diet from sterile breast milk to frequently contaminated solid food, low in protein, high in bulk (e.g. bananas, sweet potato). Cultural practices often add to the weanling's problem: in different societies milk is considered unfit for boys, or girls, or both; the child may be unable to consume enough food because only one meal is presented per day; high protein foods may be reserved for working men. Generally malnutrition is less common and less severe in school-age children and adults (except for pregnant or lactating women). The problem here is lack of calories as much as or more than a lack of protein. Except when the diet is based on bananas or cassava, whenever energy is adequate protein usually is too, but when energy is inadequate, some protein is used for energy production, producing a secondary protein deficiency. However, a primary adaptive mechanism in the malnourished adult appears to be a reduction in physical activity; there is growing evidence that low energy intake is a limitation on the capacity for physical work in developing countries. This can be critical at the end of the dry season when low food reserves occur at a time of the peak labour demand necessary to establish the next crop. Malnutrition is most prevalent in depressed rural areas and slums of great cities where a major factor is the lack of income to buy food which is available in adequate quantity and quality. Even in the Sahelian famine people starved to death while food was available, at a price. There was not an actual shortage of food (Miller and Holt, 1975). In these circumstances, reduction in cost is more important than increase in production or availability of food in alleviating malnutrition or under-nutrition.

Statistics on nutrient intakes are as imprecise as mortality figures. Total food production minus seed, stock feed and wastage is sometimes used to calculate food consumed. However, since taxes and landlords' receipts may depend upon production this is consistently under-reported by the farmer; wastage may be as high as 40% but can only be guessed at, and subsistence food production is so hard to estimate that it is often ignored. Detailed surveys of individuals in communities provide the only accurate data on distribution of food within a community family but such studies are expensive in time and money. Hence, small-scale investigations are extrapolated over large and varied populations. Despite this imprecision, the data are sufficient to show two-fold differences between energy and protein intakes of the best and worst-fed countries, while intake of animal protein has a six-fold range. Variations within each country increase these ranges of differences between the well fed and the malnourished. The requirements for nutrients are also not known precisely for any age or sex; for example, protein requirements for adult males, the easiest to study, have been revised downwards from 150 g/day to 40 g/day on diets adequate in energy. Apparent racial differences exist but may be the consequence of previous nutritional history rather than genetic differences. The differences between individuals from the same population in efficiency of energy retention is commonplace knowledge but little understood.

One of the most important sources of variation in nutrient requirements is variation in health status. Infections or parasitic diseases of the alimentary tract increase the need for protein and minerals, especially if frank diarrhoea is occurring. Stress of any sort results in increased catabolism of tissue proteins which act as substrates for hepatic gluco-genesis; loss of appetite, a frequent occurrence in many disease states, compounds this problem. A diet of starchy tubers or cereals, which may be adequate to support the healthy, is totally inadequate to supply the high concentration of high quality protein in a palatable form so necessary during illness and convalescence. Young children in developing countries are particularly subject to intestinal, respiratory and other infections which increase nutrient requirements, due to the poor hygienic and sanitary standards of their environment. Since at the same time they frequently have a poor quality diet, they are particularly likely to develop nutritional deficiencies. Thus it is impossible to define accurate values for recommended daily allowance (RDA) for individuals or populations. RDAs for healthy adults of small stature may be estimated with reasonable accuracy from body weights. But for the child, whose weight indicates the combined effects of repeated illness and malnutrition, feeding to standards based on weight will merely perpetuate the stunting of growth. The RDA must be guessed from the child's age, if known, or from weight plus a liberal allowance for compensating 'catch-up'

Chapter 12 references, p. 283

growth. This extra allowance for protein may be up to 100% of the normal RDA for a child of that weight while for energy an extra 20—25% may be needed, so that the protein/energy ratio must increase. Scrimshaw (1979) provides a thorough review of this entire area, and states, 'It cannot be stressed too much that growth is as much dependent upon adequate supplies of dietary energy as on protein and the other essential nutrients'. It is obvious from this review that provision of a diet of the necessary nutrient density and protein quality for the recovery of starved, diseased children would be much easier if animal products were readily available, although animal protein is not actually essential: milk is the nutritionally optimal source if lactose intolerance is not a problem. McLaren (1974), in a challenging article entitled 'The Great Protein Fiasco', proposed that it is rare that protein is the primary limiting nutrient even in very young children, and that concentration of research on protein deficiency has obscured the real need, which is quite simply for more food. His view is receiving wider acceptance, as nutritionists re-emphasize the 'protein-sparing' effects of carbohydrates and develop an 'energy and protein' orientation in their view of malnutrition. Nonetheless, restoration of protein adequacy may be the more difficult, due to limitations on availability and cost.

The current population of the absolute poor was estimated by the World Bank to be 770 million in 1975. 'Absolute poverty is a condition of life so characterized by malnutrition, illiteracy, disease, high infant mortality and low life expectancy as to be beneath any reasonable definition of human decency' (McNamara, 1978). The refugees, the famished and the chronically poor, the majority of whom are young and often ill, these embody the current world food problem.

3. THE 'POPULATION PROBLEM?'

> 'People at present think that five sons are not too many and each son has five sons also, and before the death of the grandfather there are already twenty-five descendants. Therefore people are more and wealth is less; they work hard and receive little' (Han Fei-Tzu, c. 500 B.C., cited by Meadows et al., 1972).

'Necessity is the mother of Invention'

Long before the Reverend Thomas Malthus (1798) postulated in his often quoted first 'Essay on the Principle of Population' that population growth would outrun food supply unless checked by war or disease, and many times since, there has been pessimism over the world's population and food situation. This has been followed by optimism as the cause of temporary shortages has been eliminated, or a technical advance has resulted in increased food production although, in some areas, the pessimists have been vindicated and famine has indeed occurred. Shortly after Malthus' dire prediction, the Agricultural Revolution resulted in intensification of English agriculture and this technological revolution continues to spread throughout the world. North America and Australia became granaries for Europe, and global famines did not occur. Similarly, predictions by Sir William Crooks to the British Academy of Sciences in 1898 of food shortages in Britain within 30 years due to lack of cereals did not eventuate. Predictions of shortages in the nineteen-sixties were followed by the 'Green Revolution' which was so successful that in India by 1970 there was concern over grain surpluses. Mankind has always survived in precarious balance with his food supply, with the difference between adequacy on a world basis and a crisis situation being only 4% or 5% of the world's grain production (Drosdoff, 1979). Bennett (1963) claimed that the average nutritional status of humankind had probably improved in the previous half century and would continue to do so for the next 20 years. Wortman and Cummings (1978) confirm this from 1955 to 1975, when growth of food production at 3%/annum exceeded the population growth of 2.7%/annum. What is the truth of the Malthusian doctrine that population must outrun the means of subsistence? Malthus states as a criterion 'on few subjects can any theory be pronounced just, that has not stood the test of experience'. His first postulate, 'Population, *when unchecked* (my italics) increases in a geometrical ratio' he immediately refuted: 'in no state that we have yet known has the power of population been left to exert itself with perfect freedom'. The next: 'Subsistence increases only in an arithmetical ratio' is substantiated only by 'it is impossible to suppose that the produce could be quadrupled. It would be contrary to all our *knowledge*. The very utmost *that we can conceive* etc. etc.', presupposing no significant change in knowledge. A third assertion 'That population does invariably increase, where there are means of subsistence, the history of every people that have existed will abundantly prove' is certainly not true of modern Europe. Table 12.1 shows the poor relationship between increase in subsistence and population in developing countries, and on a world basis there is no such simple relationship. Only in populations ignorant of the immediate local consequences of increased population (i.e., a reduced per capita share of all currently available resources) does population increase automatically to absorb an increase in limiting resources and maintain a constant standard of living. Rarely does a human population suffer from this degree of ignorance. Malthus under-rated the ability of humankind to further its own self-interest when it can generate the power to do so. In his later, less quoted writings, Malthus concedes that most populations have understood the consequences of population growth and have limited population, when necessary. He also admitted that there was a 'natural lack of will on the part of mankind to make efforts for the increase of food beyond what they could possibly consume'. Under most circumstances, an increase in population means an increased workforce which can produce more, as well as an increased demand which will cause the use of the reserve capacity which existed even at previous labour and technology levels. This applies in the long run but in the short run, imbalances of supply and demand for food and employment do occur, causing human misery.

Malthus' stimulation of ideas on population is of great significance. But his major contributions, which are the mathematical formulae so frequently quoted, are so qualified, so simplified and so widely extrapolated, that they were generally disregarded by demographers after the mid 19th century.

Nonetheless, in a finite world, the parallel advance of population and food cannot continue indefinitely. This obvious concept, which predates Malthus, has been frequently attributed to him in the recent flood of publications on the 'Population problem' which followed the realization, in the nineteen-fifties, that the population growth rate in developing countries exceeded anything which preceded it, and that the number of people in poverty was enormous. What is less obvious is that, in an economic sense, there is dispute as to whether this really is a 'finite world' since the supply and price relationships do not react as would be predicted in such a case. Simon (1981) examines the argument and demonstrates that man's ingenuity and capacity for invention consistently make so-called 'limited resources' more available as his demand for them increases. Indeed, 'Necessity *is* the mother of invention'. Thus it is not possible to predict a 'Sustainable Population' at any

Chapter 12 references, p. 283

TABLE 12.1

Annual rate of change of food (crops and livestock) production in relation to population growth for selected developing countries, 1970–1979.

Food production (%)	Population (%)					
	1.5 and below	1.6–2.0	2.1–2.5	2.6–3.0	3.1–3.5	3.6 and above
−3.0 and below			Kampuchea Dem.			
−2.9 to −2.0	Barbados					
−1.9 to −0.1	Trinidad and Tobago		Mozambique, Congo	Morocco, Gambia, Namibia, Mauritania, Ghana	Algeria	
0.0–0.9	Uruguay, Gabon, Jamaica, Suriname, Cyprus	Samoa	Ethiopia, Lebanon, Angola, Guinea, Egypt	Togo, Peru, Somalia	Honduras, Iraq	Jordan
1.0–1.5		Chile, Yemen Ar. Rep.	Haiti, Nepal, Chad, Sierra Leone			
1.6–2.0		Cuba, Fiji, Mauritius	Lesotho, Madagascar, Bangladesh, Laos	Tanzania, Zaire, Uganda, Burma, Benin, Mongolia	Dominican Rep., Nigeria	Kenya
2.1–2.5		Guinea Bissau	Buthan, Guyana, Vietnam, Central Afr. Rep., India, Indonesia		Rhodesia, Malawi, Liberia	
2.6–3.0			Upper Volta, Burundi, Papua New Guinea	Afghanistan, Niger, Botswana, Mali		

3.1–3.5	China	Reunion	*Cameroon, Yemen Dem.*	Swaziland, *Rwanda,* Paraguay	Mexico, *Pakistan* Venezuela
3.6 and above	Argentina	*Sri Lanka,* Korea Rep.,	Tunisia, Costa Rica, Colombia	Bolivia, Ecuador, Panama, Sudan, Turkey, Malaysia, Iran, *Senegal, El Salvador,* Brazil, Philippines, New Hebrides, Bahamas, Korea Dem., Thailand	Nicaragua, Zambia, *Guatemala,* Saudi Arabia, Syria · *Ivory Coast,* Brunei, Libya

From: FAO (1979).

Note: Countries in each group are listed in ascending order of the annual rate of change in their food production. Most seriously affected countries are italicized.

Chapter 12 references, p. 283

given standard of living at any time, due to this unpredictable relationship between utilization of resources and technological change. Additionally, feedback mechanisms for population control are slow, so that, even when fertility level drops to the replacement rate of 2.1 babies per woman, a stable population will not be achieved for about 70 years unless death rate increases (Guilmot, 1978). It is feared that the sustainable population will be identified in hindsight as we recognize the cost of exceeding it, when death rate increases. Even if there is no theoretical maximum population for mankind, there may well be a maximum rate of increase, above which the development, organization and application of new technologies may not be able to prevent human disaster in specific areas.

The writings of Malthus, Meadows, Erlich and others are not predictions of what will occur, but of what may occur if nothing is done. They can be regarded as describing the necessity for invention and may have served to initiate it at an earlier date and accelerate it.

The implication in Meadows et al. (1972) of 'one world with one fate' has been challenged in 'The Second Report to the Club of Rome' in which Mesarovic and Pestel (1975) divide the countries of the world into 10 classes. However, an examination of Table 12.1, which groups countries on the basis of food production and population increases, shows the difficulties in any such classifications. The positions of countries in this Table are not consistently associated with their agricultural or other resources, size, climate, political or religious systems, or colonial history, especially if recently developed ex-colonies such as Singapore, Japan, Canada and New Zealand were to be included in the comparison. Each country is unique. The 'world collapse' will not occur. Instead individual poor countries may degenerate into a state of anarchy, famine and epidemic while more affluent developed countries survive. The uneven distribution of resources, wealth and power make the 'spaceship Earth' analogy a weak one in the short-term consideration of food supplies regardless of its long-term appropriateness.

Predictions in 1966 of the world population in 2000 A.D. were 7500 million if current growth rates continued, while the UN 'medium prediction' was 6130 million. This latter included a growth for Europe of 24% between 1960 and 2000. But Guilmot (1978) showed that in Western Europe 13 countries had fertility rates below replacement level and predicted population growth of 5% between 1970 and 2000 A.D. These unexpectedly rapid declines in fertility are being repeated in many developing countries.

Increase in the human population can occur only if birth rates exceed death rates. Since there is no evidence of a recent increase in human fertility, the population increases of the last two centuries must be due to reduction in death rates, of which the main components are nutritional inadequacy, disease (often acting together) and violent death (including infanticide, abortion, war). Even during war, a major cause of death is a combination of disease and malnutrition among non-combatants. It follows that the 'unprecedented' increase in human population has been permitted as a consequence of an unprecedented degree of nutritional adequacy, good health, peace and social harmony on a world-wide basis, although the improvement of nutrition, health and peace does not necessarily cause population increase. The current situation is that, despite the existence of '770 million people in abject poverty', never before have there been as many healthy, well-fed people on the earth, relatively or absolutely. Population growth thus represents economic success and human triumph. However, this is scant comfort for those 770 million.

In developing countries, in the nineteen-sixties and the early nineteen-seventies, population growth rate overall was 2.3%/year due to declining death rate (Demographic Stage 1) while in developed countries the rate was 0.9%. Fig. 12.1 shows the strong relationship between GNP/head and birth rate, a relationship little affected by race, religion, culture or political systems. Among the poorest (below $500/year), Latin American countries have a higher birth rate at a given income than Asian countries, with African countries intermediate, but these differences are trivial compared with the primary relationship. Recent studies show that developing countries are advancing to the 'second demographic stage' of declining birth rate more rapidly than predicted. The world average birth rate per woman fell from 4.6 in 1968 to 4.1 in 1975, but in Indonesia, it fell from 6.5 to 4.6 (Drosdoff, 1979). In 15 other developing countries the birth rate per woman currently aged 48–49 had ranged from 6.8 to 7.2 but this has recently declined to 5.4 in Thailand, 5.3 in Peru, 4.2 in Colombia, and only 3.8 in Costa Rica (Anonymous, 1979). Only in Bangladesh, Nepal and Pakistan had this drop not occurred. An increasing number of women desired no more children and did not wish for their current pregnancy or last child. This desire was substantial in all socio-economic groups, equally in urban or rural women. Also, the age of marriage is rising, thus increasing the generation interval.

The key to these changes occurring so rapidly is the acceptance of family-planning programmes and an increase in knowledge of contraception, as well as the increase in efficiency of contraception methods. Three-quarters of women surveyed knew of at least one method and only in Nepal was this knowledge not widespread (Anonymous, 1979). The Pill and IUD were in common use, but the only permanent method of contraception, surgical sterilization, was surprisingly popular in both sexes. The incidence of currently married fecund women, or their husbands, having been sterilized

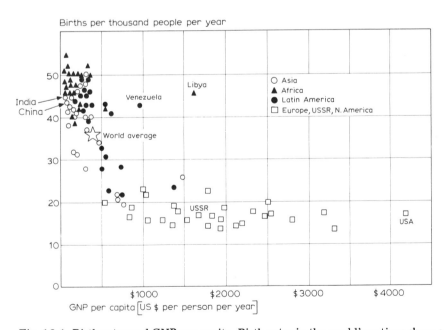

Fig. 12.1. Birth rates and GNP per capita. Birth rates in the world's nations show a regular downward trend as GNP per capita increases. More than one-half of the world's people are represented in the upper left-hand corner of the graph, where GNP per capita is less than $500 per person per year and birth rates range from 40 to 50 per thousand persons per year. The two major exceptions to the trend, Venezuela and Libya, are oil-exporting nations, where the rise in income is quite recent and income distribution is highly unequal. Source: U.S. Agency for International Development (1970).

Chapter 12 references, p. 283

ranged from 11% to 26%. Despite reputed cultural obstacles, by 1976, 63 developing countries had official family-planning programmes. Consequently, world population predictions for 2000 A.D. have been reduced to 6250 million by the UN Fund for Population Activities and to 5500 million by the U.S. Office of Population (using the same base data).

The estimated population of the world, of 2000 million in 1930, will have trebled within one human life span. The estimated population of the absolutely poor will have been reduced by 180 million by 2000 A.D. but it is likely that 600 million will remain in that category (McNamara, 1978). As population increases from 3500 million in 1970 to about 6000 million in 2000, the ratio of population in developing countries to population in developed countries increases from 2.8:1 to 4.0:1. For the developed countries, the possibility of assisting the developing countries will become more difficult. Within developing countries, increasing urban drift means that the rural population, which in 1970 produced food for half as many city dwellers, will be out-numbered by them two or three times by 2020 A.D. As the rural population is also increasing, albeit more slowly, the area per family is declining, resulting in increasing under-employment as well as under-production of food for the family. The need for employment in rural and urban areas will become more acute. The prospects for 'solving' the world food problems will remain remote during the remainder of this century, but there are still several factors which justify greater optimism than is frequently expressed.

(i) Developing countries now accept that they will have to produce most of the food they need, and would prefer to produce all of it, to eliminate dependence upon foreign governments' policies and foreign agricultural capacity (Sisler and Blandford, 1979). They are beginning to invest in agriculture at an extremely rapid rate. FAO (1980) indicated that, of 27 countries examined, four had a declining government investment in agriculture while six countries increased government expenditure at 0—10%/year, 10 countries increased 10—20% and seven increased more than 20%/year, in real terms. The experience of India in stimulating grain production in the nineteen-sixties is an encouraging example which many countries (Pakistan, Algeria, Malaysia, Philippines) are following with some success.

(ii) There is greater understanding that technology and genetic resources normally cannot be transferred to different environments without drastic alteration, but must be developed at the site of utilization. The understanding of other aspects of national development is continually improving.

(iii) An increasing number of trained and experienced technical and managerial people and an increase in research and training institutions, extension services, fertilizer factories and farm capital-forming institutions mean that as new knowledge is obtained and new techniques develop, they can be put into practice. For Latin America, and in tse-tse fly infested countries in Africa, there are large areas of land not yet developed for farming or grazing, which await the development and application of these new techniques. (Asia, however, does not have much under-utilized land.)

(iv) The realization is slowly becoming more widespread that poverty, hunger and under-employment constitute the basis of social unrest which may be the most dangerous aspect of the 'population—food' problem, leading to civil strife and revolution, from which few will benefit.

'...on any reasonable calculation the developing countries are going to have to make massive investments if these individuals (the poor and the urban unemployed) are to find productive employment. It is sometimes argued that the cost will be too high; that the world just cannot afford it. But the truth is really the other way around. What the world cannot afford is procrastination and delay while dangerous social pressures build' (McNamara, 1978).

(v) It now appears that the prospect of an ever-increasing population absorbing all increases in production (of food, employment, all resources) may be a more short-lived phenomenon than previously predicted. The world population demands are at the limit of mankind's capacity to satisfy, as they have usually been historically, and will remain so for many years, but a significant downturn in growth rate is occurring. Additionally, it is contended that the increased population first demand and then generate further resources. 'The ultimate resource is people, skilled, spirited and hopeful people, who will exert their will and imaginations for their own benefit and so, inevitably, for the benefit of us all' (Simon, 1981).

4. PARTIAL AND SHORT-TERM ANSWERS

A number of 'solutions' for the food supply problem involving import of food from developed countries, major changes in agricultural systems or the development of new production systems or new foods have been proposed in the last two decades. Wortman and Cummings (1978) reject some of these as 'non-solutions', despite their unquestioned ability to supply food, because they do not address the basic causes of the food problem by reducing poverty and under-employment. Many peasant farmers are engaged in tasks of low profitability due to inadequate resources for some or all of the year: these under-employed (usually rural) and the unemployed (usually urban) both require jobs to provide satisfaction of social and psychological needs as well as physical requirements of food. Additionally, countries with food deficits want security as well as food: talk of 'resource-diplomacy', 'lifeboat ethics' and an appraisal of the function of the international food trade make some developing nations view self-sufficiency in food as their highest priority (Sisler and Blandford, 1979). Self-sufficiency is not essential for national survival, particularly if industrial exports are being generated, and indeed few nations are self-sufficient, while even fewer have an actual exportable surplus of food (Wortman, 1976). However, an assured supply is necessary for social and political stability.

(a) Increased production of food by developed countries will not contribute directly to the solution of the world food problem unless developing countries can afford to pay (as many can, particularly the oil-rich countries) since rich countries cannot afford to give on a scale of ever-increasing generosity, except for *short-term* famine relief. The potential for increase in production of beef, if the production per beast in developing countries (minus India and China) could be lifted to that of Australia and New Zealand (where grain feeding is of little significance), is about 10 million tons (Table 12.2). That is 10 times the overseas sales of the world's biggest exporter, Australia, and more than twice the world's total exports of beef, obtained without any increase in cattle numbers. In the long term, carrying capacity and national herd size could increase along with the improvements needed to double productivity. However, initially a reduction in herd size will be essential in many areas to reduce over-grazing and soil erosion, which are already severe in much of Africa and South Asia, and permit pasture renovation and improvements. Similar calculations for pig meat show that if developing countries attained the efficiency of production of developed countries without increasing numbers, the increase in production, 4 million tons, would be nearly twice the total world export for 1979 (AMLC, 1980). Calculations for other agricultural products yield the same result: there appears to be great potential for increased food production in some developing countries by boosting production to levels closer to those *already achieved*

Chapter 12 references, p. 283

TABLE 12.2

A comparison of productivity of cattle herds in developed countries, with grain feeding
or grazing systems, and developing countries

	Cattle nos. (millions)	Beef and veal production (million tons)	Production per beast in herd (kg/year)
Developed countries			
Grain feeding			
U.S.A.	111	9.83	89
Canada	12	.96	80
France	24	1.79	75
West Germany	15	1.52	101
Grazing			
Australia	26	1.79	69
New Zealand	8	0.49	61
Developing countries			
Argentina	60	3.00	50
Brazil	93	2.10	23
Columbia	27	0.60	22
Developed countries[a]	405	29.65	73
Developing countries[b]	303	9.64	32
India	241	0.19	0.7
Developing countries (minus India) at Australia and N.Z. level	303	19.70	65

[a] U.S.A. and Canada, Europe, U.S.S.R., Australia, New Zealand.
[b] All others bar India and China.

in developed countries. It seems more sensible to attempt this than to rely
on solving the food supply problem by achieving, in developed countries,
totally unprecedented levels of both agricultural productivity and generosity
in food aid.

The amounts of under-utilized land appear to be much greater in developing countries: one analysis shows that in Africa 22% of 'potentially arable
land' is farmed, 574 million ha lie fallow, and in South America the estimates
are 11% utilized, 602 million ha 'unutilized', while in Europe, North
America and Oceania combined, only 480 million ha are still 'available'
for development (Wortman and Cummings, 1978). Although some of this
land is currently being developed, technical and economic constraints
preclude the development of some areas now and perhaps permanently.
However, in Asia there are countries with virtually no under-utilized arable
land and overall utilization is 88%. Van Bamweke (1979) analyzed more
conservative figures calculated by Buringh et al. (1975) and Buringh and
Van Heemst (1977) for availability of land, constrained by its suitability
for labour-oriented agriculture. Africa and South America contained over
300 million ha each, but in Asia the amount being farmed exceeded the
suitable area by 71 million ha and in these areas poor production occurs
and soil erosion is severe. The sustainable population under these assumptions with labour-oriented agriculture can increase 130% in South America,
30% in Africa, but Asia is currently 'over-stocked' by about 1500 million
people. Here food imports to sustain life will be essential until more productive farming techniques are available. These regional groupings and national
averages conceal great differences in the utilization of land, which may be
partly due to historical or cultural rather than purely agricultural factors;
e.g., Indonesia on average appears crowded, at 72 people/km², but actually
Java is very crowded at 600/km², Sumatra and Sulawesi are less crowded at

$48/km^2$, while the enormous islands of Kalimantan and Irian Jaya, respectively four and three times the size of Java, have 10 and 2.4 people/km^2 (1970 figures). Since the people of Java continue to survive and there is a lot of unused land elsewhere, it is obvious that Indonesia can support a much larger population even with available technology, if time is available to develop this land!

(b) Fish flour, single cell protein, extracted leaf protein and various mixtures of plant proteins have been widely acclaimed in the nineteen-sixties (Parpia, 1968) but have not yet realized their apparent potential in alleviating protein deficiencies. Tannenbaum and Mateles (1968), who carried out much of the research into single cell protein, stress that there can be no single simple solution to the world protein-supply problem, and single cell protein, oilseed meal protein, leaf protein and high lysine corn are only components which may contribute to the solution. Pirie (1975) rightly challenges the conventional claim of the conservatism of food tastes in hungry people in developing countries. Their tastes sometimes change inconveniently quickly as far as nutrition is concerned, as evidenced by the ubiquity of Coca Cola, Kentucky Fried Chicken and other expensive snack foods, and even the staple energy source can change, although less rapidly, as exemplified by the recently established dependance on imported rice grain and wheat flour in Papua New Guinea. Widespread acceptance of these products depends upon convenience, palatability and particularly marketing as a sophisticated 'Western' product, rather than as a 'poor man's food'. Market development for the protein concentrate products has not yet resulted in widely consumed food except in institutions and for famine relief. But even when such acceptable protein-rich foods are available they will still suffer from the disadvantage of being factory made. This requires technology, usually large scale, which may not be available in some countries and requires that the food be sold: for the poorest, any price will be too high. Additionally, the production of single cell protein from petroleum wastes, an attractive idea in the nineteen-sixties, is no longer being considered as a long-term contribution to the human diet in developing countries since petroleum is a 'finite' resource with a rapidly and unpredictably increasing price. There is a role for these high protein flours, where the requirements of price, high technology, high-pressure marketing and ease of incorporation into the diet present few problems. This is in the food-processing industries of North America and Europe, where these colourless, odourless products can be incorporated into bread, sausages, 'extended' meat products, canned and packeted foods on a large-scale industrial level. The labels of such foods in the U.S.A. divulge the wide range of unsuspected ingredients added.

The role of protein concentrates in developing countries will probably occur first in foods manufactured on a large scale, such as bread but will only occur as gifts of aid in the lowest economic strata where people have little money and must produce their own food. The situation here is not the same as the introduction from another area of a cereal or vegetable for which recipes and processing technology are already available. There is no existing consumption technology for extracted plant protein, etc., which can be transferred or modified. Nevertheless, fish flour, extracted leaf protein and vegetable protein mixtures can make a contribution to the nutrition of poor people when recipes, preparation techniques and marketing are developed, if the price is appropriate.

(c) The direct transfer of large-scale Western farming technologies will also supply the food but not the answer to the food problem. Highly capitalized and mechanized farmers in Australia and U.S.A. may produce

Chapter 12 references, p. 283

food for 40 times their number, using technologies devised in those countries, not imported unchanged from Europe. Similar large-scale farming techniques could eventually be developed in the tropics. Areas are available in Africa and Latin America, but not in much of Asia where it would be necessary to displace small-scale farmers to obtain appropriate areas. Produce from large farms, if used to feed the urban poor, may cut into the markets of the rural poor. In terms of production/man-hour, the mechanized farmers are more efficient, but when labour is not limiting such a statistic is meaningless. Production/ha is greatest for smallholders using well-developed complex systems involving intercropping, multiple cropping, hand-weeding and pest control and hand-feeding of weeds, stubbles and processing residues to appropriate domestic livestock. In cases where the smallholder is less efficient, this is usually in a recently instituted production system where the process of technology transfer and adaptation is not yet complete, as in the cattle industry of Papua New Guinea.

5. CONVENTIONAL ANIMAL PRODUCTION

The Provisional Indicative World Plan for Agricultural Development (FAO, 1969) defined as the four major objectives of agriculture the provision of (i) food, (ii) employment to eliminate poverty, (iii) capital for rural development, and (iv) foreign exchange (generated or conserved) for national development. These objectives are in conflict, e.g., a given piece of land (and other resources) can be used to produce (i) food or (iv) rubber for export; land can be used in a subsistence fashion to provide (i) food and (ii) employment for peasants, or farmed extensively to provide (i) food for cities, (iii) cash from sales for development, but little employment. No one farming system or development programme can hope to meet all objectives but programmes need to be evaluated in the light of those objectives so that the combination is optimal.

In 1969 the need was seen for an increased availability of protein, i.e., objective (i) was to supply the human dietary need for animal protein: meat, milk and eggs were to become a larger part of the peasants' diet. Since the estimates of protein requirements have been reduced, objective (i) is now seen as the need to provide food in its widest sense, including mainly grain and vegetables. Objective (ii) is paramount, to provide cash to purchase food. Due to the high price and elasticity of demand for animal products, animal production is now seen as a way of generating cash for the peasant to buy grain for his own consumption.

5.1. Intensive animal production

FAO (1969) advocated the development of grain-fed pig and poultry industries in developing countries as the only way of accelerating the production of meat fast enough for the anticipated demand. It was assumed that grain for these industries would come, in part, from surpluses generated by the 'Green Revolution'. In their more recent review, 'Agriculture: towards 2000', FAO still advocates this course, amongst a range of programmes. Although such industries produce meat, the grain, surplus to human needs, is not currently being produced in the developing countries, which as a group have not been net exporters of grain since 1947. In 1964, net imports were 15 million tonnes, in 1977, 42 million tonnes despite the 'Green Revolution', and FAO projections (Anonymous, 1981) indicate gross imports of 95 million tonnes for 1980–1981. Locally-grown feed grain supplies seem more remote than ever for developing countries, although their share of the international grain trade is growing. Resource-rich developing countries do import feed grains from developed countries, although fears of dependence upon other countries for animal feeds are voiced, due to possible uncontrollable variations in cost and availability. This influences national policy to minimize such imports by achieving self-sufficiency or minimizing grain feeding practices to reduce vulnerability to outside political pressure (Charles et al., 1980).

The subject of feeding grain to domestic animals has brought forth much comment upon 'the competition of man and animals for food' (Blaxter, 1970) and emotional outbursts about the immorality of feeding grain to animals while humans starve. It must be understood that there is no competition between domestic animals and man for food: such competition does occur in hunter—gatherer societies, but the penned pig, the caged bird and the feedlot steer are permitted no competitive ability. They are entirely dependent upon man for their food. The competition is between rich man and poor man for a limited amount of resources (McClymont, 1976). If there is immorality it is the immorality of uneven distribution, of wealth in the midst of poverty. Grain feeding is only one index of this inequality. If it is immoral, it is no more (or less) so than the possession of large houses and cars, the enjoyment of extravagance in food, clothing and travel, the use of agricultural land to produce tobacco, wine and flowers, the maintenance of pet dogs and cats on prepared diets; in short, any blatant over-indulgence in and over consumption of the world's resources (usually defined as 'more over-indulgent than the moralizer himself'). This is characteristic of a wide range of affluent lifestyles found in developed countries and enthusiastically emulated by the recently affluent in developing countries. 'We have met the enemy — and the enemy is us' (McClymont, 1976).

Grain feeding of ruminants on a large scale is a post-World War II phenomenon brought about, to a considerable extent, by U.S.A. stockpiles of cheap surplus grain. The removal of these surpluses in the early nineteen-seventies has resulted in less emphasis on grain feeding to ruminants and more fattening of cattle on grass so that the system is not of the same significance as it was 10 years ago. Since grain feeding cannot be as easily rejected on moral grounds as first appears, without rejecting at the same time the economic structure of the developed countries, what are the possibilities for using the rapid reproduction and efficient feed utilization by pigs and poultry especially, to satisfy some of the 'four objectives'?

In 1979, FAO published *Agriculture; toward 2000*, in which trend projections and possible projections of agricultural production are developed for the rest of this century. Their 'Normative Scenario' is defined as 'the maximum FEASIBLE extent of agricultural growth with priority, in the general case, to the basic food commodities for increasing self-sufficiency'. It is emphasized that these are not predictions of what will happen, but projections of the upper-limit of what can happen. Their analysis is extensively quoted below.

The technology of intensive pig and poultry raising consists, in part, in isolating the animal as much as possible from its environment, in respect of disease, climate and nutrient availability. These systems are reputedly more readily transferred between environments than are systems of grazing animals management, since an entire package of genetic material, feed and equipment can be purchased 'off the shelf'. But this view ignores the need for trained manpower to make the systems work, and this lack is apparent in the low turnoff projected. Additionally in their developed country form they are labour-extensive and overseas capital intensive, requiring the

Chapter 12 references, p. 283

importation into developing countries of equipment, feed and genetic material, if these industries are to expand at the rates projected by FAO. The objective of generating employment and cash for the smallholder and landless rural community will not be met by establishment of large-scale, highly mechanized 'factory farms'. Increasing attention is needed towards small and medium-sized farms. This imposes constraints upon the scope and nature of services supplied by feed manufacturers, veterinary services and suppliers of ancilliary equipment. 'Scale — negative' or 'scale — neutral' technology may have to be developed: current Western intensive farming is 'scale — positive'.

The FAO projections of possible egg and poultry meat production (Table 12.3) show the capacity for increase of these industries. It must be

TABLE 12.3

Poultry meat and egg production structure

Region	Birds (millions)	Off-take (%)	Carcase wt. (kg/ animal)	Production (thousand tons)	Growth rate 1980—2000	
					Animal nos. (% p.a.)	Production (% p.a.)
Poultry meat						
1980						
90 Developing countries	2416	178	1.1	4480		
Africa	489	137	0.9	600		
Far East	739	138	1.0	980		
Latin America	847	223	1.2	2240		
Near East	340	213	1.1	670		
Low-income countries	552	112	0.9	580		
2000						
90 Developing countries	7437	215	1.2	18 898	5.6	7.5
Africa	1446	171	1.2	2860	5.5	8.1
Far East	2398	173	1.1	5880	5.7	8.8
Latin America	1912	273	1.3	6220	4.1	5.2
Near East	1682	246	1.2	4540	8.4	10.1
Low-income countries	1194	136	1.1	2730	3.8	8.1

	No. of laying hens (millions)	Eggs wt. (kg/laying hen)	Production (thousand tons)			
Eggs						
1980						
90 Developing countries	918	5.3	4900			
Africa	241	2.4	582			
Far East	248	5.3	1588			
Latin America	296	7.5	2119			
Near East	133	4.6	611			
Low-income countries	239	2.3	581			
2000						
90 Developing countries	2292	7.5	16 440	4.7	6.2	
Africa	531	4.8	2360	4.0	7.3	
Far East	643	7.9	5340	4.9	6.3	
Latin America	643	9.4	5510	3.9	4.9	
Near East	476	7.3	3170	6.6	8.6	
Low-income countries	454	3.8	1920	3.3	6.2	

remembered that these rates of increase have occurred previously; from 1963 to 1975, poultry meat production increased 7.4%/year, compared to 2%/year for beef and mutton. Commercialization of egg production usually precedes that of meat production. The projected increase is due to 150% increase in hen numbers, 41% increase in egg production/bird. The unequal rates of egg production between regions should be noted as an indication of the amount of improvement possible. Intensive poultry meat production is one of the most efficient methods of producing high quality protein, on a feed conversion basis. Again, the greater part of the increase projected is due to numbers, not production/bird.

The production of pig meat is projected to rise more rapidly than beef and mutton, less rapidly than that of poultry in all areas, but in the Near East it will continue to be of minimal importance. In the Far East, where the possibilities of expansion of grazing animal production are more limited, there is expected to be some substitution of white meat for red. The major constraints are feed supply and disease control; if these can be overcome, a rapid rise in pig numbers (3.9%/annum) and total production (5%/annum) is predicted for the '90 developing countries' (Table 12.4). These projections assume a steady growth in small-scale, traditional pig production, but much of the increase will come from medium and large-scale commercial farms.

By-products of small-scale food-processing units supply much of the feed for poultry and pigs, especially for the smaller units. Large-scale production relies mainly upon cereal concentrates, as in Thailand and Malaysia. Most processing of food for human consumption involves the extraction of the more digestible fractions of protein, fat and carbohydrates, leaving the more fibrous residues which are of similarly low value for man, pig and poultry. By-products alone cannot support a large, highly productive industry. Devendra (1981) demonstrates the preponderance of fibrous residues produced by Malaysian agricultural processing industries. He shows that even if his plea for pre-formed protein to be fed only to monogastric animals is granted, there is no possibility of supporting the present pig and poultry industries from by-products of current feed processing industries without

TABLE 12.4

Pig meat production structure

Region	Animals (millions)	Off-take (%)	Carcase wt. (kg/ animal)	Production (thousand tons)	Growth rate 1980—2000	
					Animal nos. (% p.a.)	Production (% p.a.)
1980						
90 Developing						
countries	138.1	53.7	57.0	4232		
Africa	6.3	90.5	41.5	235		
Far East	50.0	72.8	45.2	1644		
Latin America	81.4	44.1	64.9	2328		
Near East	0.5	100.0	52.6	25		
Low-income countries	15.9	49.1	37.7	293		
2000						
90 Developing						
countries	300.3	60.1	61.9	11 155	3.9	5.0
Africa	13.7	118.2	54.0	873	4.0	6.8
Far East	126.5	70.4	53.3	4744	4.7	5.4
Latin America	159.3	49.7	69.3	5481	3.4	4.4
Near East	0.8	100.0	68.6	57	3.0	4.2
Low-income countries	36.9	54.5	48.8	979	4.2	6.2

Chapter 12 references, p. 283

either growing or importing grain. However, the output of fibrous residues of oil palm processing and other industries can easily supply ruminant needs for energy until 2000 A.D. Over the '90 developing countries', FAO (1979) estimates that 95% of inputs for livestock will be for purchase or production of feeds, with grain and milling by-products making up 44% of high-energy feeds (48 million tons of feed grain) in 1975, 54% (190 million tons of feed grain) in 2000 A.D. This implies a growth rate of use of feed grains of 6.2%/annum from 1980 to 2000, slightly above the 1963—1975 growth of 5.4%/annum.

Table 12.5, from FAO (1979) illustrating projections from 'current trends', indicates that per caput consumption of grain does not change markedly over this period, the main changes being an increase in the grain surplus to direct human consumption for 'petroleum importing countries' and 'high-income countries'. Even these modest increases result in a decrease in self-sufficiency for grain from 92% in 1975 to 80% in 2000, with African overall self-sufficiency falling to 61%. Under the FAO 'Normative Scenario', production of cereals by the '90 developing countries' could rise from 360 million tons in 1975 (with net imports of 33 million tons) to 775 million tons in 2000, achieving 90% self-sufficiency (and requiring 88 million tonnes net imports). These 'net imports' conceal very low levels of self-sufficiency for some poor countries. On self-sufficiency as a political concept, FAO (1979) warns

'The decision as to what degree of food self-sufficiency to aim for is — within moderately broad limits — made more on political than on economic grounds. Countries should, however, be aware of the implications of undue insulation from international trade and competition. Imports of a larger share of competing products would undoubtedly be cheaper than domestic supply in some instances and, with the incentive to improved production coming from dependable market availability, could be a dependable source. Food and agricultural policies should benefit the consumer and not only the producer.

As regards food, greater collective self-reliance on the part of developing countries is a goal needing careful interpretation. There is a good deal of natural production complementarity between the temperate zone and relatively extensive agricultures of the developed countries and (frequently) tropical and labour-intensive agriculture of developing countries. Furthermore, larger agricultural import flows from developed countries could advantageously also be linked with expanded markets for manufacturers from developing countries.'

As an example of the complex issues, Charles et al. (1980) analyze the consequences of attempting to achieve regional self-sufficiency in animal feeds within the Caribbean Community. They show an uneven distribution,

TABLE 12.5

Cereals demand per caput (kg p.a.)

Group	1963	1975	Trend projections	
			1980	2000
Direct feed demand				
90 Countries	141	146	152	153
All users				
90 Countries	185	200	207	213
Petroleum exporters	167	195	204	212
Other developing countries				
Low income	164	168	175	179
Medium income	202	217	219	223
High income	273	312	316	326

between countries, of the effects on unemployment, foreign exchange and internal self-sufficiency; no country received an advantage in all three. Import substitution was feasible but advantages and disadvantages appeared in different categories which were difficult to summate. Even for a single country, there is often no simple, uniformly advantageous answer to the question of whether to strive for self-sufficiency in grain or to use the relatively greater efficiency in producing other products to generate foreign exchange (Sisler and Blandford, 1979). Certainly every effort must be made to increase the efficiency with which by-products are utilized within the country for pigs and poultry, rather than exporting these relatively low-value products at ridiculously low prices to compensate for shipping costs, as occurs with flour-milling products in some Southeast Asian countries.

5.2. Meat and milk from ruminants

Projections of cattle and buffalo meat production (Table 12.6) show a different picture from the poultry and pig industries. Production increase is 178% (poultry meat 322%, pig meat 164%) since numerical increase is only 45% (poultry 208%, pigs 117%). However, offtake proportion increases 37% and yield/animal 14%, while the equivalent increases are only 27% and 9%, respectively, for meat birds and 12% and 9% for pigs. The slow increase in numbers is due, in part, to the biological constraint on reproduction, and in part due to lack of new grazing land. This latter reason is especially true throughout Asia, although pockets of under-utilized land do exist, while in Latin America there is considerable apparent potential for increases in herd size, in many areas. In Africa, there may appear to be much under-utilized land, but in the Sahel, a reduction in grazing pressure is currently being suggested for the immediate future, if at all possible, to allow the over-utilized grazing lands to recover. In the subsequent few decades, reduction of the trypanosomiasis problem could release several million km^2 for cattle production.

This improvement in offtake requires, first, the reduction of wastage due

TABLE 12.6

Cattle and buffalo meat production structure

Region	Animals (millions)	Off-take (%)	Carcase wt. (kg/an.)	Production (thousand tons)	Growth rate 1980—2000 Animal nos. (% p.a.)	Production (% p.a.)
1980						
90 Developing countries	852	10.8	137	12 583		
Africa	132	11.6	113	1732		
Far East	376	4.3	105	1711		
Latin America	292	14.6	191	8132		
Near East	52	15.0	129	1009		
Low-income countries	447	4.8	101	2185		
2000						
90 Developing countries	1238	14.8	156	28 668	1.9	4.2
Africa	190	15.4	128	3756	1.9	3.9
Far East	500	7.2	116	4147	1.4	4.5
Latin America	482	18.3	209	18 417	2.5	4.2
Near East	65	24.2	150	2349	1.2	4.3
Low-income countries	589	7.9	111	5190	1.4	4.4

Chapter 12 references, p. 283

to disease: calf mortality and infertility of cows are to be reduced and the slaughter of pregnant stock prevented. Subsequently, an improvement in range management and nutritional status will not only result in improved performance of the indigenous stock, but, association with disease control, may permit the introduction of less hardy but more productive breeds, if such exist. All three changes will contribute to greater slaughter weight at younger ages, with the bonus of higher price for this higher quality beef. One characteristic of these improved management systems may be an increased stratification of the industry into 'breeder' and 'fattener' components: the first an extensive grazing system, the second utilizing fibrous agricultural wastes and perhaps some grain or molasses and urea in areas closer to markets. Again, better quality beef may be a bonus.

Dairy production from cattle and buffaloes (Table 12.7) is projected to increase 107%, due mainly to the 67% increase in numbers, while yield/ animal increases 24%, considerably greater than the equivalent increase in the beef sector. The increase in numbers is due to a greater proportion of all cows being milked, while increase in yield is due more to changes in disease and nutritional status. The most critical factor is the quantity and quality of feed available: the combination of poor quality tropical forages, high temperatures which depress appetite if feed quantity is not limiting, and poor dairy genetic constitution preclude efficient dairying so that new systems must be developed, although beef production may be possible under such conditions. Specialized feed production and improvement in the nutritional quality of by-products are necessary although, if specific fodder crops can be produced, many developing countries would opt for attempts to grow grain or export crops in these areas.

Sheep and goats are mainly important as meat and fibre producers although in some Near East countries they provide up to 25% of the milk. The fibre is of particular significance as the basis of cottage industries. Sheep and goats are of significance in drier or mountainous country, or as 'poor man's cattle' in situations where little or no grazing is available and most feed must be 'cut and carried'. Over-grazing in the Near East precludes great increases

TABLE 12.7

Cattle and buffalo milk production structure

Region	Animals milked (millions)	Yield (kg/an.)	Production (million tons)	Growth rate 1980—2000	
				Animal nos. (% p.a.)	Production (% p.a.)
1980					
90 Developing countries	132.4	638	84.4		
Africa	18.2	326	6.0		
Far East	66.0	497	32.8		
Latin America	32.7	1088	35.5		
Near East	15.5	657	10.2		
Low-income countries	83.6	456	38.1		
2000					
90 Developing countries	221.1	792	175.1	2.7	3.7
Africa	30.8	448	13.8	2.7	4.3
Far East	109.7	604	66.2	2.6	3.6
Latin America	58.7	1289	75.6	3.0	3.8
Near East	21.9	888	19.4	1.8	3.3
Low-income countries	136.8	552	75.5	2.4	3.5

in numbers there; overall the potential for increased meat production is much less than for beef cattle, mainly due to a lower potential for increased offtake (18% vs. 37% for beef) (Table 12.8) although yield/animal may increase 17%. However, their role as the grazing animal of the poorest people must be given considerable weight.

Although the pig and poultry industries are capable of increasing production more rapidly than the ruminants, they have an initial capital cost in housing and an ongoing feeding cost which limits their significance for the poorest of men. Those who are unemployed and whose only asset is their labour can engage in ruminant production by herding on commons or wasteland, or by cutting vegetation from within plantations, on paddy bunds, etc., for stalled animals, while they could not hope to be involved in intensive pig or poultry production. The importance of meat and dairy production from ruminants by the poorest levels of society is not in the production of food for themselves. Meat and milk are luxury products, priced too high for consumption by the poor if markets can be found. It is to their nutritional advantage to exchange most, if not all, of these products for grain. For example, using prices I obtained in an Indonesian market: If a cow was sold for meat, and rice purchased with the proceeds, not only did the rice yield many times the energy value of the cow, and several times the crude protein, it yielded three times the weight of essential amino acids, so low is the price of rice relative to meat. The most economical source of milk for infants among the poverty stricken is obviously human breast milk, with its associated advantages of cleanliness, lack of adulteration, availability and even the fertility-reduction caused by lactation. Cow or buffalo milk is better sold for cash. Animals provide many things for the poor: fuel, fertilizer, a cash reserve, prestige, etc., as detailed by McClymont (1976), but their direct contribution to the diet may be low. There appears to be a place for even smaller herbivores, and rabbits and guinea pigs are used on

TABLE 12.8

Sheep and goat meat production structure

Region	Animals (millions)	Off-take (%)	Carcase wt. (kg/ animal)	Meat prod. (thousand tons)	Growth rate 1980–2000		Share of sheep in total nos. 1975 (%)
					Animal nos. (% p.a.)	Production (% p.a.)	
1980							
90 Developing countries	791.5	30.4	12.6	3037			58
Africa	222.8	31.9	10.5	747			50
Far East	196.6	34.1	10.5	705			37
Latin America	154.1	21.0	12.8	413			75
Near East	218.0	32.8	16.4	1173			69
Low-income countries	357.9	32.1	11.3	1299			..
2000							
90 Developing countries	1172.9	36.0	14.7	6214	2.0	3.6	
Africa	347.3	40.5	12.7	1787	2.2	4.5	
Far East	297.3	39.1	12.4	1441	2.1	3.6	
Latin America	225.5	24.2	15.8	862	1.9	3.8	
Near East	302.8	37.9	18.5	2124	1.7	2.0	
Low-income countries	549.6	38.0	13.3	2781	2.2	3.9	

Chapter 12 references, p. 283

a significant scale in some countries, e.g., the annual slaughter of domestic guinea pigs in Peru is about 70 million; with 250 g carcasses these provide 1 kg meat/head of population in a convenient poor-family sized package. Unfortunately both these species are more demanding in their nutritional requirements than the ruminants, as well as being more susceptible to predation. Thus they require a high standard of management and nutrition and are unlikely to achieve great importance in the hotter tropics where forage quality is poor.

For both pig and poultry meat industries, the projected increase in technical efficiency of these imported technologies is very low for turnoff % and production/animal, over the next 20 years. That there is room for increased efficiency is indicated by turnoff from piggeries in developed countries four times greater than in developing countries ostensibly using the same systems. A similar situation is found with poultry meat although with eggs, production/bird increases rapidly. The real increases in meat production are mainly due to increases in numbers of animals. This is a gloomy reflection upon the efficiency of transfer of managerial and technical skills to developing countries, indicating that the role of skilled manpower is the limiting factor in boosting production. Among the pastoral industries, which are not as heavily reliant upon imported knowledge and management, offtake and production/animal are projected to increase more rapidly. Offtake is low in relation to that in pastoral industries in developed countries such as New Zealand and Australia, and carcases much lighter. The improvement in turnoff and carcase weights will initially be a response to improved health status using current technology such as vaccination, dipping and parasite control. The feed constraints will be harder to remove unless fibrous by-products, produced locally, can be processed or modified to provide adequate nutrition, since costs preclude transport of roughages long distances and feed grains will be in high demand for the pig and poultry industries.

Throughout the animal production industries, the projected rates of increase in production indicate a disappointingly slow rate of development of new production technologies or alternatively, of successful adoption or adaptation of introduced technologies. Even these slow rates of increase in the grazing industries are almost double the historical rates of 2% increase/annum from 1963 to 1975. This casts doubts upon the possibility of achieving this 'maximum feasible' level of production.

5.3. Draught

Perhaps the greatest contribution of animals to the well-being of man in developing countries is as draught, pack and riding animals. Their contribution to the power requirements for crop production alone is estimated at 27.9 billion man-day equivalents, about 30% of the entire power needs in 1980 (Table 12.9). Although there is debate about the significance of draught animals as an index of poverty, it is concluded by FAO that their number will increase slightly over the next two decades (Table 12.10). The power input from human labour will increase at about the rate of increase of the human population, 2.6%/year, and number of tractors will increase four fold. Power requirements rise proportionately with increase in cropped area but by less, 0.2—0.7%, per 1% increase in yield. Power demand is increasing most rapidly in Latin America, which is already the most mechanized, and Africa, which uses the greatest proportion of human labour. Only Africa (especially West Africa) and the Far East will experience an increase in draught animals but the declines elsewhere are predicted to be

TABLE 12.9

Percentage share of different sources in total power output[a]

Region	1980			2000		
	Labour	Draught animals	Machines	Labour	Draught animals	Machines
90 Developing countries	66	29	5	67	20	13
Africa	81	16	3	81	11	8
Far East	64	34	2	67	25	8
Latin America	56	25	19	49	13	38
Near East	63	25	12	68	13	19
Low-income countries	63	35	2	68	25	7

[a]Power requirements are set equal to power output in those calculations.

TABLE 12.10

Number of draught animals and tractors and rate of increase 1980—2000

Region	1980		2000		Rate of increase 1980—2000	
	Draught animals (millions)	Tractors (millions)	Draught animals (millions)	Tractors (millions)	Draught animals (% p.a.)	Tractors (% p.a.)
90 Developing countries	190	2.3	208	9.9	0.5	6.9
Africa	21	0.2	24	1.1	1.3	8.0
Far East	137	0.5	154	3.3	0.6	8.1
Latin America	19	1.1	18	4.2	—0.3	6.7
Near East	13	0.5	11	1.3	—0.7	4.5
Low-income countries	137	0.4	152	2.8	0.6	10.5

very small. The bulk of power in all developing areas will still be human labour in 2000 A.D.; this is partly because increased intensification of cropping and improved yields generate more weeding and harvesting operations where animal power is less used. Machines are estimated to cultivate 24% of the area ploughed in 1980, rising to 50% by 2000 (FAO, 1979). The remainder 50% include much of the cropping area of the poorer levels of rural society, those least able to change, due to financial or educational limitation or limits of scale of farming operation.

Goe and McDowell (1980) review the use of animal power and identify 16 species employed, including horses, asses, buffalo, cattle and camels as the major sources of power; they point out that these animals provide meat, milk, hides, fuel, and fibre as by-products of their major function. Little is known of the power outputs or necessary nutritional inputs of draught animals. Data which are available differ in load, speed, duration, number in team, age and weight investigated and climatic conditions and time of day. The general picture that emerges from their review is as follows.

Horses are used as pack animals and for riding in mountainous areas, and for carting in all areas due to their superior speed. This speed is even apparent in field work in lowland rice fields but horses are not used under these conditions due to higher purchase price, greater susceptibility to injury and higher nutritional demands.

Chapter 12 references, p. 283

Asses and mules are used for packing and riding, but to a lesser extent for draught in cropping.

Swamp buffalo are the main animal used for draught in South and Southeast Asian rice production, where their more flexible foot joints and larger hooves allow better performance than cattle. They are slow for road traction, but are capable of a larger pull on a dead weight. Their ability in areas other than rice paddies is limited by their poor thermoregulation if wallowing is not available. Cockrill (1974) has extensively reviewed usage.

Cattle are used extensively in India and Africa for dry-land cultivation as well as in rice cultivation. Their speed is intermediate between horses and buffalo and their capacity for work under hot, dry conditions exceeds that of the buffalo. Excessive weight is given to the 'problem' of relatively humpless crossbred *Bos taurus* × *Bos indicus* cattle as draught animals. Since the beef breeds of Europe were developed also as draught animals, and *Bos taurus* were used extensively in Europe, U.S.A. and Australia, the lack of a hump is not a real problem. Yoke designs must be different, but these are available from the above countries. Their lack of heat tolerance provides a more valid objection. Cows are occasionally used as draught animals but the available research shows that this has adverse effects on lactation, reproduction and growth, given the present level of livestock management and the resources available.

Camels, of which there are 12 million in Africa and 5 million elsewhere, are used mainly for pack and riding purposes, but also find use in ploughing, operating mills and pumps, and carting. They are clearly superior under dry conditions, since horses and cattle have to haul their own feed as well as the payload in areas where camels can subsist on desert browse.

Goe and McDowell (1980) conclude that considerable improvements can be made in the efficiency of performance of these animals by changing harness design. Knowledge is inadequate on performance, due in part to the draught animal falling between animal production and crop production and being ignored by researchers in both areas. Consequently feeding standards and potential outputs are not known with any precision. As it is highly unlikely that the number of draught animals, presently almost 200 million, will become insignificant in the predictable future, there is a great need for this information.

6. UNCONVENTIONAL ANIMAL PRODUCTION

The utilization of wildlife has been discussed by Ledger in Chapter 9 of this volume. Between wild animals and the conventional farm animals previously discussed are a group of animals at various stages of domestication, ranging from hypothetical to small-scale economic production. The requirement with the highest priority is that the animal fills a perceived need; secondly there must be an understanding of the basic biology, and finally a management system must be developed which fills the need at an acceptable price.

As an example of an animal at the beginning of the domestication process, the manatee (*Manatus americanus*) is an aquatic mammal up to 4 m long and over 1 tonne weight. Only the first requirement is met: the perceived need which is based on this animal's appetite for water hyacinth, one of the worst of aquatic weeds in many inland waterways; the manatee may aid in its control as well as producing meat. Some of the basic biology of this creature is known, but there is no technology available for managing it or harvesting apart from hunting. Its potential productivity can only be guessed but may be very large.

Further along the domestication process, the eland (*Taurotragus oryx* Pallas) and other African ruminants have been the subject of domestication attempts in Kenya and elsewhere, as meat animals. Data are available on their biology, and management systems are being developed. The production data of these mammals show that their advantages over their competition, i.e., cattle, sheep and goats, are often not very great, especially when the management system eliminates an advantageous characteristic. For example, free-ranging animals may grow rapidly but when restrained by fences and even more by corralling at night, the limitation in grazing time may be so great that their apparent advantage is reduced. The oryx (*Oryx beisa callotis*) is able to go for long periods without water, but the provision of water points may eliminate this advantage over cattle and may be easier than domesticating the oryx. However, these ruminants often have the advantage that they are not susceptible to diseases which make cattle production impossible. In any case the total potential productivity of such almost desert lands is extremely low and it is unrealistic to expect to produce large amounts of meat from an environment offering such limited food resources.

In extreme cold environments, many other ruminants have been used; the yak (*Bos grunniens*) in Asia, the llama (*Lama glama*) and alpaca (*Lama pacos*) in South America, reindeer (*Rangifer tarandus*) and others in Arctic Europe and Asia. They supply a wide range of needs — meat, milk, draught, hides and fibre. Their contribution is of fundamental importance in these regions where other farm-animal production would be almost impossible and human habitation equally so without them. Like the hot, dry desert and near deserts, the potential productivity of these areas is very low, and these animals have failed to show advantages when transferred to other environments (e.g., the alpaca in southern Australia).

An animal which may offer more potential production is the capybara (*Hydrochoeros hydrochaeris*), an aquatic rodent, weighing up to 75 kg, found in South American swamps (González-Jiménez, 1977). This animal has been partly domesticated for over 400 years, although it has not been used extensively. Its potential lies in the extent of its habitat, the tropical swamps, and the high feed production of this well-watered environment. In this aspect the capybara exemplifies the ideal new domesticated species in that it can utilize a large feed and environmental resource which is not being used as efficiently, if at all, by any other domestic animal. However, within the capybara's area of distribution, the swamp buffalo is being introduced in Brazil to utilize some of this environment, suggesting that the capybara does not have the economic production characteristics required, or that it suffers from another disadvantage. This latter may be its role as a reservoir of cattle and horse diseases.

Closest to the conventional system of animal production is deer farming, currently undergoing rapid development in New Zealand, Scotland, Australia and elsewhere. Several species, including Fallow (*Dama dama*), Red (*Cervus elaphus*), Rusa (*Cervus timorensis*), and Sambar (*Cervus unicolor*), are being farmed successfully. In most cases the reason for farming deer is not to take advantage of an environment which is not utilized by any conventional domestic species, but to take advantage of a speciality market for venison, hides and 'velvet' antlers. Although deer are farmed in some areas where they out-perform cattle and sheep by a wide margin, their use does not lead to a great increase in meat production on a world basis.

Man domesticated his cattle, sheep, goat, horses, buffalo, pigs and poultry thousands of years ago, and has migrated throughout the world with them. During these thousands of years, he has adapted them to the new environments he has colonized, from Highlands of Scotland and the Great Plains

Chapter 12 references, p. 283

of the U.S.A. to the deserts of Central Australia and the humid tropics of Java. With the exception of the aquatic and semi-aquatic environments, the physical environments in which novel species have an advantage and which defy amelioration to make them suitable for conventional farm animals are the harshest and least productive environments on earth. The temperatures, moisture and soil fertility limit feed production so that animal production must be at a low level. This production is of critical importance to the sparse human population it supports, but on the world scene it is of little significance. The only place where a novel form of animal production can have much potential for augmenting world meat supplies is a situation where use can be made of a large feed supply currently under-utilized by domestic stock or harvested wildlife. The tropical swamps and waterways are the major such environments. Apart from these areas, even assuming that utilization can occur, animal production on a large scale means production by conventional livestock during the rest of this century. It has been argued that selective breeding and improved management systems, when and if devised, may increase the apparent advantages of some new species even further. But it is known, as detailed in Section 5, that selection and improved management of conventional livestock can improve their production several fold. New species will continue to join the ranks of our livestock, and new management systems will increase production. But to displace the highly evolved animals and systems already utilized will take many years, as is evidenced by the slowness with which the Red deer, the eland and the capybara make their way into the agricultural scene.

7. CONCLUSIONS

(i) Malnutrition and under-nutrition occur on an enormous scale throughout the developing world; the actual numbers afflicted depend mainly on the nutritional criteria adopted. The major nutritional deficiencies are energy and protein. Intercurrent disease boosts needs for proteins and minerals, especially iron. The most seriously affected are small children and pregnant or lactating women. The current 'appalling' situation must be a considerable improvement upon conditions when famine, disease and war suppressed the ability or the desire of the population to increase, which were widespread until about two centuries ago.

(ii) The cause of nutritional inadequacy is poverty, the inability of individuals and families to produce or purchase food, rather than overall lack of agricultural land or food production on a world basis. Social and economic disorganization during and after decolonization contributes to this poverty.

(iii) The contribution of population increases to the problem of nutritional inadequacy is not of the overriding significance which was stated in the Doomsday books of the nineteen-sixties and nineteen-seventies. More people mean not only more appetites, but also more economic demand and more producers. The current rate of population expansion appears to be slowing down.

(iv) Agriculture can provide the food requirements of populations in excess of present levels but this will require major reorganization of the present economic order. If production is by the developed countries, without generating employment of the poor, the problem of poverty is untouched.

(v) The main contributions of animal production to the poor are the provision of employment, draught power and fertilizer, and generation of a cash income to permit purchase of cereals. Thus the argument about the relative capacities of animal and plant production to supply food for the poor is irrelevant: for them, animals are often not direct food producers.

(vi) While conventional production systems can be modified to increase production several fold, radical alterations are not readily accepted or adopted. Development can succeed best by enhancing what already exists.

(vii) Grain feeding of livestock is an inevitable consequence of uneven distribution of wealth and high elasticity of demand for animal products. The optimal source of grain is a complex issue in which political values may override agricultural and economic considerations. There is no uniformly applicable answer.

(viii) Unconventional animal species and systems and unconventional sources of food such as 'single cell protein' will be utilized in the long term, but within this century, the vast majority of animal production will continue to be from conventional livestock.

(ix) Generalization, amplification and extrapolation, across countries or years, from Malthus to Meadows and beyond are found to have low validity. Each country is unique, as is each year of its development, so that the optimal answers to the food problem vary from country to country and year to year. There are no simple answers, no long-term master plans, and no substitute for detailed, on-the-ground, ongoing analysis, assessment and application of human imagination and hard work.

8. REFERENCES

AMLC, 1980. Statistical Review of Livestock and Meat Industries. Australian Meat and Livestock Corporation.

Anonymous, 1979. World survey detects a sharp decline in fertility. Res. Reprod., 11 (4): 13.

Anonymous, 1981. Record trade seen in Cereals. Ceres, 14 (1): 13.

Bennett, M.K. 1963. Longer and shorter views of the Malthusian Prospect. Food Res. Inst. Stud. Agric. Econ. Trade Dev. (Stanford), IV (1): 3—11.

Blaxter, K.L., 1970. Domesticated ruminants as sources of human food. Proc. Nutr. Soc., 29 (2): 244—253.

Boerma, A.K., 1969. Cardinal O'Hara Memorial Lecturer. University of Notre Dame, Indiana, U.S.A., 24 October.

Buringh, P. and Van Heemst, H.J.D., 1977. An Estimation of the World Food Production Based on Labour-Oriented Agriculture. Centre for World Food Market Research, Wageningen, The Netherlands.

Buringh, P., Van Heemst, H.J.D. and Staring G.J., 1975. Computation of the Absolute Maximum Food Production of the World. Agricultural University, Wageningen, The Netherlands.

Charles, C.F., Blandford, D. and Boisvert, R.N., 1980. Economic feasibility of import substitution of livestock feed in the Caribbean community. Cornell Int. Agric. Mimeogr. No. 80. Ithaca, NY, July.

Cockrill, W.R., 1974. The Husbandry and Health of the Domestic Buffalo. FAO, Rome.

Devendra, C., 1981. The feed requirements for animal production in Peninsular Malaysia: an assessment of the feed resources, implications and strategies for efficient utilization. Conference on Exotic and Crossbred Livestock Performance in Malaysia, 11—12 September 1981. Genting Highlands, Malaysia.

Drosdoff, M., 1979. Perspectives of the World food situation. World Food Issues Series, No. 1. Centre for Analysis of World Food Issues, Cornell University, Ithaca, NY, 8 pp.

FAO, 1969. Indicative World Plan for Agricultural Development. FAO, Rome.

FAO, 1979. Agriculture: toward 2000. Conference of the Food and Agriculture Organization of the United Nations, 10—29 November 1979. FAO, Rome.

FAO, 1980. The state of food and agriculture 1979. Food and Agriculture Organization of the United Nations, Agric. Ser. No. 10. FAO, Rome.

Goe, M.R. and McDowell, R.E., 1980. Animal traction: guidelines for utilization. Cornell Int. Agric. Mimeogr. No. 81. Ithaca, NY, December.

González-Jiménez, E., 1977. The Capybara. World Anim. Rev., 21: 24—30.

Guilmot, P., 1978. The demographic background. In: Population Decline in Europe: Council of Europe. Edward Arnold, London, pp. 3—53.

Malthus, T.R., 1798. An essay on the principle of population. In: G. Hardin (Editor), Population, Evolution and Birth Control. W.H. Freeman, San Francisco.

Mayer, J., 1976. The dimensions of human hunger. Sci. Am., 235 (3): 40—49.

McClymont, G.L., 1976. Animal production in a grain-hungry world — or competition between man in a resource-limited world. S. Afr. J. Anim. Sci., 6: 129—137.

McLaren, D.S., 1974. The great protein fiasco. Lancet, ii: 93—96.

McNamara, R.S., 1978. Address to the Board of Governors of the World Bank (25 September, 1978). I.B.R.D., Washington, DC.

Meadows, D.H., Meadows, D.L., Randero, J. and Behrens, W.W., 1972. The Limits to Growth. Universe Books, New York, NY.

Mesarovic, M.D. and Pestel, E., 1975. Mankind at the Turning Point. The Second Report to the Club of Rome. Hutchinson, London, 210 pp.

Miller, D.S. and Holt, J.F.J., 1975. The Ethiopian famine. Proc. Br. Nutr. Soc., 34: 167—172.

Morley, F.H.W., 1969. Challenges to animal production. Proc. 2nd World Conf. Anim. Prod. American Dairy Science Association, St. Paul, MN, pp. 23—30.

Parpia, H.A.B., 1968. Novel routes to plant protein. Sci. J., 4 (5): 66—71.

Pirie, N.W., 1975. Some obstacles to eliminating famine. Proc. Nutr. Soc., 34 (3): 181—186.

Scrimshaw, N.S., 1979. Protein energy requirements under conditions prevailing in developing countries: current knowledge and research needs. United Nations University World Hunger Programme. Food Nutr. Bull., Suppl. 1, July, 73 pp.

Simon, J.L., 1981. The Ultimate Resource. Martin Robertson, Oxford, 415 pp.

Sisler, D.G. and Blandford, D., 1979. 'Rubber or Rice?': the dilemma of many developing nations. Paper No. 13 in World Food Issues. Centre for the Analysis of World Food Issues, Cornell University, New York, NY, 11 pp.

Tannenbaum, S.R. and Mateles, R.I., 1968. Single cell protein Sci. J., 4 (5): 87—92.

U.S. Agency for International Development, 1970. Population Program Assistance. Aid to Developing Nations by United States, Other Nations and International and Private Agencies. Office of Population, Bureau of Technical Assistance. A.I.D. Government Printing Office, Washington, DC.

Van Bamweke, A.R., 1979. Land Resources and World Food Issues. Paper No. 2 in World Food Issues. Centre for the Analysis of World Food Issues, Cornell University, Ithaca, NY, 4 pp.

Wortman, S., 1976. Food and agriculture. Sci. Am., 235 (3): 30—39.

Wortman, S. and Cummings, R.W., 1978. To Feed This World. John Hopkins University Press, Baltimore, MD.

Chapter 13

Animal Products and Their Competitors

M.V. TRACEY

1. TRADITIONAL FOOD USES AND TECHNOLOGY

A necessary consequence for adult man of maintaining a stable body weight and normal activity is the daily loss of about 10 MJ as heat (which can be provided by the metabolism of about 0.6 kg of protein or carbohydrate or of 0.3 kg of fat), 5 g of combined nitrogen, 1 l of water, and small but significant amounts of compounds the body is unable to synthesize. These range in amounts from about 4—5 g of the essential amino acids (corresponding to less than 1 g of the combined nitrogen loss) to as little as 1 μg of vitamin B_{12}. All these requirements can be met from a diet of entirely animal origin. All, save the requirement of 1 μg of B_{12}, can be met from a diet of entirely plant origin — and since it is difficult to exclude all microbial and invertebrate sources of B_{12} from such a diet, B_{12} deficiency in vegans is far less common than might be expected.

The value placed on meat in the diet of most people is the result of many factors. The availability of leaves, seeds and fruits in temperate climates changes greatly with the seasons, while the population of the larger food animals is relatively constant. However, meat cannot be stored for long in temperate or hot climates in its natural state. (In 1423 the Worshipful Company of Butchers in London prohibited 'the sale of any flesh that had been killed above three days in winter and two in summer'.) Plant seeds, on the other hand, because of their very low natural water content, can be stored without difficulties provided that visible pests such as insects and rodents are excluded. Plant products susceptible to microbiological attack commonly contain large amounts of carbohydrate and little protein, while animal products contain little carbohydrate and much protein. The result of microbiological degradation in the first instance is fermentation to alcohol and subsequently acetic acid; in the second, putrefaction. On the whole, the products of fermentation are harmless while those of putrefaction are often harmful — perhaps this is why we find fermentation odours generally pleasant but are disgusted by the volatile nitrogenous products of putrefaction. It is also important in considering man's motivation to the use of animals as food, that while food plants produce few useful by-products other than straw, the animal is a peculiarly rich source of them. The hide alone is of enough value as covering and as a raw material for containers, footwear, harness, thongs and many other uses, for animals to have been slaughtered even had they been inedible. It is, perhaps, for reasons such as those set out, that although man's needs can be met by an exclusively plant diet, most people prefer one with a substantial animal component and are prepared to pay for it. This is shown by the increase in meat intake with

Chapter 13 references, p. 306

attainment of enhanced economic status within a community, and by the increased consumption of animal products that invariably follows a growth in disposable income in developing countries.

1.1. Use of the animal for food

While most species of plants are probably toxic in varying degrees to man, no mammal or bird is so. The qualifications inevitable to such a sweeping statement are trivial, including as they do the livers of polar bears, Arctic foxes and some seals and whales, which may literally represent too much of a good thing in that they contain toxic amounts of vitamin A, and occasional instances of toxicity arising from the unusual food eaten by the animal concerned. Instances of the latter include quails that have fed on hemlock and the milk of cows that have eaten the weed snake root (*Eupatorium rugosum*) (the cause of the virtual depopulation of initially settled areas in

TABLE 13.1

Foods, the consumption of which French people would like to increase if their financial status were improved (% of total replies)

	All replies	Professional people	Skilled workers	Rural workers
Meat	53.3	31.7	71.7	46.8
Processed foods	12.9	31.7	2.6	4.1
Fruit	7.4	9.8	2.6	18.4
Vegetables	6.9	—	12.8	12.3
Fish, shell-fish	4.9	14.6	—	3.0
Other foods	14.6	12.2	10.3	15.4
Total	100.0	100.0	100.0	100.0

Source: Claudian and Serville (1968).

TABLE 13.2

Taiwan: indices of production 1965—1970

	Crops	Livestock
1965	100.0	100.0
1966	101.7	102.4
1967	105.0	117.7
1968	108.5	122.9
1969	103.8	131.5
1970	104.4	145.0

Source: Brown (1972).

TABLE 13.3

Changes in protein and calorie intake head^{-1} day^{-1} in Japan

	1949	1953	1958	1963	1968	Ratio 1968/1949
Total protein(g)	68	60	71	71	77	1.13
Animal protein(g)	16	22	24	29	32	2.00
Calories	2087	2041	2108	2090	2214	1.06

Source: Oiso (1971).

the Old Northwest of the United States in the early 19th century). While in general no part of an animal or bird's body is poisonous, much of it is, of course, unrewarding as food as are, for example, feathers, hair and, to a lesser extent, hide and bone. In Table 13.4 I have attempted to set out the yield of edible products from a steer of moderate fatness. In looking at this Table, it must be remembered that in advanced societies there is a tendency to eat less of the animal than did our ancestors; in part, because we can afford to divert the less appetizing parts of the carcase to our pets or as components of stockfood, and partly because we may value the pharmaceuticals such as insulin and thyroid extract highly enough to direct the pancreas, thyroid and other glands to therapeutic uses. Table 13.5 tabulates the weight of individual products from an average beef animal at the same liveweight during the processes of slaughtering and butchering. Table 13.6 lists typical offal weights from sheep and lambs.

Both plants and animals contain structural materials (proteinaceous in animals, polysaccharides in plants) which are not digestible by man. In the plants he eats, every cell is surrounded by a coat of cellulose, save in the endosperm of grains, and willy-nilly he must ingest this indigestible material (a technology by which this may be evaded is described later). The structural and protective indigestible materials of an animal are, however, localized in tissues such as skin and bone, and can readily be rejected by dissection of the carcase. Thus animal food is characterized by low residues after digestion and by a very low content of carbohydrate. Milk, however, is an exception in that it has a high content (40% of dry matter) of lactose, a sugar occurring in no other food, and also has the distinction (since it is a secretion and not a tissue) of containing only traces of nucleic acids; these, however, are derived from bacteria, lymphocytes or cellular debris from the udder walls.

The fat of ruminants is more saturated than that of monogastric animals such as the horse, pig and chicken, and this has become of interest to nutritionists in relation to susceptibility to heart disease. The universal occurrence of cholesterol in edible animal tissues (including eggs and milk) is also significant and contrasts with the absence of cholesterol in plant foods.

TABLE 13.4

Composition of a 500 kg steer (medium fat) and yields of edible protein and fat

Beast at slaughter				
40 kg gut contents				
293 kg water	63.7%			
75.75 kg protein	16.5%			
68 kg fat	14.8%			
23.25 kg ash	5.0%			
Protein			Protein normally eaten	
Keratin	6.75 kg	8.9%		
Collagen	20.7 kg	27.3%	1.5 kg	3.6%
Good quality protein	48.3 kg	63.8%	40.25 kg	96.4%
	75.75 kg		41.75 kg	(55.3% of total protein)
Fat			Fat normally eaten	
Non-carcase (offal) fat	17 kg			
Carcase trimmed before sale	30 kg			
'Visible fat' of retail meat	13.5 kg		21 kg (31% of total fat: this excludes	
'Invisible fat' of retail meat	7.5 kg		fat recycled via cooking fats, marg-	
	68 kg		arine, and suet)	
Carbohydrate				
Negligible				

Chapter 13 references, p. 306

TABLE 13.5

Estimated yield of offals and meat from cattle (liveweight 500 kg; carcase weight 280 kg)

Offal	(kg/Body)
Tail	0.950
Small intestine (hand-stripped)	6.800
Large intestine (hand-stripped)	5.350
Oesophagus and trim	3.057
Tripe: rumen and reticulum	8.200
bible and reed	10.400
Caecum	0.376
Thick skirt (diaphragm)	0.800
Cheeks	2.250
Liver	5.850
Heart	1.300
Tongue	1.700
Kidneys	0.850
Pancreas (sweetbread)	0.325
Lungs	3.100
Brain	0.330
Pituitary gland	0.0022
Adrenal glands	0.024
Thyroid gland (throatbread)	0.028
Thymus gland (heartbread)	0.400
Ovaries	0.020
Uterus: uterine horn to 5 cm below cervix	0.900
Udder	4.500
Testes	0.700
Penis	1.300
Spleen (milt, lien)	0.650
Forehocks	3.750
Hindhocks	4.300
Horns	1.300
Head	13.100
Spinal cord: with skin	0.320
without skin	0.200
Bladder contents	0.300
Bile: liquid (9% solids)	0.390
Hide: salted	25.000
Ears, face and lug pieces	2.766
Blood: liquid	10.000
Paunch contents: undried	38.000
Caul fat	3.850
Other internal fats	15.416

Carcase weight	280 kg
Retail boneless meat	155.1
Bone in retail cuts	16.9
Mince and sausage (butcher)	35.5
Bone from butcher	46.6
Fat	25.9
Cooked weight of retail cuts less bone	117.0

With an increase in the standard of living there is a tendency to regard less and less of the slaughtered animal as edible, and the words used to describe organs we now consider fit only for pet food are becoming unfamiliar to the general public. Thus 'pluck' described the contents of the forward end of the body cavity (heart, liver and lungs) when used as food, and 'fry' referred mainly to the intestine, though in some localities it was used specifically for liver. 'Lights' is used for lungs when sold as food, and the Victorian working class often ate 'faggots', described as 'a sort of cake, roll or ball made of chopped liver and lights, mixed with gravy and wrapped

TABLE 13.6

Estimated yield of major offals from sheep and lambs
  Sheep: Liveweight 40 kg; carcase weight 21 kg
  Lambs: Liveweight 30 kg; carcase weight 16 kg

Offal	Sheep (g/Body)	Lambs (g/Body)
Tongue	155	110
Liver	700	450
Heart	200	135
Brain	110	100
Spleen	90	50
Lungs	400	250
Skirt	95	70
Pituitary gland	0.9	0.5
Adrenal glands	3.2	2.8
Kidney	105	75
Gall	25	15
Thymus gland		30

in pieces of pig's caul'. In this usage 'caul' was a name for the omentum. The shift in public taste towards an almost exclusive use of skeletal muscle is of some sociological interest in that it appears to have been more evident in the lower-income groups, for products of the slaughterhouse such as kidney, brain and tripe remain more strongly represented on the menus of expensive restaurants than in the mass diet. It was summed up in 1801 by Nisbet, 'The chief part of quadrupeds preferred in food is the muscular part, as being the most nourishing, the best flavoured, and the easiest digested'.

This muscular part of a steer with a liveweight of 500 kg (carcase weight 280 kg) weighs at the retail level only about 190 kg and will, of course, include not only visible fat present after trimming but also intramuscular fat. Boneless beef, that is all the beef muscle and associated fat from a dead animal after removal of blood, head, feet, skin, viscera and organs except for the kidneys, as sold to the consumer, may contain, according to the state and age of the animal at slaughter, from 5.7—41.0% of fat (data from 31 animals selected as a wide cross-section of those coming to slaughter (Callow, 1947). The corresponding figures for 29 lambs were 13.25—45.23%. It is probable that under current marketing conditions the upper extremes for these meats as sold would be lower, as more trimming would occur at the retail level. A current Australian estimate is that about 12% of the boned weight of a carcase would be trimmed as fat and not remain with the retail cuts. Thus about 74% of carcase weight is retailed as bone-in cuts, mince and sausages, or 68% of carcase weight excluding bone. This latter figure is equivalent to 34% liveweight (Table 13.5). These figures are of some importance to the nutritionists in that official statistics for consumption of meat are given in terms of carcase weight or carcase weight and offal moving into consumption. Even if it is assumed that the entire gut is eaten as casings and tripe and that the heart, liver, tongue, kidneys and tail are also consumed, the yield of meat is still only about 75% of carcase weight, i.e., about equivalent to the yield of bone-in retail cuts alone. The figure of meat consumption per head of 80 or 90 kg/annum in carcase equivalent weight common for the heavy meat-consuming countries is likely to be equivalent at most to 60—68 kg at the retail level (164—185 g/day) and less than 45—51 kg on the plate (125—140 g/day) — a far more comprehensible figure than the carcase equivalent weight/day of 219—247 g! In other words, the yield of meat on the plate from carcase weight of beef is 56% of the

Chapter 13 references, p. 306

reported carcase weight consumed — others, e.g., the National Live Stock and Meat Board, Chicago, suggest a much lower figure of 35%, in part due to the apparent exclusion of offal (which would result in a 51% yield) and perhaps in part due to cooking losses approaching 50%. Calculated yields from veal, pork and lamb are similar to the figures given. In countries consuming at this level (Australia, U.S.A., New Zealand and the Argentine), the total consumption/head is equivalent to about 0.4 of a bovine or four to five sheep or lambs per head per annum.

In 1848 Lawes and Gilbert began an extraordinarily detailed series of dissections, weighings and analyses of domestic farm animals in different conditions and after periods of controlled diets. They calculated from their very extensive results the proportions of the total nitrogenous compounds and fat of the bodies of these animals which would be consumed as human food. They considered that all the carcase save the bones, together with head flesh, tongue, brains, heart, liver, pancreas, spleen and diaphragm would be eaten, but not most of the intestine. Their general conclusions appear in Table 13.7.

A comparison of the figures in Table 13.7 relating to the eighteen-fifties with those in Table 13.4 shows clearly that it is now customary to eat directly much less of the total fat of a beef beast than was eaten a century ago.

The animal as a whole coming to slaughter can be regarded from the point of view of human nutrition principally as a source of protein and fat. Of the protein, about 9% is keratin and more than a quarter is collagen, a protein of little nutritional value in itself. In all, about a third of the total protein of a bullock is inedible or of poor nutritional quality, nearly 95% of this total in the form of hide, hair, horns and bones being rejected before ingestion. Collagen, principally from hides and hide scraps unsuitable for the production of leather, is processed to form gelatin for food. Only the highest quality gelatin is used in this way, more degraded products finding industrial outlets described later. For a long time it was believed that gelatin shared the high nutritive value of other proteins. In 1815 a Gelatin Commission was appointed by the Academy of Science in Paris to determine whether the gelatinous extract prepared by boiling bones could replace meat in the diet, and it commented, 'everyone who is familiar with broth knows that its nutritious qualities are due primarily if not entirely to gelatin'. The second Commission of 1842 reported that gelatin alone could not replace meat in the diet of animals but was not willing to say that it was useless nutritionally if combined with other nutrients. At this time there was public concern about the diversion of food products to manufacture, which has a curiously modern ring. A government pamphlet distributed in Paris at the

TABLE 13.7

Estimated total nitrogenous compounds and total fat consumed as human food

	% Consumed as human food	
	Of the total nitrogenous compounds of the body	Of the total fat of the body
Calves	60	95
Oxen	60	80
Lambs	50	95
Sheep	50	75
Pigs	78	90

Source: Lawes and Gilbert (1859).

time of the second Commission report, for example, contained the slogans 'A bone is a tablet of bouillon formed by nature' and 'A box, a knife handle, a dozen buttons are so much meat stolen from the poor'. With the discovery of the essential amino acids and the demonstration that some are not present in collagen and hence absent from gelatin, the reason for the inadequate nutritional properties of gelatin became clear. Its obvious value as a supplement to other proteins containing the missing essential amino acids but themselves low in some contained in gelatin, was finally demonstrated by Chick in 1951.

All of the fat can be used as a dietary energy source, though in countries with abundant food supplies it is increasingly regarded as desirable to reduce the percentage of total energy present in the diet as fat from current figures in the region of 40% down to 30% or less. This will increasingly result in diversion of animal fat to manufacturing purposes or, more sensibly, in the marketing of animals less fat than has been customary.

The consumption of beef in nine EEC countries in 1973 averaged 25 kg per head per annum. If we assume that unpreserved meat would on average throughout the year have to be consumed within a very few days of slaughter, one beast of 500 kg liveweight would provide meat, to be eaten in 2—3 days, for about 600 people. Obviously it would not be easy to arrange that each beast slaughtered be eaten totally by 600 people without wastage from decomposition or a system of glut and famine. Much advantage could be achieved if the meat or parts of it could be preserved for a period. Until the advent of refrigeration in the last century, this involved either the subtraction of water or the addition of preservative materials or, preferably, both processes.

Drying in air requires the meat to be in thin layers or strips, and is dependent on dry weather and adequate air movement, and assisted by high temperatures. Among the traditional air-dried raw meat products are biltong (South Africa), charqui (South America), both produced in hot climates and salted, and pemmican from the colder regions of North America and Europe. Biltong is produced from high quality lean meat while both charqui and pemmican have a high fat content. More reliable drying can be secured by fibre and smoke and has the added advantages of the preservative action of the condensable vapours of smoke from wood.

Preservation by immersion in strong brine (a process equivalent to the subtraction of water) was one of the most successful means not involving drying, since salt can diffuse from the surface into considerable thicknesses of meat and may be assisted in doing so by cuts and pricks. The need to conserve scraps of meat and the observation that the effect of a preservative was from the surface only, particularly if it included insoluble spices, must have led early to the attempts at preserving meat in a chopped or minced form, and the simultaneous availability of hollow visci such as the gut, bladder and uterus made the sausage inevitable. A length of gut stuffed with comminuted meat and fat with or without added salt and herbs gave a product that lasted well in temperate climates if it were dried or smoked to reduce the water content. In winter, in high latitudes, microbiological attack was delayed and sausage with a higher water content could be kept for short periods, while a longer life could be assured if a lactic fermentation were encouraged which, producing a lower pH, discouraged the growth of putrefying organisms.

Each division of the gut and the bladder and uterus gradually developed a specific use as a container of preserved meat. The rumen of the sheep became the sine qua non of the haggis, though the Romans preferred the uterus of a sow for a similar product; without the bladder the mortadella would not be possible, and in fact the traditional diameters of the innumerable varieties of dried, smoked or fermented sausage developed primarily in Europe, as their names derived from the original centres of production indicate, depended on the characteristics of different parts of the gut of sheep, cattle and pigs used in their preparation. The rumen and reticulum of cattle were seldom used in this manner, however, since they were valued more as a food in their own right, the first yielding plain tripe and the second honeycomb tripe, while the fourth stomach of the calf (known as the 'vell') is the source of rennet. The best grade of rennet is from calves that have fed only on milk, the next grade is from calves that have had mixed feed in addition to milk, while the poorest is that from animals fed on feed and grass. Grass-fed calf abomasum is worthless as a source of rennet. The vells are preserved by freezing, salting or drying after squeezing out the contents, washing the outside and trimming off fat.

Since fat is insoluble in water and melts below the temperature of boiling water, it can readily be prepared in a water-free state by boiling fatty tissues in water and removing the solid fat from the surface after cooling. The fat of the pig (lard), being less saturated, is of particular value as an ingredient in cooking, and was traditionally packed in the bladder of the animal. The fat found round the kidney and in the omentum of ruminants is even more highly saturated than that of the depots beneath the skin. Its high melting point allows it to be shredded or chopped finely in a form which can be kept for long periods without consolidation by melting and with little development of rancidity; it is sold as suet instead of joining the fat trimmings and being rendered for the preparation of tallow and the manufacture of margarine.

In current jargon, the milk of animals is their only renewable resource as food. While available daily throughout the year, it is very perishable and has a relatively high water content (about 690 g water/100 g solids vs. 220 g water/100 g solids in meat) and much was to be gained by transforming any surplus to a convenient, stable and less weighty form by the abstraction of water. The principal products were, of course, butter (18 g water/100 g solids) and cheese (59 g water/100 g solids for cheddar). Of the two, cheese is by far the more stable product at normal temperatures in that the fat it contains is far less susceptible to oxidation. It is of interest that while both have been known to man since primitive times, only cheese has a long history as a food. In Roman times butter seems to have been used principally as an unguent or as a cooking oil, and until the last century or so was normally marketed as an evil-smelling, rancid, semi-liquid material in which oxidation was often encouraged by efforts to bleach it by exposing it in thin layers to the sun. The product as we know it today, as a bland spread, only became widely available after improvements in technology enabled it to be rapidly transported to market after hygienic preparation and to be kept chilled at all times during transport and marketing. A simple method of processing butter that results in a product which can be kept for months was developed by the nomadic Aryans some 4 millennia ago — ghee. Butter is heated until clarified and all water driven off and then allowed to solidify to pure butter fat. If carefully done there is little loss of vitamins. It was, and is, used as a cooking medium or poured melted over food. In India it became, and remains, a most important product, for it is regarded by the brahmins as having the property of purifying, in a religious sense, other components of the diet. Consumption in India is of the order of half a million tons a year. A product made by the controlled hydrogenation of vegetable oils (conceptually analogous to margarine) with a coarsely crystalline texture, called vanaspati, has been produced as a ghee substitute. It has achieved a similar market penetration to that of margarine in Western countries.

2. FOOD TECHNOLOGY IN THE INDUSTRIAL AGE

2.1. Preservation with minimal change

Preserving animal foods by drying in the sun and wind, or in front of a fire, by smoking and drying in kilns, by lactic fermentation applied to both milk and ground meat, by pickling in vinegar and by salting, were all methods well explored before the industrial age. Though preservation by cooling in cellars and with ice when available was also exploited, it was of limited value since subsoil temperatures are higher and stored natural ice hard to come by in the areas where cooling was most needed. Preservation at low temperature was, however, regarded as very desirable as it was, and still remains, the only method available that may completely avoid the changes — often very marked — in taste, appearance and texture caused by other methods of preservation. Custom and the necessity of using these methods in the past led to the acceptance of the new and changed properties in the products with the result that the new qualities became valued for themselves. A leg of pork preserved in the form of a ham, for example, is considered as a desirable alternative to the joint in its natural state, and indeed commands a higher price. Today, however, the consumer expects food to be preserved in a state as close to the original, traditional preserved, or the cooked form as possible, and this is achieved only by the methods of freeze-drying, maintaining low temperatures during storage or canning, all of which have been made possible in the last two centuries.

Beef was cooked, placed in tin containers which were then filled with hot fat and gravy and sealed against contamination, in Holland in the late 18th century, and sent for use by Dutch troops in Surinam, and a similar process appears to have been applied to fish in Holland at the same time. Nevertheless, it was not until Appert in France demonstrated the efficacy of processing in sealed containers in 1809 that the foundations of the canning industry were securely laid. Wide application of the process, however, had to await the manufacture of high quality tin plate, which began in England in the eighteen-twenties. It was fostered first by the needs of mariners and then by those of burgeoning urbanized populations of Europe. While the product was not preserved in the fresh form, it nevertheless sufficiently resembled the cooked form to be widely acceptable.

International trade in natural ice began at the time of Appert's early experiments, when Mr. Tucker of Boston sent a cargo of ice cut from local ponds to the West Indies and lost $4500 in the enterprise. His second attempt in 1815 was successful, and by 1833 he had expanded his ambit to include India. The natural ice trade in the United States developed very rapidly, for by 1848 the ice houses of New York and New England had a capacity of 325 000 tons, and by 1875 the Hudson River ice crop alone amounted to about 2 million tons. During that time the first commercially successful refrigeration machinery capable of producing ice in bulk and of refrigerating storage space had been developed by Harrison in Australia in 1856. In 1860 a number of his machines were in operation in Victoria and New South Wales. It was not, however, until 1880 that the arrival in London of the s.s. Strathleven with 40 tons of frozen beef, mutton and lamb from Sydney signalled the real origin of the international frozen meat trade.

To be effective, cold-rooms should be reasonably air-tight, and this requirement meant that it was feasible to consider the possibility that meat, particularly chilled meat, might keep better over extended periods in an atmosphere other than air. In 1933 both Australia and New Zealand exploited a British observation that beef had a longer shelf-life in an atmosphere

Chapter 13 references, p. 306

containing 10% carbon dioxide by modifying shipping cold-rooms so that such an atmosphere could be maintained during transport to markets in the Northern Hemisphere. More recently, the availability of plastic films with suitable gas permeability characteristics and the growing market acceptance of cuts of meat rather than whole carcasses has led to the preservation of meat by chilling cuts sealed into plastic pouches after evacuation. Oxygen in the remaining air is rapidly used by post-mortem respiratory processes in the meat, and the residual atmosphere is nitrogen and carbon dioxide. Carbon dioxide commonly rises to 20—30% of the total gas with time, in part due to microbial activity.

In recent years almost 10% of world production of beef entered international trade, together with about 7% of pig meat production and about 20% sheep and goat meat production. Since production of the first two accounts for more than 90% of the total, the proportion of all meat produced that enters world trade is a little less than 10%, amounting to rather more than 6 million tons carcase equivalent, of which most is frozen. Of Australian meat exports (some 15% of world totals), 90% at least is frozen, almost 3% is canned, less than 1% dried, salted, smoked or pickled, and the rest (3—5%) chilled.

In the mid nineteen-fifties, a new process for the preservation of foods by drying from the frozen state was developed commercially. The process had originally been developed during the second world war as a means of preserving blood serum and other pharmaceuticals. It depends on the sublimation of ice from the frozen product, and from the point of view of application to foods has the advantage that no heating occurs and bacterial multiplication is totally inhibited by the low temperature used during drying. Additional advantages are that the product, having a very low water content, is stable microbiologically at room temperature and, if packed in an inert atmosphere, does not deteriorate in storage as a result of oxidation. After rehydration, moreover, the product is in most instances very little altered from the original state. The process is best adapted to liquids but can be used successfully with many vegetables. Its application to meat is limited, as the product — particularly in large pieces such as steaks — suffers a serious loss of quality on rehydration, probably as a result of irreversible effects on the cellular membrane structure of the meat. The process is expensive energetically and has been restricted to high-cost products. The only application to meats has been by the defence forces of a number of countries which are willing to pay the considerable costs involved in the preparation of a product that is stable, low in weight, and of high quality. In view of the rising costs of energy, it is very unlikely that freeze-drying will become quantitatively important as a means of preserving meat.

2.2. New products

In the past, parts of the meat animal which are now relegated to the renderer for tallow and meat meal production, were directly used as food. As implied earlier, these parts were mainly products such as trotters, chitterlings, heads, lungs and some other offals. In addition, the necessity for high throughput in the abattoir has led to a considerable amount of trimmings (tissue that is predominantly fat), together with meat scraps adhering to bones, going to the renderer. These meat scraps may amount to 5 kg/beef carcase, and represent a potential source of high quality meat protein retaining, in addition, the virtues of flavour and texture that characterize meat.

A dressed weight carcase of 250 kg may yield on cutting, 20—21% bone, 63—66% meat for sale, and 14—16% trimmings, of which a fifth is lean tissue.

Recovery of this tissue by dissection would be intolerably expensive in terms of flavour, but much of the protein can be extracted, giving a yield of 1.5—2.0 kg dry protein which can be reconstituted to a liquid which, like reconstituted dried egg, can be fried, producing a patty-like product of acceptable meat texture and taste. Meat can also be recovered from the surface of bones by mechanical means. The finely comminuted product is suitable for forming into sausage and hamburger-like products, but may contain unacceptably large amounts of finely subdivided bone resulting in a high calcium and phosphate content.

Meat in pieces weighing 25 g or more can be processed into a coherent homogeneous form. In one process the pieces are tumbled by being placed in a rotating drum in the presence of salt amounting to 0.6% or more of the weight of the meat. Salt-soluble proteins, of which myosin is the principal component, transfer to the surface under the influence of the massaging action of the impact of other pieces of meat during the tumbling (or mechanical stirring may be used) and coagulate when the compressed mass of pieces is heat processed. In this way, blends of meat in blocks or cylinders of good texture and coherence can be made. The process can conveniently be combined with curing, which is accelerated as a result of more rapid penetration of the curing agent during the mechanical working of the mass.

It is also possible to reshape normally boned cuts or intact muscle such as beef rump under pressure so that a mass of uniform cross-section is produced which may be sliced into portions of equal area, thickness and weight. The meat is warmed from the frozen storage temperature to about −4°C ('tempering'), pressed in a mould at some hundreds of pounds square inch, cut into slices of the desired thickness, wrapped and packed and cooled again to the frozen storage temperature. If desired, enzymic or mechanical tenderizing can be carried out on the reformed boneless cuts before the pressure treatment. A similar pressure-moulding process may be used with meat flaked mechanically from the frozen state. At this stage, meat of differing fat content may be blended and tougher finely flaked material mixed with more coarsely flaked more tender meat. The blend of flaked meat is then mixed to a doughy consistency with 0.75% of salt, moulded into an approximation of the desired final shape, frozen to less than −10°C, tempered to −5°C, moulded under pressure, sliced into 'steaks' or 'roasts', packed, and frozen again for storage.

The advantage of these processes lies in the production of uniform portions of meat which are ideal for mass-catering. The flaked products can, moreover, be closely controlled in fat content and, of course, provide an opportunity for upgrading the meat used, for blending meats of different origin, and for adding flavours.

The most interesting new forms of ruminant meat products, however, are those in which the composition of the fat has been substantially modified by dietary means prior to slaughter. By feeding sheep or cattle for some weeks before slaughter with a supplement of polyunsaturated fat protected from saturation by rumen organisms, the level of polyunsaturated fat in both milk and the edible fat of the carcase can be raised from the usual figure of about 3% to 20% or more. Medical authorities in many countries are increasingly advising the public to increase the ratio of polyunsaturated to saturated fat in the diet as a health measure likely to decrease the incidence of heart disease. Since ruminant fats have a low and nearly constant level of polyunsaturated fat which is independent of the degree of polyunsaturation of the fat the animal ingests, it has been a necessary feature of diets recommended to reduce the incidence of heart disease, that to varying degrees they place severe restrictions on the amounts of ruminant meats,

Chapter 13 references, p. 306

milk and milk products contained therein. The process which was developed in Australia by CSIRO depends on coating polyunsaturated seed fats such as sunflower seed oil in the form of minute globules with a protein skin which is treated with small quantities of formaldehyde. The treated protein coat is unaffected by conditions within the rumen and is also resistant to microbial attack, thus protecting the enclosed fat from the microbial hydrogenation which is the normal fate of dietary polyunsaturated fats in the rumen. When the protected material leaves the rumen, however, it encounters the lower pH of the abomasum, the protein formaldehyde link dissociates, and both fat and protein coating become available to the normal digestive enzymes of the animal. Thus the ruminant digestion becomes in relation to the encapsulated fat and its protective protein layer that of a monogastric animal, in that the fatty acid content of the meat and milk reflects that of the diet of the animal, and the utilization of the protein employed is more efficient. In addition, protection of the fat avoids the limitation on dietary fat levels necessary to retain satisfactory rumen function. As a result of this, the feeding of protected fat, whether polyunsaturated or not, presents an effective way of greatly increasing the energy density of the diet in circumstances in which it is profitable to increase the weight of an animal as rapidly as possible.

It has been shown that ruminant meat and milk products may routinely be produced in this way with a linoleic acid content of at least 20% of the total fatty acids. Cholesterol levels in the products are unaffected.

It has been found, as a result of extensive development work, that the problems posed by the increased level of polyunsaturation, and hence susceptibility to oxidative change with time, are not serious. The duration of frozen storage of the meat is reduced to below that of normal beef and mutton but only to the level of pork and chicken, which also contain appreciable levels of linoleic acid. Chilled storage time is unchanged, being governed by microbial growth as with the conventional meats. Polyunsaturated milk, however, is very susceptible to oxidation and the development of off-flavours, and this tendency is greatly increased in the presence of copper even at levels less than 0.1 ppm. Oxidation can, however, be held in check for 14 days or more by the use of an antioxidant (butylated hydroxyanisole) at a level of 0.02% of total fat, even in the presence of added copper. If polyunsaturated milk with a low copper content is used for cheese-making, oxidation is not a problem, being controlled apparently by the activity of the bacteria involved in the process.

The presence of 20% polyunsaturated fat in the products leads naturally to some change in the physical properties of the fat which has a lower melting point and is softer than that of the conventional products under comparable conditions. In addition to changes in resistance to oxidation and in physical properties, there are detectable changes in flavour. This is perhaps most pronounced in lamb in which the mutton odour disliked by many consumers is much reduced. This desirable change can be produced at levels of linoleic acid less than the 20% aimed at for medical purposes. Cheddar cheese, surprisingly, was found to lack flavour rather than to have acquired a different one when made from polyunsaturated milk. It can, however, be made with a completely satisfactory flavour by using *Lactobacillus bulgaricus* in addition to the normal cheese starter. In the manufacture of butter from polyunsaturated cream, churning has to be performed at low temperature (0—2°C) in order to achieve the necessary working, for at normal temperatures the churning time is very short. The butter is of good quality and maintains its grade well during storage. The process is in the course of application, both for the production of polyunsaturated products and as a high energy supplement in finishing cattle in many countries, including the United States, Canada and the United Kingdom. Polyunsaturated meat and dairy products are successfully marketed in South Africa, and polyunsaturated cheese has achieved a steady, though small, sale in Australia.

2.3. Plant substitutes and diluents

Plants are the primary source of food for man and for his domestic animals. Consequently animal food from domestic animals is almost invariably more expensive than equivalent foods from plant sources unless the animals are using plant food not directly available to man, as is the case with fish, game and domestic animals at free range on land not suitable for, or not needed for, crop production. Moreover, most people place a higher value on animal products as food. As a consequence there is much to be gained by the production of food resembling animal products from plant sources. It was first realized that this might in principle be possible in the early 19th century, when chemistry had developed to a stage at which the first chemical analyses of foods were done, and it was realized that both plants and animals had very similar components in terms of fats, carbohydrates and proteins. Indeed, at first it was thought that the similarities were so great that only four or five proteins existed which occurred in both plants and animals, and that they could be incorporated, essentially unchanged by digestion, into the animal from its food whether plant or animal in origin. If 'casein', for example, occurred in both milk and plants, then in principle milk could be made from plant protein. However, it rapidly became plain that such an approach was a great over-simplification, but it was realized that fats of plant and animal origin had much more in common both in composition and properties than the other components. In the then state of knowledge it seemed reasonable to suppose that by exposing a fat to the transforming virtues of the tissues of a cow or of an extract therefrom, butter could be produced.

In the eighteen-sixties, Mège-Mouries (a food technologist then working for the French Navy) reasoned that milk fat must be formed by a transformation of body fat. He therefore heated suet in warm water with chopped cow's udder and a little milk. The resultant product had too high a melting point and the lower melting fraction was squeezed out between heated plates and was marketed as margarine. By 1880, fifty factories in Holland were producing it. In the eighteen-nineties it was found that vegetable oils were an even cheaper raw material, and 20 years later the technique of achieving a desirable consistency by hydrogenating vegetable oils was discovered. The desirable flavour necessary to the product was achieved by the addition of small quantities of cultured or sweet skim milk.

The penetration of margarine into the traditional market for butter initially depended solely on its low cost, since for a long time its flavour could not compete with that of butter. The flavour was, however, progressively improved, and with social changes it became possible to exploit new advantages it had in relation to butter. It was found that margarine could be formulated in such a way that it remained spreadable over a wider range of temperatures than butter. The ever-increasing use of domestic refrigeration made this a very real advantage and finally led the dairy industry to consider seriously the possibility of processing the natural product to resemble, at least in this respect, the substitute more closely. The second unanticipated advantage was that a proportion of the plant fats and oils used as raw material could be highly unsaturated, and margarine of satisfactory taste and spread-

Chapter 13 references, p. 306

ability could be formulated with high ratios of polyunsaturated fatty acids to saturated fatty acids. Since opinion in the medical profession over the last few years has increasingly suggested nutritional and health advantages might lie in an increased intake of polyunsaturated fats in relation to saturated fats, the margarines with high ratios of polyunsaturated fats to saturated were seen to be more desirable as a dietary component than butter. Therefore, attempts are being made by the dairy industry to develop butter blends in which the ratio is increased by blending butter with highly polyunsaturated seed oils.

The enormous success of margarine as a butter substitute depended not only on its great price advantage but also on the fact that milk, and consequently its products, has no cellular structure and is homogeneous. Obviously it is simpler to simulate a product with no apparent gross structure than one which derives some of its properties from its structure. It might be expected that the widespread introduction of margarine would have quickly led to the development and marketing of simulated milk and cheese, but this did not occur for very many years, in part because of effective legislative measures initiated in most advanced countries to protect these products from competition — particularly milk, which would have been almost certainly the next product to be simulated successfully.

'Filled milk', in which animal fat is replaced by cheaper plant fat, has also been developed and enjoys some success in those states of the U.S.A. where it can legally be sold. However, in the foreseeable future its penetration of the milk market is likely to remain less than 10%. It is technically possible to produce a milk-like product using not only plant fat but also plant protein, but it has not been commercially successful in the United States, and the necessity to use emulsifiers and stabilizers militates against its public acceptance. In India, milk 'extended' by the use of plant seed proteins and plant fats has, however, been successful in reducing the shortage of milk for children. Further inroads into the traditional market for milk and cream have come from the production of 'whiteners' for tea and coffee, whose acceptance is based largely on their stability and convenience.

The problem of simulating meat was very much more difficult technologically than that of simulating milk and its products. Meat is a fibrous tissue and recognized by the consumer not only by its taste but by its visibly fibrous structure and feel in the mouth. As a consequence, the first efforts were restricted to the simulation of processed meat products in which the fibrous structure had become no longer apparent to the eye. Products such as sausages and pastes were the obvious target and those first chosen by people attempting to vary a vegetarian diet with meat-like products. Those first produced were successful only with consumers motivated by vegetarian beliefs to accept them in spite of their failings in flavour and texture to come near the product simulated. Acceptance as meat substitutes by those who enjoyed meat and its products as a constant component of their diet had to await the achievement of greatly improved texture and flavour. Two approaches to the problem were possible. If the substitute were to be limited in its use as a diluent for comminuted meat products, then provided it had no marked flavour of its own, it could rely on that of the meat product it accompanied, provided also that its texture in terms of chewiness (or 'mouth feel') was sufficiently similar to the cooked meat component as not to be obvious. The second approach to a total simulant of meat is much more demanding — the product has to be fibrous, the fibres must be well oriented and bound together, of similar physical properties to those of cooked meat, and it must have added to it a flavour close enough to that of the simulated product to be acceptable.

The first serious attempts to process seed proteins to a form in which they could simulate an animal protein were directed to providing an alternative fibre to wool. The Ford Company in the United States supported research into making a satisfactory fibre for car upholstery from soybean protein. Earlier substitutes for animal proteins such as milk had included the use of regenerated cellulose, or its derivatives, and simulated wool (Lanital), or glues with improved properties were made from animal proteins, such as casein, which were relatively cheap at that time. There were many efforts to use other proteins to make substitutes for wool and they were often successful technically but never commercially. Nevertheless, it became obvious that fibres could be spun from proteins of widely varying properties, composition and origin, and this led to the development of a successful method of making a total simulant for meat.

The starting material is a soybean fraction containing more than 90% protein. It is dissolved at an alkaline pH and after maturing is forced through spinnerets into an acid coagulating bath. The emerging fibres are stretched, while being wound out of the bath, to about a third their emergent diameter to increase their strength and elasticity and to impart the necessary 'chewy' quality to the final product. To form the meat simulant, an edible binder, such as egg albumin or wheat gluten, is added together with flavouring or colouring materials, and the impregnated material is compressed as a 'tow' of oriented fibres, and then heat-treated to ensure adhesion. The product can be diced into the desired shapes. Variables that can be exploited to impart variety to the product include fibre diameter, properties of the binder used, stretching of the fibre, and compression of tow, in addition to flavouring or colouring to achieve the desired product properties. On a dry matter basis the final product, such as simulated bacon, may contain 55% spun soybean isolate, 16% protein binder, 21% edible oil and 8% flavouring. These additives increase the price of the product. The structure of simulated meats may vary widely; for example, it can be made to appear like chicken, scallops, beef or ham.

The cost and problems associated with the production of meat 'extenders' are much less than those associated with those of the true simulants. They are made by the extrusion of a paste of defatted soybean flour (containing about 30% protein) through a die at temperatures above 100°C. A high pressure is necessary and the paste expands very rapidly as it leaves the die as a result of some of the water turning into steam. This sudden expansion of the hot paste results in stretching and orientation of the protein which gives a chewy open texture to the product.

2.4. 'Single cell protein'

During the nineteen-fifties and nineteen-sixties a widespread belief grew up that the most important single deficiency in world diet was protein. It was true that many suffered, and still do, from an inadequate intake of protein, but in general this was a secondary inadequacy in the diet, the primary deficiency in most instances being of total metabolizable energy. Except in limited areas in which the diet is almost restricted to cassava, yams, sweet potato or plantain, it is probably very unusual for an adult diet to be deficient in protein, provided that it is adequate for needs in energy content. If it is not adequate in this respect, then available protein may be lower than desirable and this deficiency may be compounded by the body's necessity to metabolize the protein to satisfy energy requirements, thus diverting it from tissue repair and growth. Needs for dietary protein are higher than normal in pregnant and lactating women and in children suffering chronic infections, and in some situations where total intake of food is restricted (or in the case of infants where the 'protein density' of the diet is low so that repletion occurs before satisfaction of protein requirement). It is these groups that are most exposed to the likelihood of protein shortage. This description of the problem of protein shortage is markedly different from that current in the two post-war decades when it was said with great authority that a third or even a half of the world population was short of protein. In that climate of opinion it was not surprising that many proposals for filling the 'protein gap' were advanced. The general recognition in recent years that the protein requirements of man had been greatly over-estimated — which abolished the conceptual protein gap — has not, however, affected the interest in the production of protein from unusual sources — principally microbiological — but, for the near future at least, has resulted in a redefinition of their end use to stockfeed rather than human food.

Food or feed products of microbiological origin are commonly described with evident inaccuracy as 'single cell protein'. They contain in addition to protein, fat, carbohydrate, minerals and vitamins and are normally the dried remains of organisms grown in bulk culture which are not necessarily unicellular. As a class they are not a completely new article of diet, for fungi have always been a part of the diet of some peoples, as have the multicellular marine algae. Single celled algae have also been a traditional component of the diet in both Africa and Central America, and yeast was used in the diet by the ancient Egyptians. Bacteria, however, have never been a conscious or deliberate food. 'Single cell protein' in the commercial sense of the phrase is prepared at present by the large-scale submerged culture of yeast grown on hydrocarbon fractions or ethanol (derived from hydrocarbon fractions or fermentation); bacteria grown on methanol or methane as principal substrate, or a fungus (*Fusarium* sp.) grown on a variety of carbohydrate sources. Large plants for the production of 'single cell protein' have been built — for the production of yeast from hydrocarbon substrate these are at present of 4000, 20 000, 60 000 and 100 000 tons/annum capacity. There are two of the 100 000 ton capacity in Italy and it is said that total installed capacity in the U.S.S.R. is of the order of 300 000 tons. In England a plant of 60 000 tons/annum capacity is being erected to grow bacterial products using methanol as substrate. The products in production and development have been rigorously tested on a number of species of animals and no evidence has emerged that they are in any way harmful as major components of the diet of animals nor that they are likely to be so to man. Yeast products contain 60 ± 5% crude protein, bacterial products a little more than 70%. The nucleic acid content, about 6 and 11%, respectively, is high but poses no problem for animals which metabolize the purines beyond the uric acid stage. It may be advisable to reduce it before their use in human diets. In practice their main outlet is seen in the feeding of monogastric animals such as pigs, poultry and fish, since they are unlikely to be economic in the feeding of adult ruminants, though they may have application as starter feeds for the pre-ruminant calf and lamb. Their application in humans seems to be feasible but not likely for some time to come.

2.5. Protein from leaves

The principal source of protein for ruminants is forage. Conversion of this protein to meat and dairy products involves major losses. If it were possible for man to substitute leaf protein for animal protein in his diet, then the amount of protein available to him would be greatly increased. Work over the last few decades has shown the feasibility of preparing leaf protein as an acceptable though unusual component of the diet both on the small (village) and large scale. Since leafy crops are usually available throughout the year, small-scale production and immediate consumption as a quantitatively small supplement to the diet does not involve problems of preservation or sophisticated processing. The process has been shown to be feasible but has not yet been adopted on a significant scale. It remains, however, as an option that in the future may well be exercised as a means of increasing the protein content of low protein diets.

2.6. Synthetic foods

The term 'synthetic' has become vague in its application to foods and is used with different meanings. It has been widely applied, for example, to meat substitutes made from plant protein; in this usage 'artificial' might be a more appropriate adjective. It is also applied to mixtures of individual natural components such as sugars, fats, amino acids and vitamins, many of which are natural products, derived directly or indirectly by hydrolysis from foodstuffs, though some, such as the amino acids and vitamins, may be chemically synthesized from starting materials not derived from foodstuffs or by micro-organisms growing on natural and inorganic components. Finally, in its strictest sense it may be applied to synthetic products such as 1, 3-butanediol (an energy source) or to glycerol synthesized from petrochemicals rather than prepared by the hydrolysis of natural fats.

'Synthetic foods' in the first sense, such as margarine and meat extenders, have already been discussed, and in this section those foods composed of mixtures of defined single chemical compounds of biological or organic synthetic origin will receive brief attention.

The discovery in the early years of this century that some components of the diet present in small amounts were essential to the health of man had been foreshadowed in 1876 when Escher showed that gelatin as the sole protein source in the diet of a dog led to loss of nitrogen, while supplementation of gelatin with tyrosine restored the nitrogen balance. The recognition of the existence of vitamins not only rendered suspect most previous work on the requirements of animals for individual constituents of foods, but also led to the need to make sure that all essential trace components in the diet were identified. This in turn led to the necessity to formulate diets made up ab initio from known pure components for use in feeding trials. As a result of endeavours in this direction it has proved possible to maintain man in health for many weeks on liquid diets of rigorously defined composition and, in principle, it is now possible to formulate a completely synthetic diet which, however, is not only very expensive but has little attraction for the consumer. The practical benefits from this work at present and in the near future appear to be confined to medical uses. The absence of polymers from these diets avoids the need for digestion before absorption and is useful in treatment of patients with a non-functional pancreas, malabsorption, and severe allergic reactions to dietary protein. They can be fed, moreover, without any fibre content or other indigestible material, and in this form are useful in the treatment of intestinal fistula and ulcerative colitis, and also are an excellent preparation for gastrointestinal surgery, since their use for a few days reduces the microbial population of the intestine by several orders of magnitude. In patients with chronic renal failure the nitrogen content of these diets can be controlled accurately and, if necessary, the essential amino acids of the diet can be replaced by the keto and hydroxy analogues.

Chapter 13 references, p. 306

TABLE 13.8

The principal by-products with their source and uses

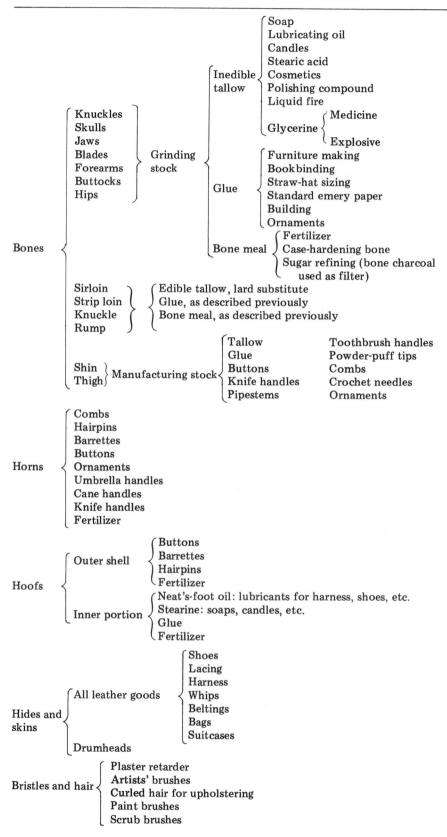

Bones	Knuckles, Skulls, Jaws, Blades, Forearms, Buttocks, Hips } Grinding stock	Inedible tallow { Soap, Lubricating oil, Candles, Stearic acid, Cosmetics, Polishing compound, Liquid fire
		Glycerine { Medicine, Explosive
		Glue { Furniture making, Bookbinding, Straw-hat sizing, Standard emery paper, Building, Ornaments
		Bone meal { Fertilizer, Case-hardening bone, Sugar refining (bone charcoal used as filter)
	Sirloin, Strip loin, Knuckle, Rump } { Edible tallow, lard substitute; Glue, as described previously; Bone meal, as described previously
	Shin, Thigh } Manufacturing stock { Tallow, Glue, Buttons, Knife handles, Pipestems — Toothbrush handles, Powder-puff tips, Combs, Crochet needles, Ornaments

Horns { Combs, Hairpins, Barrettes, Buttons, Ornaments, Umbrella handles, Cane handles, Knife handles, Fertilizer

Hoofs { Outer shell { Buttons, Barrettes, Hairpins, Fertilizer
Inner portion { Neat's-foot oil: lubricants for harness, shoes, etc.; Stearine: soaps, candles, etc.; Glue; Fertilizer

Hides and skins { All leather goods { Shoes, Lacing, Harness, Whips, Beltings, Bags, Suitcases
Drumheads

Bristles and hair { Plaster retarder, Artists' brushes, Curled hair for upholstering, Paint brushes, Scrub brushes

TABLE 13.8 (continued)

Wool: All woollen goods

Intestines
{
Sausage containers
Lard containers
Snuff containers
Gold beater's skins
Fancy bottle caps
Drum snares
Tennis strings
Ukulele strings
Harp strings
Violin strings
Other kinds of musical strings
Surgical ligatures
Ox-gall { Setting dyes, used in paints, dyes, laundries and dry cleaners
Activate bile secretion
Gall-stones: for ornaments
}

Fats
{
Oleomargarine
Soaps
Glycerine .... { Medicinal purposes
Liquid fire
Candles
Cosmetics
Polishing compounds
Stearic acid: explosive
Red oil lubricants
Ointments
Special oil for leather dressing
Torpedo lubricating oils
Heavy lubricating oils
Illuminating oil
}

Meat scrap and blood
{
Meat meal: stock feed
Blood meal: stock feed
Tankage: fertilizer, feed
}

Glands—18 preparations issued by the medical fraternity to treat various diseases such as rickets, goitre, mental backwardness, influenza, also in obstetrics, surgery, menopause, etc.
{
Pituitary: base of brain
Pineal: surface of brain
Thymus: neck
Pepsin: stomach
Spleen: spleen
Thyroid: neck
Uretic: abdomen
Suprarenalin: kidney
Ovaries
Corpus luteum: part of ovaries
Mammary
Pancreatin: pancreas
Parotid: neck
Parathyroid: near tongue root
Kephalin: brain
Lecithin: brain
Thromboplastin: brain
Red bone marrow: tonic
}

Source: Tomhave (1925).

3. TECHNOLOGY OF FOOD ANIMAL BY-PRODUCTS

Less than 40% of the dry weight of a beef animal is normally used as food today in advanced societies; rather more was eaten a century ago and is eaten still in countries where the demand is greater. Man has shown remarkable ingenuity in developing uses for all inedible slaughter products, and this realized its peak of diversity and ingenuity in the United States in the first half of this century before the burgeoning of the plastics industry. The claim of the Chicago slaughterhouses that 'we use all of a pig except its squeak' was largely justified not only for pigs but for beef animals too. Table 13.8,

Chapter 13 references, p. 306

compiled by Armour and Co. in 1925, shows the thoroughness with which every possible economic use of non-food products of the slaughter-house was identified and exploited. At the time it was compiled, competition for these uses had just begun as cellulose nitrate and, later, acetate began to compete in the market with bone, horn and hoof in the fabrication of products such as combs, buttons and handles of various kinds. As synthetic plastics appeared in later years in ever-increasing variety and at more and more competitive prices, the diversity of the market for many animal by-products was steadily reduced. It is largely true that the uses of inedible animal by-products which were dependent on their bulk physical properties were the first to suffer from this competition. Inedible products with a microstructure significant in determining their final useful properties, and difficult or impossible so far to duplicate economically in synthetic materials, have survived competition well. Wool, in spite of decades of the most active competition, still remains a salable product and, in spite of a reduced share in a vastly expanded market, still enjoys sales not very different quantitatively from those which it had in the days of its dominance. In spite of the invasion of plastics in the footware market and the virtual disappearance of other markets, much the same can be said of leather. It cannot be said of those inedible products depending for their use on bulk properties, as a glance at Table 13.8 will show under the headings for bone, horn, bristles and hair.

Products depending on their nutritional qualities for man and beast have not suffered, and will not suffer in the near future, in the same way from synthetic competitors, except for those fats and oils which had alternative uses as lubricants or raw materials for the synthetic chemical industry. In the last two respects, competition from products derived from crude oil by fractionation or synthetic routes has been very severe, particularly in the once very considerable markets for illuminating oils, candles, light and heavy lubricating oils and, to a lesser extent, soap. In the previous section it has been said that truly synthetic sources of food are unlikely to be of significance in the foreseeable future for man, and this is just as true in relation to animal feeds, though it appeared a few years ago just possible that a synthetic energy source such as 1,3-butanediol might be used. Recent changes in the price of crude oil and natural gas appear to have disposed even of this remote possibility. Apart from the production of high grade gelatin from some primary slaughterhouse products, outlets for protein containing edible by-products are confined to stockfeeding (mainly poultry) with a minor component going to other animal food uses such as fish farming and mink farming. In this area they encounter competition from plant seed products (soy meal and oil seed cakes such as linseed) and will meet increasing competition from single cell protein produced as a result of growth on crude oil fractions or bulk products such as methanol from natural gas. It is likely that these sources of energy for single cell protein production will be progressively replaced by food-processing wastes — even including liquid wastes from abattoirs. While as a result the profit margin on meat meals may shrink, they are likely to remain significant when costs of alternative means of disposal of the raw material as waste are taken into account.

The fats not sold for human consumption through the meat trade may, if derived from beasts fit for meat, be used after rendering as raw material for the manufacture of margarine, cooking fat and for other human food purposes. Fats from rendering of material declared unfit for human consumption, together with low grade fats from the first category, still find a market in the manufacture of soap and glycerol and as raw material for detergent manufacture. They are also, of course, a component of stockfeed either as

TABLE 13.9

Relative vulnerability of animal by-products to innovations in technology

Inedible products
Very vulnerable, have already lost many markets either because market has vanished as a result of social change or because competing products have cost advantage or are superior in properties. The least vulnerable are those such as wool and leather in which the microstructure contributes significantly to their properties though they have suffered and will suffer from competition from substitutes inferior in properties but lower in cost.

Edible products
Essentially invulnerable as competition can only come from plant sources and single cell protein. It seems certain that at worst it will in most circumstances be more profitable to sell these products as stock food than to dispose of them in conformity with increasingly rigorous environmental regulations.

Pharmaceutical products
Many — insulin, for example — appear invulnerable but their high cost and limited availability against a demand likely to steadily increase makes them vulnerable to replacement via biosynthesis, chemical synthesis or, in some instances, analogue synthesis.

TABLE 13.10

Average yields of the more important glands and tissues used for pharmaceutical production

Raw material	No. of glands/ animal	Yields (g)			Product
		Bovine	Ovine	Porcine	
Pancreas	1	250—400	30	50—130	Insulin, glucagon, pancreatin, trypsin, chymotrypsin
Pituitary	1	2.0—3.0	Sheep 0.8 Lamb 0.5	0.4—0.8	ACTH[a], thyrotropin, vasopressin, growth hormone, oxytocin and others
Pineal	1	0.3	0.04—0.12	0.1	Possible use in treatment of schizophrenia
Adrenal	2	15—30	2—3	3—5	Adrenaline[a], cortex extracts in Addison's disease
Thyroid	1	25—30	3—9	5—10	Thyroid extract for treatment of thyroid insufficiency[a]
Parathyroid	2—4	0.5	0.2	0.15	Parathyroid hormone
Ovary	2	10—20	1.5	5.0	No significant uses
Testes	2	500—1000	100	150	Hyaluronidase
Spleen	1	500—900	Sheep 90 Lamb 55	90—150	No significant uses
Spinal cord	1	100—150	40	50	Cholesterol, raw material for synthesis of active steroids
Gall (fresh)	1 bladder	300—400 (9% solids) Calf 10—15 (8% solids)	Sheep 25—30 (11% solids) Lamb 15—20 (12% solids)		Once used as detergent, now as source of bile acids used as raw materials for synthesis of corticosteroids
Vells[b]	1	390 green 20 dry			Rennet
Raw blood		4% of carcase weight	7% of carcase weight		Bovine serum albumin; foetal calf serum (tissue culture)

[a]Now being produced by synthesis, or synthetic analogues can replace product.
[b]Abomasum of calf. Rennet used in cheese-making. Now suffers increasing competition from analogous enzymes from fungi.

Chapter 13 references, p. 306

present in meat meal or blended with it during final formulation. In addition to their use in feeding monogastric animals, they may now play an increasing role in the fattening of ruminants in the protected form referred to earlier.

In Table 13.9 the by-products have been classified as inedible, edible and pharmaceutical. The last is minute in volume compared to the others, but the products in some instances can command high prices. Table 13.10 lists average yields of these glandular by-products from cattle, sheep and pigs. Fifty years ago these by-products represented the only source of the pharmaceutically active compounds listed, apart from adrenaline for which a satisfactory synthetic route existed. In the intervening years more and more valuable pharmaceuticals have been identified in these by-products, and during the same period more and more of those known to be present have become available by synthesis, or suffered from competition from synthetic analogues or products from alternative sources. This trend can be expected to continue with the probable result that the glandular and other by-products listed will become less and less worth while collecting individually and will either become more available as food (pancreas) or enter the renderer to emerge as meat meal. The least likely to suffer this fate are, perhaps, as far as can be seen at present, the pituitary and possibly the pineal. Much will depend on how straightforward it proves to code micro-organisms for the synthesis of polypeptides such as pituitary growth hormone and proteins such as insulin.

4. ACKNOWLEDGEMENT

I am indebted to members of the Industry Section of the Meat Research Laboratory, CSIRO Division of Food Research, for the data in Tables 13.5, 13.6 and 13.10.

5. REFERENCES

Brown, L.R., 1972. The green revolution and world protein supplies. P.A.G. Bull., 2 (2): 25—33.

Callow, E.H., 1947. Comparative studies of meat. 1. Chemical composition of fatty and muscular tissue in relation to growth and fattening. J. Agric. Sci., 37: 113—129.

Chick, H., 1951. Nutritive value of vegetable protein and its enhancement by admixture. Br. J. Nutr., 5: 261—265.

Claudian, J. and Serville, Y., 1968. La menagère devant la viande. Cah. Nutr. Diet., 3 (3): 29—32.

Escher, T., 1876. Über den Ersatz des Eiweisses in der Nahrung durch Leim und Tyrosin, und deren Bedeutung für den Stoffwechsel. Vierteljahrschrift der naturforschenden Gesellschaft in Zürich, 21: 36—50.

Lawes, J.B. and Gilbert, J.H., 1859. Experimental inquiry into the composition of some of the animals fed and slaughtered as human food. Philos. Trans. R. Soc. London, 149—494.

National Live Stock and Meat Board, Chicago, 1977. Meat Board Rep., 10 (1), 12 pp.

Nisbet, W., 1801. A Practical Treatise on Diet. Phillips, London, 429 pp.

Oiso, T., 1971. Ann. Rep. Natl. Inst. Nutr., Tokyo, 46—48.

Tomhave, W.H., 1925. Animal by-products and sausage making. In: Armour & Co. (Editor), Meat and Meat Products. J.B. Lippencott, Philadelphia, PA, pp. 237—267.

Chapter 14

Livestock in Economic Development

H.E. JAHNKE

1. INTRODUCTION

Almost by definition, livestock production and productivity are very low in developing countries compared to the levels which could be achieved, as demonstrated by the performance of the sector in the developed world. Unfortunately, however, these low levels of production do not in themselves suggest what type of intensity of effort would be required to improve the performance of the livestock sector. The situation is further complicated by the fact that present and potential levels of livestock production vary widely in different developing countries.

Livestock production can play an important role in agricultural and economic development, though this role will vary from country to country according to the baseline situations and the specific characteristics of the development process. In this chapter, the major factors which influence the development of the livestock sector will be analyzed within a theoretical framework. The broad patterns of development will be described, which can be expected given specific factor constellations, and the history of livestock development in dryland Australia, the tropical and subtropical regions of South America, and Germany will be described. Hopefully, this will shed some light on the basic policy dilemma facing planners in the developing countries, as described by Pino (1975):

> 'They can't simply say it's good to produce livestock. They have got to know under what conditions it's good to produce livestock ... or under what conditions it's necessary not to.'

A comparative historical approach will be followed, as a complement to theoretical discussions and cross-sectional analyses. It is felt that current theories of agricultural development are not sufficiently refined to explain the role of livestock production in the development of an economy. Cross-sectional analyses, by themselves, are difficult to interpret, due to differences in the natural environment or similarities in the level of national development which make it impossible to distinguish the different roles played by the livestock sector. However, a comparative historical approach will not produce unambiguous results with immediate applicability: rather the aim is to formulate a number of hypotheses which can be tested in real situations in the developing world.

The focus on the role of livestock production in economic development is justified on two counts. In the first place, livestock is unique as an agricultural commodity because animal foods differ from other foods, both in

---

Chapter 14 references, p. 328

terms of their nutritional value and in terms of their elasticities of demand. In the second place, the role of livestock is by no means limited to the production of animal foods. Livestock also provide draught power, fertilizer and other by-products, and domestic animals often play an important social and cultural role. In some agricultural systems, these other functions of livestock are more important than the production of animal foods.

Throughout this chapter, the term 'livestock' is used in a broad sense to refer to all agricultural animals, though the emphasis is on ruminants. This is because ruminant production plays a central and changing role throughout the process of economic development.

2. PATHS OF AGRICULTURAL DEVELOPMENT

Economic development is a process characterized by the growth of aggregate variables, such as gross national product and national income, as well as profound changes in the structure of the economy. Such changes are the introduction of technical innovations into the production process, increasing economic differentiation and specialization, the relative — and later an absolute — decrease of the labour force in agriculture, an increasing shift of emphasis from the primary to the secondary and tertiary sectors of the economy and, in general, the increasing importance of external economic links and international trade.

Agricultural development is linked with general economic development in several ways. [For two authoritative descriptions of these interconnections, see Malassis (1975) and Johnston and Mellor (1961)]. Industry supplies capital equipment and other inputs to agriculture, and agriculture supplies raw materials to industry, as well as labour and capital to more productive sectors of the economy. Agriculture may also play an important role in earning foreign exchange which is then invested in industrialization. Overall economic development, in turn, determines domestic demand for agricultural products, in particular, food.

It is argued here that three factors are crucial in determining the development of the agricultural sector, or of a subsector such as livestock, within the context of general economic development. These factors are: (1) the expansion of domestic demand, (2) the export situation, and (3) the stage of production technology.

Starting by assuming a closed economy, the expansion of domestic demand for agricultural products is determined by two factors: population growth and the increase in per capita incomes. Population growth is generally assumed to lead to a symmetrical increase in demand for food, i.e., all commodities are equally affected[1]. Thus a 3% increase in total population would lead to a 3% increase in the demand for food. Increased per capita income, on the other hand, is assumed to lead to an asymmetrical increase in the demand for different types of food, depending on the income elasticities of demand[2]. This phenomenon will be discussed in more detail in the next section.

[1]The simplifying assumptions are that the accompanying change in the age structure of the population has no effect on the structure of demand and that the interdependence between population growth and the growth of per capita income can be ignored.
[2]The income elasticity of demand is the percentage by which the amount consumed increases with a given percentage increase in income. Typically in a low-income country, the income elasticity of demand for food is about 0.8, or perhaps somewhat higher. This means that with a 1% increase in income the demand for food will increase by 0.8%.

Leaving the assumption of a closed economy, it is also clear that agricultural development is significantly influenced by the export potential of a country's agricultural products, which in turn depends on world-wide demand and transport costs. Thus, the emergence of rich, urban industrial centres earlier in this century, together with the development of transport technology and an accompanying dramatic fall in transport costs, had a profound effect on world agricultural development.

Internal and external demands go a long way in explaining the rate of growth and overall volume of agricultural production. But these factors reveal little or nothing about the nature of agricultural production or the way in which development takes place. In a 'pre-development' situation — before widespread industrialization, with per capita incomes near subsistence level, and before economic development has become a major government policy — there are still profound differences in the agricultural and livestock production systems of different countries, different regions and different continents. Two factors appear to be largely responsible for these differences — the natural environment and density of the population.

Looking first at the natural environment, the most important factor affecting land use, and thus agricultural and livestock production systems, is climate. By considering biological and economic factors, along with climatic conditions, it is possible to define ecological zones in greater or lesser detail according to the needs of a particular study. In this chapter, the ecological zones discussed will be subtropical and tropical arid areas, as exemplified by large parts of sub-Saharan Africa and Australia; subtropical and tropical humid areas, such as large parts of South America and also sub-Saharan Africa; tropical highland areas, as in Eastern Africa; and temperate areas, for which Germany will serve as an example.

Population density and population growth are the second major factor determining the nature of agricultural development. Herlemann and Stamer (1958) propose that the most important factors determining the nature of structural change in a developing economy are the density and absolute growth rate of the population and the reserves of arable land available at the beginning of industrial development. These two authors go on to hypothesize that with different initial population densities development takes different paths, and they propose models for development in densely and sparsely populated countries. The primary distinction is that, at a given level of technology, in areas of low population density agricultural production can be significantly expanded by bringing new land into production, whereas in densely populated areas this is not possible because all the land is already being used.

Basically, they hypothesize that the pattern of agricultural development is determined by the initial balance between population and natural resources. For a densely populated country, they claim that agricultural development typically follows the following pattern: first an increase in population density is followed by an intensification of agricultural production, which in turn is followed by mechanization, which leads to consolidation and expansion. In sparsely populated areas, on the other hand, agricultural development tends to pass through the stages of expansion, then mechanization, and finally intensification (Table 14.1).

These models are based on two production factors — land and labour. Before the significant use of capital in agricultural production or the introduction of major technological innovations, it is enough to consider the interactions between labour, land and climate, as well as transport costs, in order to explain the differences, in economic terms, between production systems and intensities of production. (This is the basis of Thünen's theory

Chapter 14 references, p. 328

of agricultural production, which is taken up again later in this chapter.) In
the course of economic development, however, the use of material inputs in
agriculture increases, and yield and net income are increasingly determined
by capital, along with labour and land. Maximization of net income is no
longer determined only by the intensity of production and the value of the
product (as a function of market factors and accessibility), but is also
increasingly dependent on the choice of the best combination of inputs.

A farmer will choose a combination of production inputs according to
their relative prices. In other words, the price—cost structure in the economy
(which is an indicator of the general technological level) determines the
relative quantities of production factors used in the agricultural sector. This
means that for any combination of initial resource endowment (basically,
land and labour) and level of economic development (expressed in terms of
the level of industrialization and ultimately the relative costs of capital
inputs), there is an optimum mix of production inputs for the underlying
assumptions, such as the 'law of diminishing returns', see Herlemann (1961).

These hypotheses provide a general framework in which to view the
pattern of agricultural development. The rest of the discussion here will
focus on a more specific consideration of the livestock sector.

3. FACTORS AFFECTING THE ROLE OF LIVESTOCK PRODUCTION IN ECONOMIC DEVELOPMENT

3.1. Internal demand

With increasing per capita income, the proportion spent on food tends to
decline, as illustrated in Table 14.2. (This statement is based on two economic theorems referred to as Engel's law and Gossen's law. An elaboration
of these would go beyond the scope of this discussion.) For the purpose of
the argument here, this is equivalent to saying that the income elasticity of

TABLE 14.1

Evolution of agricultural production in the course of economic development

Development phase	Change in the cost structure associated with:	Substitution of ... by
(a) Densely populated countries		
Densification	Increasing land values declining wages	Land ... labour
Intensification	Increasing land values declining cost of capital	Land ... capital
Mechanization	Declining cost of capital increasing wages	Labour ... capital
Expansion	Increasing wages	Labour ... land
(b) Sparsely populated countries		
Expansion	Increasing wages declining land values	Labour ... land
Mechanization	Declining capital costs increasing wages	Labour ... capital
Intensification	Increasing land values declining costs of capital	Land ... capital

Source: Herlemann and Stamer (1958).

TABLE 14.2

Per capita consumption and proportion spent on food: international comparison[a]

Per capita consumption (U.S. $/annum)	Percentage spent on food[b]	No. of countries in sample
Under 250 (148)[c]	48.8	12
250—499 (405)	39.5	6
500—749 (557)	34.0	4
750—999 (849)	32.3	4
1000—1249 (1113)	27.0	4
1250—1499 (1386)	26.0	4
over 1500 (1928)	22.4	5

Source: FAO (1972).
[a]Figures for 1965.
[b]These are unweighted averages which are only illustrative and not statistically valid.
[c]The unweighted averages for the group are given in brackets.

demand for food declines with increasing per capita income[3]. In fact, the level of income is probably the most important factor in determining the income elasticity of demand, as shown in Table 14.3. Income elasticities of demand are different for different types of food, specifically for foods of animal and plant origin, as shown in Table 14.4. This is an important factor in explaining differences between the pattern of livestock development and the development of crop farming.

The effect of growing per capita incomes on the demand for agricultural products has been described by Grigg (1974) in a historical perspective.

TABLE 14.3

Per capita gross national product and income elasticities of demand for food: international comparison, 1957—1959

	Per capita GNP[a]	Income elasticities of demand for food[b]
Asia and Far East (excl. Japan)	165	0.9
Near East and Africa (excl. South Africa)	260	0.7
Latin America (excl. Argentina and Uruguay)	491	0.6
Japan	910	0.6
Mediterranean Europe	575	0.55
European Economic Community	1235	0.5
Other Western Europe countries	1440	0.2
North America	2190	0.16

Sources: Mellor (1970), after FAO, *Agricultural Commodities Projections for 1970*; FAO, *Commodity Review 1962*, Special Supplement.
[a]In U.S.$ at 1955 prices according to purchasing power parity.
[b]In terms of farm-gate prices.

[3]Total household expenditures are taken here to be synonymous with total household income, but for a more detailed analysis this relationship would need to be qualified (FAO, 1972). Similarly, the difference between demand parameters in terms of farm-gate prices and consumer prices is neglected here.

Chapter 14 references, p. 328

TABLE 14.4

Income elasticities of demand for major food groups[a]

	Plant foods[b]	Animal foods[c]	All foods
North America	−0.03	0.23	0.16
Oceania	−0.06	0.14	0.10
European Economic Community	0.10	0.57	0.47
Mediterranean countries	0.18	0.90	0.55
United Kingdom	0.01	0.28	0.24
Japan	0.20	0.94	0.58
Argentina, Uruguay	0.04	0.22	0.17
Latin America (excl. Argentina and Uruguay)	0.03	0.60	0.47
Near East and Africa (excl. South Africa)	0.40	1.16	0.68
Asia and Far East (excl. Japan)	0.62	1.49	0.89

Sources: Mellor (1970), after FAO, *Agricultural Commodities Projections for 1970;* FAO, *Commodity Review 1962*, Special Supplement.
[a]Expressed in terms of quantities.
[b]Cereals, starchy pulses, sugar, vegetables, fruit and nuts.
[c]Milk, milk products (excluding butter), meat, eggs and fish.

'Real income per capita grew steadily in western Europe and the European areas overseas in the second half of the nineteenth century, and this led to a growing demand for meat, milk, fruit and vegetables, as well as products such as coffee, cocoa and tea which had hitherto been luxuries.... Nor did economic growth affect food products alone, for early industrialization was based upon textiles, and this prompted not only the expansion of cotton growing in the United States — and during the American Civil War in many other parts of the world — but also wool production in Australia and Argentina and jute in India. Later other agricultural products entered international trade. Vegetable oils for soap, candles, lubricants and margarine began to be important in the later nineteenth century, prompting the increased cultivation of groundnuts, oil-palms and coconuts, and towards the end of the century demand from the motor car and electrical industries led to spectacular growth of rubber plantations in South East Asia.' (Grigg, 1974 pp. 47—48).

Both historical and cross-sectional data indicate that income elasticities of demand for food decline with the growth of per capita incomes and that there are basic differences in the demand for plant and animal products. A model can be put forward of an economy in which per capita income is increasing (i.e., which is experiencing growth) and in which the demand structure for food is changing in the typical way.

The basic assumptions of this model are that at the lowest income level 60% of income is spent on food (calculated in terms of farm-gate prices), with 10% of this spent on livestock products. At the middle income level, only 35% of income is spent on food (though in absolute terms this is a larger amount than that spent by the lowest income group), with 33% of this spent on livestock products. These hypotheses are illustrated in Table 14.5.

In terms of absolute quantities, per capita meat consumption for the lowest income group would be about 10 kg/annum, and annual milk consumption about 25 kg. For the middle-income group, per capita meat consumption would be about 25 kg/annum, with a correspondingly increased milk consumption. With these basic consumptions, plus an annual growth rate of per capita income assumed at 5% and a population growth rate assumed at 2%, the income elasticities of demand can be calculated, as well

TABLE 14.5

Hypothesized values of factors determining the demand for food in less developed countries[a]

	Lowest-income group	Low-income group	Medium-income group
Population (million)	10	10	10
Per capita income (U.S.$)	200	500	1000
Proportion of income spent on			
all food	60.0%	50.0%	35.0%
livestock products	6.0%	10.0%	12.0%
other food	54.0%	40.0%	23.0%
Income elasticities of demand[b] for			
all food	1.0	0.6	0.4
livestock products	2.0	1.2	0.8
other food	0.9	0.5	0.2
Growth rate of demand[c] for			
all food	7.0%	4.3%	2.8%
livestock products	12.0%	8.5%	5.7%
other food	6.5%	3.2%	1.4%
Actual demand (million U.S.$) for			
all food	1200	2500	3500
livestock products	120	500	1200
other food	1080	2000	2300
Increase in demand[d] (million U.S.$) for			
all food	84.6	106.5	99.0
livestock products	14.4	42.5	68.0
other food	70.2	64.0	31.0

Source: Compiled by the author.

[a]The proportion of income spent on different food items is interrelated with the elasticities of demand for the individual items, and these interrelationships are accounted for in this Table. However, the figures are rounded so cross-checks may not add up. All figures are based on farm-gate values.

[b]The elasticity coefficients (e) calculated here measure the response of demand to changes in income from one year to the next, defined as

$$e_i = \frac{dy_i}{y_i} : \frac{dx}{x} = \frac{dy_i}{dx} : \frac{y_i}{x}$$

This represents the percentage change in expenditure (y_i) on the ith product in response to a 1% change in income (x). No assumptions are made about the form of the Engel curve implied.

[c]A growth rate of per capita income is assumed at 5% and a population growth rate at 2%/annum. These are simply added for an estimated growth rate of demand.

[d]This is the increase from the year represented in the Table to the following year. Further increases would be higher because of the compounding effect.

as the absolute demand for all food, for livestock products and for other foods.

Table 14.5 reveals the potential for growth in the demand for livestock products. At the same time, it points out the fact that in the early stages of development the importance of livestock products in overall domestic consumption is very small. This model is of a closed economy, so no account is taken of external trade factors. In addition, two important deviations occur from this pattern of domestic demand which should be noted.

In the first place, in some societies at the beginning of the development process the proportion of animal products in total food consumption is much higher than indicated in the model. The pastoral peoples of Africa are a particularly extreme example: there are instances of pre-development societies whose food consumption consists entirely of animal products. Dahl and Hjort (1976) estimate that an African pastoral household of six persons needs 318 g of protein and 13 800 kcal a day, equivalent to 116 kg of protein and 5 million kcal annually. This corresponds to an annual per capita consumption of 600 kg of milk and 185 kg of meat. In such a situation, development is likely to be accompanied by a reduction in the proportion of animal products in the diet, i.e., there will be a negative income elasticity of demand for animal products.

Less extreme examples of this situation are found in certain South American countries, Australia and New Zealand. In 1968, von Oven estimated annual per capita meat consumption at 83 kg for Argentina, 77 kg for Uruguay, and 54 kg for Paraguay. Guadagni (1964) and Guadagni and Petrecolla (1965) found a very low income elasticity of demand for meat in Argentina, 0.3, with a per capita income of U.S.\$ 670. This deviates significantly from the model calculations given in Table 14.5.

A second deviation from the model occurs in countries, such as Japan, where the demand for animal products increases only slightly in spite of rapidly growing per capita income. In 1975, per capita income in Japan was estimated at U.S.\$ 3600, but per capita milk consumption was only 73 kg and meat consumption 16 kg/annum.

These two deviations concern the quantities of animal products consumed in terms of per capita income. If expenditures on animal products are compared, the deviations will probably be less striking. In other words, in the pastoral societies of Africa and the beef-producing countries of South America the per capita consumption of animal products may be high, but expenditures on these products are relatively low. The converse may be true for Japan. Thus, while cultural factors undoubtedly play a role in the consumption of animal products, broad deviations may well reflect different price situations, as well as the availability of substitutes. For a more thorough analysis, demand elasticities would have to be considered in terms of quality as well as quantity, and the problems of aggregation and the mathematical specifications would have to be dealt with in more detail. An introduction to this problem complex is given in FAO (1972).

3.2. External trade and transport technology

The discussion so far illustrates the limits on the expansion of livestock production set by internal demand in a closed economy. Even in South American countries where beef production can be increased at very low opportunity costs, thus keeping prices low and allowing a high level of local consumption, production would not expand substantially if it were based on internal demand alone.

In the real world, however, production generally does not depend solely on internal demand. Nevertheless, in the first half of the 19th century, Thünen developed his theory of a spatial differentiation of agriculture as a function of distance from the market. The underlying assumptions were that the economy was a closed one with one central market only, that there were few purchased factors of production, and that high transport costs (particularly overland transport) were of such overriding importance that at a certain distance from the town commercial agriculture would stop altogether because the cost of transporting the produce to market would be prohibitive. A summary of Thünen's theory is given in Herlemann (1961). Since Thünen's time, his model has been completely superseded by the industrial revolution, substantial population growth, and the accompanying trends of national and international migration and urbanization. Thus in 1800, only 2.5% of the world's population lived in towns of 20 000 or more, but by 1900, 10% lived in towns of this size, mostly concentrated in Western Europe and the eastern seaboard of the United States. For the first time in history, a substantial proportion of the world's population did not produce its own food. One consequence of this was a great increase in international trade: in the eighteen-fifties, world food exports were less than 4 million tonnes/annum, but this figure had quadrupled by the eighteen-eighties, and by the eve of World War I annual food exports were nearly 40 million tonnes. Over this period, the trade in livestock products (beef, veal, pork, bacon, ham, mutton, lamb and butter) increased from 100 000 tonnes to 1.5 million tonnes (Grigg, 1974, p. 42).

This dramatic expansion of the world food trade was based on advances in transport technology which resulted in substantial reductions in freight costs. Ocean freight rates fell during the first half of the 19th century, and again later in the century when larger ships were built of steel, rather than wood, and powered by steam, rather than sails. Grigg writes

...Whereas the freight on a bushel of wheat from New York to Liverpool was 21 cents in 1873, it had fallen to 3 cents by 1901, and over the same period freights on wool from Australia to England were halved (Grigg, 1974, p. 48).

The fall in overland freight rates was even more spectacular, as the railways were extended after 1850: rail freight rates fell throughout the century to a quarter or less of their original levels. This made it possible to expand agricultural production in vast areas of North America, Southern Brazil, Argentina, Australia and Siberia.

Grigg points out that trade in animal products other than wool developed more slowly, being dependent on advances in canning and refrigeration:

'In the 1860s ice-boxes were used on American railways to transport meat, and refrigerated cars were common by the 1880s, so that dressed meat rather than live cattle could be moved. The meat canning industry was established in the mid-west in the early 1870s, and again reduced the weight to be moved. Refrigeration was subsequently introduced into ships, and the first chilled beef reached England from New York in 1875 and Le Havre from Buenos Aires in 1877. Frozen meat, kept at −10°C as against −1°C for chilled meat, reached London from Australia in 1879 and from New Zealand in 1882. The application of refrigeration to dairy products came a little later, but by the 1890s New Zealand butter was reaching London.' (Grigg, 1974, p. 49).

As a consequence of these developments, large areas of North America, South America and New Zealand were brought into production to meet a good proportion of the world's demand for meat. Given the growing demand for agricultural products due to growing populations and rising per capita incomes, improved transport and processing facilities meant that unused resources and resources of low opportunity costs far away from market centres could be brought into production and natural advantages could be exploited almost anywhere in the world.

No longer limited by the constraints of internal demand, livestock production, particularly of sheep and cattle, often served as the first enterprise opening up large empty spaces in the new world. At the same time, the availability of cheap grains from the areas of North America where production costs were low radically changed the nature of agricultural production on the eastern seaboard and in the European countries which did not limit imports. Developments in Denmark and Holland are typical. No longer competitive in grain production and increasingly orientated towards the expanding markets in the new industrial centres, these countries began to specialize in intensive farming systems, producing dairy products, pork, fattened beef, poultry and horticultural crops. In such economies, animal production developed as a strong and independent enterprise on the basis of cheap grain imported from North America (see Herlemann, 1961, p. 128).

3.3. Farming technologies

The expansion and development of agricultural production has been described as largely a result of the growth of world trade. The development of farming technologies and the increasing division of labour between agriculture and industry also had a major impact on the development of more intensive farming systems and on the role of livestock.

Before the onset of modern economic development, livestock supplied subsistence products which were particularly important in dry areas which could not support cropping on a sustained basis. The main food product in subsistence livestock production systems has always been milk, but the importance of non-food products — such as hides, skins, wool and horns — should not be under-estimated, particularly before the widespread use of synthetic fibres, rubber and plastic. In more humid areas, a major role of livestock has been the provision of draught power and manure for crop production. In addition, livestock can utilize fallow land, crop residues and slack labour to produce food and other products, which can be looked upon as by-products of a crop-farming system.

Grigg breaks down the development of farming technology into four main areas of innovation — the application of mechanized power to agricultural production (first steam and then fossil fuels), the introduction of new implements, the use of chemical fertilizers, and improved plant and animal breeding. With a growing division of labour between agriculture and industry, agricultural products have been increasingly processed in factories, while industrial production has supplied farms with fertilizers, coal for steam-driven implements and by-products for animal feeds. Between 1860 and 1910, cheese — and to a lesser extent butter — production was transferred from the farm to the factory, whilst fluid milk processing was transformed with the introduction of railways, cooling machines, pasteurization and the rise of large wholesale milk companies.

These developments in Europe, North America, Australia and New Zealand gave rise to an unprecedented increase in productivity, creating at the same time a wide gulf in production levels between these areas and much of Asia, Africa and Latin America. Table 14.6 gives an indication of the magnitude of this gap by 1975.

Table 14.7 demonstrates in more detail the differences in productivity and efficiency between developed and developing countries in terms of ruminant production. Not only have the volume and efficiency of meat and milk production increased dramatically in the developed countries, but the whole role of livestock within the farming system has changed substantially. The different roles of livestock within a traditional agricultural production system can be described as follows:

(1) supplying animal products for human consumption or use, such as meat, milk, wool, hides, skins, horns and dung for fuel, either for direct consumption or sale;

(2) balancing the farm enterprise and making it more profitable by spreading production and marketing risks and by utilizing inputs with low opportunity costs, such as residual labour and 'absolute feedstuffs' (which must be converted by animals to be of use to man);

TABLE 14.6

Meat productivity in kg/animal for 1975

Region	Cattle and buffaloes	Sheep and goats	Pigs	Poultry
World	33.9	4.9	63.0	3.5
Temperate zone	61.1	5.3	70.3	4.4
Tropical and subtropical zone	13.3	3.8	30.1	1.7
Industrial economies	76.4	6.2	102.5	7.3
Centrally planned economies	44.5	5.8	56.5	2.4
Developing economies	14.3	3.8	31.3	1.6

Source: Horst and Peters (1978), from FAO, *Production Yearbook 1975*.

TABLE 14.7

Productivity and efficiency of ruminant livestock production, 1972

Indicator	Developed countries	Developing countries	
		India	Others
Carcase meat production (kg/LU[a])	82	3	25
Milk production (kg/LU)	837	112	99
Physiological fuel value[b] (Mcal³/LU)	714	100	123
Physiological fuel value/ metabolized energy (%)[c]	11.9	1.8	2.7

Source: Winrock International (1978).
[a]LU = Livestock unit, the equivalent of 1.0 buffalo, 0.8 cow or 0.1 sheep or goat.
[b]Of meat and milk in human diet.
[c]Metabolized energy: estimated availability from arable land and permanent pastures.

(3) serving crop production directly by supplying traction for cultivation, power for other activities such as threshing, and dung; and indirectly by inducing rotation cropping systems which increase soil fertility;
(4) providing transport; and
(5) providing a means of saving and investment through the inherent storage and reproduction capacity of animals (Jahnke, 1978)

If economic development implies an increasing demand for food of animal origin, the substitution of manufactured goods for many natural products, the introduction of technical innovations into agriculture such as chemical fertilizers and mechanized power, and the creation of savings institutions, then it is clear that the functions of livestock enumerated here will change. In general, it is possible to say that with economic development the multiplicity of livestock functions is reduced, the integration of livestock production with cropping is also reduced, and differentiation and specialization develop to a point where animals are kept solely to produce a specialized product for the market. In this connection, Duckham points out:

> ...We should not forget that livestock were originally 'jacks of all trades' and that undue specialization in one aspect, in food production e.g., may have considerable repercussions on the remaining, but still important functions.
> ...We should remember that livestock still play a significant role in our personal social lives and that we must guard against confusing this perfectly legitimate 'cultural' purpose with the primary, more material and strictly dollars-and-cents functions of domestic farm animals (Duckham, 1946).

Chapter 14 references, p. 328

4. SOME EXAMPLES OF LIVESTOCK DEVELOPMENT

4.1. Sparsely populated dry areas — the example of Australia

4.1.1. Historical development [this section draws heavily on Peel (1973)]

The Australian continent of almost 8 million km² was practically unin-habited up to the end of the 18th century. Domestic livestock were intro-duced with the first European settlers:

> The live animals we took on board on the public account from the Cape [of Good Hope] for stocking our projected colony, were two bulls, three cows, three horses, forty-four sheep and thirty-two hogs, besides goats and a very large quantity of poultry of every kind (Tench, 1789, quoted in Peel, 1973, p. 41).

Today, less than 200 years later, there are over 175 million sheep and about 30 million cattle in Australia, and the share of livestock in total agricultural production is over 60% [this was the average for 1968—1970, according to Abercrombie et al. (1978)]. At the same time, the population enjoys one of the highest living standards in the world. For these reasons, Australia presents a unique example of livestock development and its role in overall economic development.

Up to 1820, the primary purpose of livestock husbandry in Australia was to produce meat for the small, but rapidly growing, population, and little attention was given to other animal products. About this time, the first wool exports were sent to England, and it was realized that this trade would be profitable for the better qualities of wool, in spite of high transport costs. At the same time, free settlement and investment were increasing in New South Wales: the nature of the colony was changing from a penal colony to an agricultural settlement, with free farmers, businessmen and emancipated convicts, as well as those still serving sentences and the military (Peel, 1973, p. 44). This change in the population was accompanied by a rapid expansion of livestock production, supported to some extent by a rapidly growing local demand for meat, but even more by the international demand for wool. In 1813, there were about 50 000 sheep in Australia, by 1860 there were 21 million, and by 1890, there were 100 million. Wool was first exported in 1807; exports reached 2.5 million lb. in 1830, 35 million lb. in 1860 and 641 million lb. in 1891 (Grigg, 1974, p. 251).

Given the magnitude of the export market, sheep and wool production became the leading sector in the Australian economy. The expansion of the textile industry in Britain sharply increased the demand for wool, and wool was also one of the few products which was non-perishable and whose unit value was high enough to bear the substantial transport costs. However, expansion of the industry in Australia was not accompanied by any major technological breakthroughs. Types and quality of wool were improved through selective breeding, and shearing equipment was mechanized.

Up until the eighteen-sixties, animals were shepherded with a minimum of labour on what was, in effect, an open range, much as in the American West. Following the discovery of gold at that time, rural labour became more expensive. As a result, sheep 'runs' were fenced, stock was improved and watering facilities extended. Far-reaching changes in land tenure were introduced and the larger graziers responded by formalizing their control over vast holdings through semi-legal interpretations of the legislation.

Wool exports benefited from advances in land and sea transport tech-nology and the resulting lower transport costs. Refrigeration also opened up an export market for Australian meat: 'With the successful shipment of

40 tonnes of frozen beef and mutton from Australia to London in 1879, the markets of the world were opened to Australian meat and dairy products' (Peel, 1973, p. 58)[4].

Although the acreage under crops increased between the eighteen-sixties and eighteen-nineties, agriculture in Australia was still dominated by animal production. Then a series of disasters struck — a rabbit population explosion, the collapse of international wool prices and severe droughts — which led to a number of bankruptcies and the subdivision of large holdings. At the same time, other long-term trends were taking place. By the eighteen-nineties, wheat production had spread across a great arc from South Australia to Queensland. Cattle and sheep ranches moved westwards, where innovations such as artesian bores allowed the expansion of settlement and grazing. Mixed farming operations also increased in the higher rainfall areas, based on technological advances such as the application of phosphate fertilizers, the use of subterranean clover to fix nitrogen, the invention of centrifugal cream separators (1883), the introduction of the Babcock test for quick determination of the fat content in milk (1892), and the development of fat lamb breeding techniques.

The role of livestock production as a pioneering enterprise in Australia parallels the development of the American West. The same sequence of land-use patterns ensued, with wheat production and mixed farming becoming predominant in the areas with higher rainfall, and ranching concentrated in the drier areas. According to Peel (1973, p. 72), the history of commercial ranching in Australia was shaped by the environment and economic and cultural factors: 'Geared essentially to distant export markets, its development reflects the abundance of land and scarcity of labour and capital on which Australia's economic growth was initially based.'

4.1.2. Present productivity levels

Ranching in Australia is concentrated in an arid to semi-arid zone with rainfall averaging 75—375 mm/annum. This zone is defined by the Bureau for Agriculture Economics (BAE). The statistics that follow are based on various surveys of the livestock industry carried out by the BAE in the nineteen-sixties and nineteen-seventies. These surveys have been analyzed and summarized in Beresford (1977). Within this zone, three regions may be distinguished: the higher-rainfall areas of Queensland Central, Queensland Inland North and New South Wales Plains, the medium-rainfall areas of Western Australia, Kimberley, the Northern Territory (Barkeley Tableland and Victoria Rivers) and Western Queensland, and the driest areas, such as Alice Springs and the northern part of South Australia. In each of these regions, sheep production predominates in the last place mentioned.

Table 14.8 gives some indications of the levels of productivity achieved in Australia's cattle producing areas. In particular, this Table illustrates the high level of labour productivity achieved, with one man looking after anywhere from 320 to over 800 cattle grazing on 3000—70 000 ha of land. Meat productivity per man-year ranges from 11 to 21 tonnes, at an average farm-gate price of U.S.$ 0.65/kg carcase weight, assuming a liveweight of 250 kg. The gross value of production per unit of labour is also very high, from U.S.$ 2500 to 2700/employee annually at the time of the surveys. These labour costs account for 20—40% of total operating costs.

Table 14.9 gives some indications of the productivity of sheep ranching

[4]Meat from Argentina had only to be chilled because of the shorter distance, and it was preferred by consumers. It was only in 1934 that chilling techniques advanced to a stage where chilled meat could be exported to Europe from Australia.

TABLE 14.8

Beef cattle productivity indicators for Australia, 1968/1969—1970/1971[a]

Indicator	Higher-rainfall areas	Medium-rainfall areas	Driest areas
Stocking rates (ha/CE)[b]	8—12	30—40	100+
Livestock—labour ratios (CE/man-year[c])	320—570	670—810	about 705
Land-labour ratios (ha/man-year)	3000—7700	13 600—29 000	52 000—69 000
Cattle offtake rate (%)	21—34	12—19	26
Cattle offtake per ME	101—106	95—173	184
Value of offtake/ME (U.S.$)[d]	8000—8500	7600—13 800	14 700
Value of offtake/ha (U.S.$)	1.1—2.8	0.3—1.0	0.2—0.3

Source: Beresford (1977), based on the BAE livestock industry surveys.
[a]Range of 3-year averages for typical holdings in the different regions as defined by the BAE surveys. All figures rounded.
[b]CE = Cattle equivalent; equivalent to one adult bovine.
[c]Man-year: Equivalent to one person working 50 weeks/year.
[d]Converted from Australian $ at a rate of Aus$ 0.8929 = U.S.$ 1.0000. An average selling price of U.S.$ 80/animal has been assumed.

TABLE 14.9

Sheep productivity in Australia, 1972—1973

Productivity indicator	Western Australia	New South Wales	Southern Australia
Total wool production (kg/annum)	38 528	20 220	18 988
Wool production/ man-year (kg)	9397	10 010	8599
Gross income/man-year (U.S.$)[a]	14 035	21 549	21 214

Source: Own compilation after Beresford (1977), based on the BAE livestock industry surveys.
[a]Converted at a rate of Aus$ 0.7843 = U.S.$ 1.0000, as of 23 December 1972.

in three regions of Australia. This Table shows that income and labour produc-tivity are higher for sheep than for beef cattle production, but these figures may not reflect the long-run situation accurately because they were collected just after an 80% rise in wool prices.

The returns to capital, including land and livestock, appear to be rather low for beef ranching. Only the family-owned ranches of Alice Springs achieve an acceptable figure of 8.8%. However, it should be noted that these calculations can be misleading for at least two reasons: (1) because capital on these ranches consists largely of cattle and land, which are difficult to value, and (2) because cattle are an appreciating asset which can normally be built up at a much lower opportunity cost than the market cost of capital. (For this analysis, cattle were valued at between U.S.$ 72 and U.S.$ 108 a head. In most ranching areas the value of land is well below U.S.$ 1.00/ha: the lowest value was estimated for Alice Springs at U.S.$ 0.19/ha, and the highest was in the New South Wales Plains at U.S.$ 13.5/ha.)

The sheep industry, benefiting from the high wool prices of the early nineteen-seventies, showed rather good returns to capital at 8.6—18.0%. The capital invested per labour unit for both sheep and cattle ranching in the late nineteen-sixties and early nineteen-seventies was between U.S.$ 80 000 and U.S.$ 100 000.

4.2. Sparsely populated areas in the humid tropics — the example of South America
[this historical account closely follows Grigg (1974)]

4.2.1. Historical development

Ranching, which has been described as the 'child of the Industrial Revolution', emerged as a major agricultural system throughout the world only in the second half of the 19th century. Although ranching activities were initiated earlier in South America than in Australia, the development and expansion of the industry in South America came as a response to the same factors which spurred the development of ranching in Australia. These factors were the growth in the demand for beef and wool in the urbanized areas of the Eastern United States and Western Europe, the reduction of ocean freight rates, the introduction of refrigeration in ships and railway cars, and advances in meat-canning technology.

The Spanish and Portuguese introduced cattle and sheep into South America in the 15th and 16th centuries, along with ranching systems developed in Southern Iberia, which at that time was the only part of Europe with a ranching economy. Cattle ranching played an important role in the agricultural economy of South America from the beginning of European settlement, particularly in the opening up of new lands.

In Brazil, the first Portuguese settlements were sugar plantations on the northeastern coast, but in the late 16th century cattle ranchers moved into the *sertao* (inland savannahs) through the lower San Francisco valley. Here, large, though poorly demarcated, cattle ranches were established, in spite of the government's attempts to limit the size of holdings. Cattle were grazed on natural pastures, tended by semi-nomadic herdsmen. Once a year, the cattle were rounded up, the calves branded and those ready for market were driven to the coast, where their meat was sold to the settlers on the plantations. Their hides, which were shipped to Europe, were perhaps an even more important product.

Further south in what is now Argentina and Uruguay, the *pampas* (plains) provided ideal grazing for cattle. Until the early 18th century, feral cattle were hunted and their hides sold in Buenos Aires, and later in Montevideo. By the middle of the 18th century, much of the pampas had been carved up into estates, and hunting was replaced by herding. Hides remained the major product, though a market for dried meat was found on the Brazilian plantations.

Much further north, the *llanos* (plains) of Venezuela were unoccupied when the Spanish arrived in the mid-16th century. The alternation of floods and droughts made it difficult for the pre-Colombian population to cultivate this area, and they lacked grazing animals. A century after the first Spanish settlement, over 140 000 cattle were herded on the open range by semi-nomadic cowboys, and by the late 19th century there were over 8 million cattle on the llanos.

By 1978, the humid tropical and subtropical zones of South and Central America (including Belize, Guatemala, Costa Rica, El Salvador, Honduras, Nicaragua, Panama, Colombia, Venezuela, Guyana, Surinam, French Guiana, Ecuador, Peru, Paraguay, Bolivia and Brazil) had a cattle population of about 155 million. Brazil alone accounted for 95 million. There were 55 million sheep (25 million in Brazil) and 23 million goats (16 million in Brazil) (FAO, *Production Yearbook 1976*). Livestock products account for 30—40% of the gross national agricultural output of the countries in this area [Abercrombie et al. (1978) give average percentages for 1968—1970 of 40% for Paraguay, 38% for Venezuela, 34% for Colombia, 33% for Brazil, 31% for Costa Rica, and 29% for Guatemala]. This is high in terms of the general level of development in these countries and in terms of per capita incomes. It indicates a rich endowment of natural grazing resources, relatively low livestock production costs and the major role of livestock production for export.

4.2.2. Present productivity levels

Von Oven (1968) studied beef production and productivity levels from Colombia and Venezuela in the north, southwards through Bolivia, Paraguay and Uruguay to Argentina. He found that cattle production was concentrated in three large ecological zones.

(1) The relatively humid plains and rolling uplands of the temperate climate zone. This region includes the pampas of Argentina, Uruguay and the southern part of Rio Grande do Sul in Brazil.

(2) The subtropical and tropical savannahs and parklands bordering the first zone to the north. This area includes Northern Argentina, Paraguay, eastern Bolivia and large parts of Southern and Central Brazil.

(3) The fluvial plains and coastal areas in the north of Colombia and Venezuela.

For the purpose of characterizing the economics of cattle production, these zones can be amalgamated into two: the temperate zone (e.g., Uruguay and the pampas of Argentina), and the tropical and subtropical zone (e.g., Northern Argentina, Paraguay, Bolivia, Colombia and Venezuela). This discussion will focus on the tropical and subtropical zone. Ranching operations can then be distinguished according to their primary production orientation, i.e., breeding, fattening or various combined operations.

Table 14.10 shows that variations are considerable in stocking rates, meat production, labour productivity and capital intensity. Van Oven (1968) distinguishes among five levels of intensity according to the husbandry techniques used, from a branding/castration/bull purchase package, to a management system with complete control of grazing, supplementary feeding and improved breeds. The more intensive forms of production can be neglected in this discussion, however, because they are found only in the temperate zone.

Productivity and labour intensity figures, compiled for Brazil and presented in Table 14.11, generally agree with von Oven's figures for tropical and subtropical South America. The calculations for Brazil further indicate the significantly higher labour intensity of dairying and pig production.

Tables 14.10 and 14.11 reveal some basic differences between livestock production in tropical South America and in the dry areas of Australia. The most striking difference is carrying capacity: in South America this varies from 1 to 4 ha/animal, while in Australia it varies from about 10 to 100 ha. In other words, the productivity of the land is much higher in the humid areas, as would be expected. The dry areas of Australia, however, show much higher labour productivity, only approached by the breeding and extensive fattening operations carried out in the somewhat drier areas of South America. However, the cost of labour in Australia is about five times higher than in South America, which indicates a higher stage of economic development as well as pressure to increase labour productivity by increasing inputs of land and capital.

4.3. Densely populated temperate zones — the example of Germany

4.3.1. Historical development [this section is based on Reinhardt (1974)]

Germany was a densely populated country at the beginning of the Industrial Revolution. Livestock production increased to keep pace with a growing population and increases in per capita income, but it did not play a role in

TABLE 14.10

Productivity and labour intensity indicators for cattle production systems in tropical and subtropical South America, 1978

Indicators	Type of production			
	Breeding	Breeding and extensive fattening	Breeding and intensive fattening	Fattening
Rainfall (mm)	1800	1000—1700	1800—1900	1700
Stocking rate (ha/CE)[a]	1.4	3.6	1.0	1.3
Offtake	25%	16%	20%	28%
Meat production (kg liveweight/CE)	95.0	73.6	115.3	123.9
Meat production (kg liveweight/ha)	70.6	20.4	114.9	163.6
Labour intensity (ha/man-year[b])	37.7	933.8	70.5	71.5
Labour intensity (CE/man-year)	28.0	273.1	70.3	94.4
Labour productivity (kg liveweight/man-year)	2660	20 100	8100	11 703
Meat prices (U.S.$/kg liveweight[c])	0.28	0.20	0.30	0.30
Labour cost (U.S.$/man-year)	473	606	490	421
Capital investment[d] (U.S.$/man-year)	7361	31 247	19 207	30 536

Source: compiled by the author from van Oven (1968).
[a]CE = Cattle equivalent.
[b]Including family labour.
[c]Unweighted averages, to be considered as rough equivalents only.
[d]Including land and cattle.

TABLE 14.11

Productivity and labour intensity indicators for livestock production systems in Brazil, 1962—1964

Indicators	State and main activities					
	Minas Gerais		Rio Grande do Sul		Sao Paulo	
	Beef	Dairy	Beef	Pigs	Beef	Dairy
Ave. size of holding (ha)	595	211	726	71	397	195
Livestock and livestock products in gross output	75%	72%	82%	65%	94%	75%
Stocking rate (ha/LU[a])	10.0	2.0	3.3	1.1	1.4	1.4
Labour intensity (ha/man-year)	113.6	37.9	147.0	6.7	70.0	36.1
Labour intensity (LU/man-year)	13.0	18.5	51.1	5.8	42.4	23.9

Source: Fundaçao Getulio Vargas (1974).
[a]LU = Livestock units; one bovine = 0.8 LU, one pig = 0.2 LU.

opening new lands, creating a surplus in the economy or laying a basis for export-led expansion. Increased livestock production came as the result of the introduction of new technologies and the changing role of livestock in the farming system.

Up until the middle of the 19th century, the most important role of livestock in the farming system in Germany was the contribution of animals to crop production, directly through the supply of draught power and manure, and indirectly through rotation with fodder crops which increased productivity. Over the past 100 years, this role has become practically non-existent. Animal draught power has been replaced by mechanical power, manure has been replaced by chemical fertilizers and modern agronomic practices, and crop rotation systems have been devised which rarely include fodder crops.

Originally, livestock also played an important role in diversifying the farm enterprise in order to balance risk. This was possible because the yields and prices obtained for plant and animal products varied independently of each other. The importance of this role decreased, however, in the process of specialization. With modern technology, the risks involved in crop production were reduced, and the larger enterprises which were established were in more secure positions in the market.

The role of livestock in respect to farm labour has also changed completely. Farm animals have evolved from a labour-saving role, to labour balancing, and eventually to providing an outlet for residual labour capacity. However, the evolution has been slightly different for dairy enterprises. In the 19th century, dairying was the most productive agricultural enterprise in Germany in terms of labour and land, taking into account the value of manure. In the course of development the relative advantage of dairying was reduced, and labour on large farms in particular could be more productively applied to crop production. For smaller farms, dairying is still an attractive enterprise, though working conditions are not favourable.

Another dramatic change occurred in the role of livestock in food production. Originally, the role of animals was to convert absolute feedstuffs (grass) into food which could be consumed by people. With increasingly intensive land use and rising incomes, however, animals have assumed a 'food upgrading' role, converting grain to animal protein. This trend has been accentuated by the fact that demand for animal products has grown much faster than demand for plant products, so that the prices obtained from animal products have increased disproportionately. Also the price of absolute feedstuffs has increased relative to the price of grain.

When farmers began feeding grain to their livestock rather than grass and fodder crops, animal production was significantly expanded. Economies of scale were also achieved as animal production units were enlarged. Both of these trends have resulted in a situation where consumers enjoy relatively decreasing prices for livestock products in terms of their rising incomes. The process of technological change and increased efficiency has been accentuated since World War II during a period of unprecedented economic expansion and rising per capita incomes. On average, agricultural wages and parity incomes have increased 10 fold over the 20 years from 1955/1956 to 1975/1976, as shown in Table 14.12. Purchasing power for agricultural products has increased four fold, for example 2.8 times for milk and 6.75 times for poultry.

4.3.2. Present productivity levels

The process of rapid economic growth is still underway, with its effects on livestock production. The livestock industry in Germany can best be described according to a number of stages which have followed each other over time but which, to varying degrees, also still occur alongside each other.

Dairy production, for example, has developed in response to a rapidly growing effective demand based on increased purchasing power. New technologies have been developed for grazing and forage harvesting, and for feeding, demanuring and milking. These technologies have been accompanied by economies of scale, demonstrated by the fact that overall cow numbers have remained constant since 1950 while the average number of cows on a holding has doubled. These developments have made possible a reduction in the average capital requirements per cow. The trends are illustrated in Table 14.13.

TABLE 14.12

Purchasing power of agricultural wages in Germany (F.R.), 1955/1956—1975/1976

	1955/1956	1965/1966	1975/1976
Wages (U.S.$/h)	0.36[a]	0.92[a]	3.15[c]
Corresponds to ... kg of			
Cereals	3.70	8.89	17.22
Sugar	21.50	48.30	87.45
Milk	5.00	9.06	13.95
Slaughter cattle	0.74	1.51	2.28
Slaughter sheep	0.87	1.71	2.67
Slaughter poultry	0.60	1.57	4.05
Eggs	0.37	1.10	3.35
Mixed cattle feed	3.92	8.25	14.57
Mixed pig and poultry feed	3.45	6.92	12.69
Parity income[b] U.S.$ p.a.	955.50[a]	2423.75[a]	9316.80[c]

Source: Reisch (1977).
[a]Conversion rate: DM 4.0 = U.S. $1.0.
[b]Statistically determined target family income for agriculture.
[c]Conversion rate: DM 2.5 = U.S.$ 1.0.

TABLE 14.13

Technical progress in German dairy production since 1950

Production factor	Typical level of technology and period in use[a]				
	A Up to 1955	B 1955—1965	C 1965—1970	D 1970—1975	E Since 1975
No. of cows on holding	5—8	15—20	20—30	40—60	20—90
Labour input[b] (man-hours cow^{-1} year^{-1})	210—250	110—150	70—90	55—70	40—50
Capital input[c] (U.S.$/cow)	2900	2500	2460	2500	2480
Milk yield (kg cow^{-1} year^{-1})	3000	3600	3900	4200	4500
Meat yield[d] (kg cow^{-1} year^{-1})	150	150	150	150	150
Labour productivity (kg milk/man-hour)	12—14	24—33	43—56	60—76	100—120

Source: Reisch (1977).
[a]Level of technology derived from housing, feeding, demanuring and milking practices.
[b]Includes labour for stall-keeping, summer feeding and winter feed production and conservation.
[c]Includes costs of buildings and equipment. Derived from 1976 prices in DM, converted at a rate of DM 2.50 = U.S.$ 1.0.
[d]From culled cows and calves.

Intensive beef cattle production has also developed rapidly over the same period, as shown in Table 14.14. Milking is a relatively labour-intensive activity, even with modern technological innovations, so that the reduction in labour input for beef production has been even more dramatic than for dairying. With increasing herd size, mechanization and modern stabling systems, the annual labour input has fallen from 100 to 13 man-hours/ animal and, correspondingly, labour productivity has increased to over 50 kg liveweight/hour worked. Similar developments have taken place in other areas of animal production, for example in pig production where enterprises have specialized in breeding or fattening with very high levels of labour productivity, but also very high management requirements.

Chapter 14 references, p. 328

TABLE 14.14

Technical progress in intensive beef fatting in Germany since 1950

Production factor	Typical level of technology and period in use[a]			
	A Up to 1955	B 1955—1965	C 1965—1975	D Since 1975
No. of animals on holding	10—15	15—25	20—30	80—200
Labour input[b] (man-hours animal^{-1} year^{-1})	70—115	50—75	20—25	10—13
Capital input[c] (U.S.$/animal)	1800	1440	1320	1500
Meat yield (kg liveweight/animal)	550	550	550	550
Labour productivity	5—8	7—11	22—27	46—55

Source: Reisch (1977).
[a]Level of technology derived from housing, feeding and demanuring practices.
[b]Includes labour for stall-keeping, summer feeding and winter-feed production and conservation.
[c]Includes costs of buildings and equipment. Derived from 1976 prices in DM, converted at a rate of DM 2.50 = U.S.$ 1.0.

Overall, the development of livestock production in Germany since World War II can be characterized by the increased size of holdings, the transition from diversified to specialized enterprises, intensification of land use, increasing capital intensity, and a declining availability of labour. In general, small farms have fallen behind in terms of income and productivity. Thus, in 1975/1976, farms of less than 20 ha with a fodder area of about 75% achieved a labour productivity level of 16 000 DM/man-year and a labour income of 12 500 DM/man-year. Corresponding figures for middle-sized farms of 20—50 ha were 26 000 and 21 000 DM/man-year, and for large farms 36 000 and 27 500 DM/man-year, which is a level of labour productivity and income 2.2 times higher than for small farms (these figures refer to farms in Lower Saxony and Bavaria).

With respect to the farming enterprise as a whole, small farms achieved an average annual income of DM 25 000 and profits of DM 20 000/man-year, while large farms achieved an average annual income of DM 78 000 and profits of DM 62 000/man-year, three times that of the smaller farms. There was no great variation in capital input/ha; this ranged from DM 9000 to 10 000 for all farms. However, there was a great discrepancy in capital input in relation to labour; the larger farms showed a capital input of DM 25 000/man-year, more than twice the capital input of the smaller farms.

5. CONCLUSIONS AND IMPLICATIONS FOR LIVESTOCK DEVELOPMENT IN AFRICA

Three examples of livestock development have been discussed: in Australia, tropical and subtropical South America, and in Germany. The low, erratic rainfall in the livestock production areas of Australia is comparable with climatic conditions in many parts of Africa and the Middle East. However, livestock development in the dry areas of Australia has been characterized by a relative abundance of land and scarcity of labour, while the situation in most of the dry areas of developing countries today is the reverse. In these areas, it is generally not technically feasible to increase livestock production by intensification. Increased production must first be brought about through expansion, i.e., the proportion of land must increase relative to labour. Since there is little unoccupied land available for such an expansion, the shift in the production factors must come about through an out-migration of the population from these areas. A relative increase in the land under production, compared to population, will allow an improvement in labour productivity which will eventually lead to further development through mechanization. To the extent that this development path is not available to African countries for social, economic and political reasons, the dry pastoral areas of Africa are likely to remain waiting rooms of development.

One of the most important developments which allowed the livestock industry in Australia to become a leading sector in the economy was innovations in the handling and transport of livestock products which, coupled with low production costs, allowed Australian products to compete successfully on the world market. This market is not readily available to the arid production zones of Africa because of stringent disease regulations governing world trade. The Sahelian countries benefit from trade opportunities with the West African coastal areas where disease restrictions are less rigid, and Ethiopia, Sudan and Somalia have similar opportunities in the Near East. Botswana and Kenya, on the other hand, have carried out expensive veterinary measures in order to gain access to the European market. Production technology consists of a fairly standard package of disease and fire control and the provision of water and dipping, which has remained much the same over time. It is unlikely that the livestock production sector in the dry parts of Africa will achieve any further major production breakthroughs.

The humid areas of Africa are similar ecologically to the tropical and subtropical areas of South America, and both areas are characterized by low population densities and a low intensity of land use. However, these areas in Africa are subject to trypanosomiasis which is the greatest obstacle to livestock development. Technical solutions to the trypanosomiasis problem are available, particularly in terms of tsetse fly control or eradication, but a livestock production system must then be introduced which generates enough income to meet the costs of these measures. In this situation, it is not feasible to pursue the type of livestock husbandry on the open range with a low level of inputs which was the basis for expansion of the industry in Australia and South America. Relatively intensive production systems based on artificial pastures may be appropriate, but experience with these systems is limited. Given the costs involved, internal purchasing power is likely to be the most serious constraint to the expansion of livestock production in the tsetse zone. Nigeria, with its income from oil production, may well be the one country in the region which can carry out this type of development in the foreseeable future.

Experience in the temperate zone is relevant to the development of livestock production systems in Africa's tropical highlands, where high-yielding animal and plant varieties and other technological innovations from the temperate zones of Europe have been adopted with success. Though small in terms of geographic area, Africa's highlands are important in terms of population and agricultural production. Recent experience in a country such as Germany, where an explosion in per capita income has been accompanied by a high degree of mechanization in the agricultural sector, is less relevant than experience from about 1850 to 1950, when per capita incomes increased by a factor of 10, corresponding roughly to an annual compound growth rate of 2.5%. The developments which occurred in German agriculture during this time, including increases in total output, innovations in production technology and the changing role of livestock in the farming system, provide interesting lessons for Africa's tropical highlands.

Chapter 14 references, p. 328

For example, experience in Europe and North America indicates that with the advent of mechanized draught power and chemical fertilizers, the role of animals in crop-production systems declines. It is doubtful, however, whether this development pattern is relevant for today's low-income countries. According to an FAO survey in 1966, nearly 85% of total draught power used in agriculture throughout the world is still provided by animals. Despite the development of mechanized power sources such as tractors, the developing regions still depend on animals for well over 90% of their agricultural power. The widespread introduction of mechanized power, while possibly desirable from the animal breeders' point of view (Duckham, 1946), is not likely to take place in the near future, and in many cases may even be undesirable in terms of employment and energy conservation. Thus livestock are likely to continue to play an important role in the crop-production systems of less developed countries for some time.

Specialized intensive livestock development may also be feasible in the tropical highlands, as demonstrated by the history of smallholder dairying in Kenya. With the development of infrastructure during the colonial period and the introduction of improved breeds and production technology, particularly for disease control, African farmers were able to take up dairying on a commercial basis. They were also encouraged by the land adjudication programme and the expansion of markets for milk. Circumstances in Kenya may be atypical, however, and it may take some time for other countries in the African highlands area to achieve a similar level of development.

Overall, it must be stressed that livestock production expands and its role in economic development changes slowly over a period of several decades. This is shown by the analytical model formulated at the beginning of this discussion and by the history of livestock development in Australia, South America and Germany. The model shows that with an annual increase in population of 2% and an increase in real per capita income of 5%, it is possible to increase the demand for livestock products by a factor of 10, but this is over a period of 40 years. Livestock development in the course of general economic development is a long-term process.

6. REFERENCES

Abercrombie, K.C., Clayton, E.S. and Ribeiro, I., 1978. Employment in Livestock Production in Developing Countries. Draft. Policy Analysis Division, FAO, Rome.

Beresford, M., 1977. Animals as Employers of Labour: Australia. International Livestock Centre for Africa, Addis Ababa.

Dahl, G. and Hjort, A., 1976. Having Herds. Stockholm, Studies in Social Anthropology No. 2. Department of Social Anthropology, University of Stockholm.

Duckham, A.H., 1946. The Functions of Livestock. British Embassy, Washington DC.

FAO, 1972. Income Elasticities of Demand for Agricultural Products. CCP 72/WP. 1, Rome.

FAO, 1977. Production Yearbook 1976, Rome.

Fundaçao Getulio Vargas, 1974. Caracteristicas Economicas das Exploracões Rurais 1962—1964. Rio de Janeiro.

Grigg, D.B., 1974. The Agricultural Systems of the World: An Evolutionary Approach. Cambridge Geographical Studies 5. University Press, Cambridge.

Guadagni, A.A., 1964. Estudio econometrico del consumo de carne vacuna en Argentina en el periodio 1914—1959. Desarrollo Economico, 3 (4): 517—533.

Guadagni, A.A. and Petrecolla, A., 1965. La Funcion de Demanda de Carne Vacuna en la Argentina en el periodio 1935—1961. Trimestre Economico, 32: p. 261 et seq.

Herlemann, H.H., 1961. Grundlagen der Agrarpolitik. Verlag Franz Vahlen, Berlin.

Herlemann, H.H. and Stamer, H., 1958. Produktionsgestaltung und Betriebsgrösse in der Landwirtschaft unter dem Einfluss der wirtschaftlich-technischen Entwicklung. Kieler Studien 44. University of Kiel.

Horst, P. and Peters, K.J., 1978. Regionalisierung und Produktionssysteme der Nutztier-haltung in Weltagrarraum. Z. Ausländ. Landwirtsch., 17 (3): 190—211.

Jahnke, H.E., 1978. Role and function of livestock. In: Overview and Course Materials from the Livestock Development Projects Course, Vol. 1. International Livestock Centre for Africa, Nairobi.

Johnston, B.F. and Mellor, J.W., 1961. The role of agriculture in economic development. Am. Econ. Rev., 51: 566—593.

Malassis, L., 1975. Agriculture and the Development Process. Education and Rural Development Series No. 1. UNESCO Press, Paris.

Mellor, J.W., 1970. The Economics of Agricultural Development, 2nd edn. Cornell University Press, Ithaca.

Peel, L.J., 1973. History of the Australian pastoral industries to 1960. In: G. Alexander and O.B. Williams (Editors), The Pastoral Industries of Australia. Sydney University Press, Sydney.

Pino, J., 1975. The role of animals in the world food situation: foreword. In: Proceedings of a Conference held at the Rockefeller Foundation. Rockefeller Foundation, New York, NY.

Reinhardt, H.L., 1974. Aufgaben und Wirtschaftlichkeit der Nutzviehhaltung im Wandlungsprozess. Bayer. Landwirtsch. Jahrb., 51 (1).

Reisch, E., 1977. Unterlagen zur Entwicklung der Arbeitsproduktivität in der Tierhal-tung in Deutschland. International Livestock Centre for Africa, Addis Ababa.

Von Oven, R., 1968. Produktionsstruktur und Entwicklungsmöglichkeiten der Rind-fleischerzeugung in Südamerika. D.L.G. Verlag, Frankfurt am Main.

Winrock International, 1978. The Role of Ruminants in Support of Man. Morrilton, Arkansas.

Chapter 15

Grazing Animals in the Next Few Decades

F.H.W. MORLEY

1. INTRODUCTION

Grazing animals serve men in many ways. They produce food and clothing, if necessary from plant sources which are not directly usable. They process energy from renewable resources to provide power for much of the agriculture and transport of the world. They harvest widely dispersed plant material and produce dung to provide an important fuel for domestic purposes. They form a mobile form of capital storage, and are hallmarks of prestige. They are the companions of man and also provide targets for spears, firearms and cameras (A1: 9).

Nevertheless, animal products are not essential for man's survival. Nutritionally adequate diets can be compounded from plant sources. Although such diets may lack appeal to some people, many find them wholly acceptable, especially if they have been accustomed to them from childhood. Shoe leather and woollen jackets can be substituted by products from non-animal sources, which already command the bulk of the market for products in everyday use, though not yet the luxuries.

Questions such as these have generated numerous speculations in the current technical, social and political literature, and a formal review of this literature would almost immediately become outdated and redundant. However, because of the importance of the questions to animal science, to crop and pasture agronomy, and to land-use planning, some basic framework of a discussion of the issues needs to be developed. Arguments will be supported by cross-referencing to other chapters in this volume and in Volume B1, rather than by a formal review. These will be given in parentheses with volume number, then chapter numbers, e.g. (A1: 3, 5) means Volume A1, Chapters 3 and 5.

A symposium at the Third World Congress on Animal Production (1973) provided a forum for this topic. It brought together many facts and viewpoints, some of them conflicting, on the future of animal production in general. Conflicts between economic and technical objectives became evident. For example, the technology exists to increase animal production dramatically by the use of crops and supplements, but such fodders usually cost much more than pasture. They are economically worthwhile only if they close gaps between animal requirements and pasture availability at critical times (B1: 13, 15, 18) or if their use is subsidized or protected by trade barriers. Industrialized countries may continue to afford high-cost systems such as feedlotting, but will the rising strength of consumerism tolerate the prices? Much of the research into these intensive techniques seems almost irrelevant to the needs of developing countries, or indeed of countries which rely heavily on exports of animal products.

It was perhaps significant that the critical role of draught animals in food production was scarcely mentioned at the 1973 Conference. The extent to which draught animals will be replaced by small tractors or human muscles was not considered.

Few new facts have further illuminated the way ahead. The relevance of much of the information available in 1973 has diminished as the world energy crisis has developed, costs have escalated, and economic problems have been given emphasis by recurring political crises as resources have become decreasingly adequate for the expectations of expanding populations. The protection of industries by trade barriers has caused prices of meat to vary by almost an order between different groups of consumers. The new wealth of oil-exporting nations has opened new markets for animal products. The high price of energy has meant restrictions on purchasing power, and the partial closure of some markets (A1: 11). It has also curtailed some well-established techniques of production such as grazing systems which require large inputs of fertilizer, especially nitrogenous fertilizers (B1: 3, 14), and the use of concentrated rations for fattening ruminants has been questioned.

This present economic and social scenario may be no more than a transitory commotion, which will subside as costs and prices adjust to new equilibria with production systems. The extent to which whole fields of technology, and priorities for research, may be modified in response to new opportunities and limitations is highly uncertain. Therefore, the thoughts advanced here must be inexact; in the hindsight of A.D. 2020 some may be seen to have been misleading. However, if programmes of research and development are to be allocated priorities in the use of scientific and industrial resources some assumptions are necessary. This chapter is an attempt to define the basis for such assumptions.

2. POPULATION PROJECTIONS

2.1. Human populations and the need for food and fibres

In the absence of catastrophes of a higher order than we have known in the past two centuries, the number of people in the world seems likely to double in the next three or four decades. This will lead to at least a doubling of the requirements for food and fibres. Opinion is divided, but these additional demands can almost certainly be met, even with existing technology, by increasing the intensity of cropping on currently cropped lands, by increased yields, and by cropping grazing lands and lands not currently used for agricultural production. The areas of land available for such expansion must be limited by topographical, climatic, and biological factors, as well as by available finance and political policies. Other limits to production may intervene, e.g., shortage of energy to support mechanization of agriculture, increased costs or depletion of fertilizer supplies, pollution of the environment and expanded urbanization, and probably financial and political crises (A1: 11, 12, 13, 14).

Although the larger populations of the near future could probably be fed adequate, although perhaps not interesting, diets it seems unlikely that this will be achieved. Limitations already mentioned, added to problems of finance and distribution, will probably be too great (A1: 12). If food production were the sole objective of agriculture, the role of food-producing animals would be greatly reduced, except where these use resources, such as extensive grazing lands or wayside weeds, which are not suitable for direct forms of food production. However, since food production is only one of several uses of grazing animals, changes, and especially rates of change, in patterns of production, must be highly uncertain.

In this chapter it is appropriate to ask where animal production through grazing is heading. Will the limits to resources such as land, water, energy and finance, coupled with increased demands for basic foodstuffs by expanded populations, mean that intensive forms of animal production will give way to crop production for food and fibre for direct use by man? Where cropping is marginal, because of topography, soil, or aridity, what part will be played by grazing animals?

2.2. Depletion of resources

In addition to the high costs of resources such as land and water, which are discussed below, foods of animal origin are usually more costly to process, transport and store than the equivalent plant products. They may also carry disease-producing organisms if processing has been lax. Therefore they usually cost the consumer far more than equivalent plant products per unit of energy or protein.

Depletion of energy reserves will bear increasingly on production systems which are energetically the least efficient, e.g. feedlotting of beef cattle, intensive swine and poultry production, and ranching in semi-arid and arid areas. Shortages of energy may also have important implications for the development and use of draught animals as sources of power. Systems of crop—pasture rotation which are less dependent on fertilizers for soil fertility than are more intensive cropping systems may be economically efficient, but any advantages could be offset by lower yields and reduced total production. Resources of phosphate rock may not be limiting in the immediate future, but price rises may limit pasture production. Water may be diverted from agriculture to other industries, or become more expensive. Production from pastures would seem more vulnerable to a reduction in irrigation potential than would crop production.

3. GRAZING ANIMALS IN WORLD AGRICULTURAL PRODUCTION

3.1. The role of grazing animals

Animal products, although not essential, are nevertheless highly attractive as food and fibre to large numbers of people. Consequently many are prepared to pay high prices for beef, wool, or butter, although cheaper ways of satisfying their basic needs for food and clothing could be found. This permits large industries to be founded on grazing animals, and the maintenance and development of such industries plays a critical role in the development and economic welfare of countries such as the U.S.S.R., Australia and Argentina.

Great increases in production from pastures, and especially tropical pastures (B1: 7, 8, 21) are technically feasible (A1: 14). However, the questions for the future concern objectives and resources. Which grazing lands might better satisfy human needs if used for some form of production other than meat? Will the technology of production of sheep and cattle for export be increasingly restricted by costs unless the demand for meat, wool, cheese and butter is sufficient to support matching increases in prices? (A1: 12, 13).

Milk seems likely to continue to be in heavy demand in 'Western' countries, although alternatives are available. Therefore milk production may continue to occupy a large share of the best land. This may include some of the best arable land in industrialized exporting countries, even where populations are high. In addition, dairy production will probably continue to be the main product of much steep and non-arable land, such as in New Zealand, where the rainfall and temperature permit the establishment of highly productive pastures (B1: 2, 4, 7); that is, in 'advanced' countries, dairying will probably be the best survivor among animal industries, whether for 'luxury' consumption or for infant welfare.

In less affluent countries few families will be able to purchase significant quantities of dairy products. Technical developments such as freeze-drying and reconstitution seem unlikely to be relevant to such families, and infant nutrition in urban populations will depend on the support of human lactation by suitable crops. Dairy products may be consumed in moderate quantities by the less impoverished families, depending on food preferences and prices. If prices are kept low by subsidies, or by low-cost production from pastures rather than from concentrates and conserved roughages, consumption could be substantial. Even in generally impoverished and densely populated countries the tourist industries, and the numbers of affluent people, are sufficient to demand large amounts of dairy products, almost irrespective of prices. Milk and meat from grazing animals will retain their importance in diets of some predominantly pastoral, and especially nomadic, populations.

The future of production from grazing animals will therefore interact strongly with the expectations and affluence of human populations, especially those in South America, Africa and Australasia. In addition the political and social consequences of the energy crisis may direct increasing emphasis on the role of draught animals. To complicate the complex, new trade ghettos will impose further uncertainties on production systems of farmers on both sides of trade barriers.

3.2. The use of arable land and water resources

Production per unit area of land, or of water applied to the land, is much greater from direct cropping than through systems of animal production, *provided* appropriate crops can be grown on such areas. This is not to say that cash returns, or returns on investment in land, capital, or labour will be higher from such cropping.

Most irrigated and well-watered lands are now used for food crop and fibre (cotton) production. In the U.S.A. and Australia, countries with large food surpluses, some are still used to produce fodder for animals in forms such as hay, fodder crops, pastures, and some grains. This may continue for a long time, but could change, if pressures to grow human food increase, to reduced animal production and more cropping.

Land and water resources are also extensively used for the production of tea, tobacco, cannabis, coffee, alcohol, vegetables and ornamental horticulture. This applies of course not only to fully arable land but also to semi-arable or steep land, which is nevertheless capable of growing olives, and many fruits, although perhaps not grain crops. In Italy, Spain and Greece probably less higher quality land is used for grazing than for the production of wine and olives. In Brazil the potential for animal production of areas which are used primarily for coffee production is enormous, as is its potential for production of many other crops which could be used for food energy and protein production. The diversion of production from coffee, wine or meat to cereals could increase the present levels of food production, but need not increase the satisfaction of human objectives, whether in terms of stimulants, sedatives, money, or enjoyment. The advisability of such a diversion might well be challenged by the rich, but perhaps not by the hungry.

Grazing could continue to play a useful, even a critical, role in multiple-enterprise and multiple-use systems of production (B1: 11, 12, 17). A pasture phase in a rotation may be useful in the production of many crops by economically stabilizing soil structure, and restoring some soil fertility. Although soil conditioners and increased use of fertilizers may underlie many feasible forms of continuous cropping, it will probably still be questionable whether these are economically preferable to ley systems in all, or even many, of the environments where crops are grown. The need for such rotations could become increasingly evident as long-term reserves of soil nitrogen, and other nutrient reserves, are exhausted in areas such as the corn belt of North America and the Ukraine, and the wheat belts of Australia and Argentina. Although some soils of these areas do not at present respond greatly, or at all, to fertilizer inputs, soil fertility cannot be expected to persist indefinitely in the face of depletion through the sale of crop products. If the costs of fertilizers continue to rise relative to prices obtained for food, as has been true in recent years, rotations which use fertilizers fully and efficiently may become widely used. However, this is an area hedged with uncertainty.

The use of many by-products of crop production, such as straw of cereals and the leaves of many vegetable crops, will still provide the basis of substantial levels of animal production through ruminants. Although much of these is at present burnt or returned to the soil, demands for high quality foods may well render their use highly worthwhile (B1: 17).

In summary, much of the best agricultural land in the world will undoubtedly continue to be used for urbanization and for growing products other than basic foodstuffs. Were pressure for food production to be intensified, food crops would take precedence over livestock. Livestock production may continue to play a useful role on arable lands through suitable cropping—grazing rotations for the maintenance of soil structure and fertility, the control of weeds, and the diversification of incomes through mixing enterprises. Although continuous cropping systems are becoming widespread, the crop—livestock farm could continue to be a highly significant form of agriculture on arable lands, especially where pressure for food production is not intense.

3.3. The use of non-arable lands

Sixty percent or more of the agricultural lands of the world are non-arable (FAO) and therefore are not capable of producing crops with present technology and forms of land use. In the lower rainfall areas there is possibly no alternative to animal production, which frequently is based upon a nomadic system of land use. This form of husbandry, although extremely interesting, has imposed hazards to the long-term conservation of land in marginal areas and is unlikely to be significant when considering food production on a world scale. This is not to overlook the fact that such systems are obviously very important to the people who depend on them. Therefore any developments that are feasible, and which increase the stability of the agriculture and the welfare of such people, should not be neglected. Of course it is essential that such developments are acceptable to the beneficiaries.

Where rainfall is sufficient for some forms of crop production the present classification of land as non-arable must contain a very important economic parameter. As is evident in China, the Himalayan foothills and parts of Africa, much 'non-arable' land has been brought into high levels of productivity through terracing, water conservation and high labour inputs. Intensification through terracing could prepare large areas of well-watered country for crop production, if pressure to do so were to be generated by shortages of food. Animal production in such areas, if it is important or even exists, must largely give way to crops, timber, vegetables and fruits. Such land would not be given a second thought in plans for development of crops for food production by people accustomed to the great plains of the U.S.A. or the steppes of the U.S.S.R., but it can support large human populations and their often numerous goats, alpacas, cattle, yaks and sheep, which browse and graze uncultivated areas and are fed residues of crops from the terraces (A1: 2,4).

Grazing of non-arable lands has been responsible for widespread degeneration and soil erosion. As human and animal population pressures have grown because of the decrease in death rates of pastoral peoples and the prestige associated with owning livestock, the dangers of further degeneration seem inescapable unless controls on the use of grazing lands are introduced. Usually controls are resented and resisted by pastoralists, whether they be nomadic tribesmen or affluent ranchers. Therefore reclamation, or even conservation, will be difficult to introduce and maintain.

If one were optimistic, but perhaps a little unrealistic, one could hope that the over-exploitation of grazing lands would gradually give way to relatively stable systems of land use. Although there is increased awareness of the problem, particularly in organizations such as UNEP and FAO, the next few decades may not see spectacular progress because soil erosion is a catastrophe less spectacular, but no less destructive, than an occasional flood or famine.

Demands for forest products, for fruits and national parks may become more vocal in the near future than the demands for animal products. If so, animal production may give way in the better-watered regions where the climate is suitable to the production of timber, fruits, paper and chemicals from trees. This will generate some opportunities for employment, but not necessarily for the pastoralists, since multiple-use systems could well call for more labour than is required for pastoral pursuits (B1: 12).

Where rainfall, temperature, violent winds, or topography make the growing of trees impossible or unprofitable, grazing may well continue as the sole form of agriculture. Measures to protect the environments concerned will need to be given high priority. However, in view of the relatively low levels of production and incomes generated from such areas, it is questionable whether adequate resources will ever be provided. The consequences need not be spelt out.

3.4. Animal production in cooperation with cropping

3.4.1. Crop—pasture rotation
Future developments in animal production in association with crops are likely to be directed to maximizing crop output from rotation systems in which a grazing phase is no longer than is adequate for cropping purposes. This must mean emphasis on rapid establishment of a productive pasture phase, and careful integration of components of production systems. Making the best use of cropping lands, of non-arable lands, and of stubbles and crop residues needs to be within the constraints imposed by scarce and costly liquid fuel. If the demands for food are to become twice as high as those of the nineteen-seventies, it is questionable whether crop—pasture rotations can be afforded on the best arable land, but they may make effective use of areas with lower potential productivity.

3.4.2. Draught animals

Draught animals probably contribute more food to man than most animal industries. Much of the rice of Asia, as well as many other crops, is grown with the aid of buffaloes and draught oxen. The ox, the buffalo, the mule, the donkey, the horse, the elephant and the llama are widely used for power and transport. These animals often live on crop residues, waste lands, and weeds of headlands and roadsides; that is, they are fed renewable resources.

Will draught animals be replaced by other forms of power? Although the advantages of liquid fuel could be substantial, especially where the 'Green Revolution' has meant some displacement of subsistence agriculture by larger scale cash cropping, nevertheless draught animals seem likely to continue to play an important, even critical, role in world food production. It is to be hoped that plans to select buffalo for meat and milk production do not result in an erosion of the merits of these animals for draught purposes.

Large-scale unemployment and the energy crisis have generated proposals for a reversion to man-and-animal-powered agriculture in countries with high levels of mechanization. Subsistence agriculture, based largely on muscle power, can produce sufficient food to support large populations, as in densely populated India and Indonesia, but it might not be possible to produce sufficient surplus food to support modern conurbations. Since mechanized agricultural production requires less than 20% of the energy used in modern industrial states, and is basic to the maintenance of urban populations and standards of living, energy for agricultural production will probably command high priorities in order to provide for a high level of mechanization, and even increased mechanization. That is, although draught animals will continue to be very important, they are unlikely to displace much of the power now provided by engines.

Unfortunately draught animals have the disadvantage, relative to the internal combustion engine, that they require energy for their maintenance, whether they are working or not. They also need to have been on a satisfactory plane of nutrition for some time before they can be of much use for draught. In many regions, especially where rainfall is a major limitation, timeliness of operations may be an important limitation to growing crops. Dry sowing, or sowing immediately after rains, may not be feasible if the animal power required is not available until the draught animals are strong enough from grazing the plants grown in response to the rains. Techniques of avoiding this impasse will almost certainly involve the use of stored fodder, which frequently could be usable in part by man. In the long term the use of such stored fodder could be highly economical if it were to permit increases in crop yield through, for example, increased timeliness of operations. These could far outweigh the quantity of supplements fed. In the short term the equivalent of cash flow in food supplies limits such procedures. Such limitations are particularly important in equatorial Africa, where at present much of the muscle power of the farm is provided by women. They are often unable to work hard at the end of a dry season because they are under-nourished.

4. NEW AND DEVELOPING TECHNOLOGIES

4.1. Domestication of wild species

The transition from hunted to domesticated animals has been successfully attempted in only a few species. The introduction of animals such as deer, eland and the capybara into animal production systems is attracting interest at present, and economically and technically viable systems of production will probably be developed. The production of venison in New Zealand and Europe is certainly contributing income to those who are producing the venison, and it also provides a popular luxury in parts of the more affluent world. Whether significant quantities of food can be produced by these systems, and by other 'new' domesticated species must be questionable. There are several limitations to be overcome which have been detailed in the literature (A1: 1, 2, 9 and B1: 1). Should the eland, wapiti, oryx, or wildebeeste become fully domesticated, it is doubtful whether in most circumstances the production of food from these would substantially exceed that from locally adapted races of cattle, goats, sheep or camels.

At present techniques of animal harvesting which rely upon hunting and portable abattoir facilities seem unlikely to make more than trivial contributions to world food production. Certainly they are unlikely to justify using land resources for food production in competition with cropping, but they may well be highly competitive in earning money from tourists, especially in countries which lack hard currency. Indeed, where hunting rights for lions and elephants may cost thousands of dollars, income from tourists armed with cameras or guns may permit landowners, whether national, individual, or tribal, to purchase far more food, and not only food, than they could possibly afford from selling the agricultural produce from their land, either from animals or crops (A1: 9).

4.2. Genetic improvement

The genetic structure of poultry, swine and dairy cattle populations in commercial production systems has moved away from the traditional 'stud stock', and 'pure' breed systems, which still prevail in sheep and beef cattle breeding. The control of animal feeding and movements, and the relative ease of measurement, coupled with artificial insemination, have made possible the use of more sophisticated aids to selection. Control of the oestrus cycle may bring further advances. Such technical advances pose questions of ethics (A1: 8) which are not easily soluble.

The main impediments to implementing advanced breeding plans in grazing animals, which are not handled daily, have arisen from the costs of the operations relative to expected gains. Some sections of the horse-breeding industry are also saddled with the additional burden of conservation, as well as an understandable reluctance to jeopardize the credibility of pedigrees, by refusal to accept pedigrees generated from artificial insemination. Perhaps disease may impose some level of acceptance of such techniques.

Research, largely in meat-producing sheep and in beef cattle, has disclosed substantial possibilities of improved performance by the use of crossing systems which use heterosis for maternal performance, as well as individual performance in matings which produce animals for slaughter rather than for breeding. The emphasis is starting to shift from breed improvement to breeding systems which may be more efficient and productive.

Systems of crossbreeding have for long been in intensive use in sheep breeding in regions where the species is stratified into extensive predominantly wool-producing flocks, dual-purpose flocks, and intensive, predominantly meat-producing flocks. In Australia the use of Border Leicester (or equivalent) rams on Merino ewes to produce a crossbred ewe for mating with a Dorset (or equivalent) for prime lamb production is widespread, and at present is apparently fully competitive with the use of dual-purpose breeds. It depends heavily on a large wool trade and the consequent availability of large numbers of surplus Merino ewes which are of limited value for other purposes.

Systematic crossbreeding need not be based on such stratification of an industry. The time and cost required to develop and maintain inbred lines rule out 'Hybrid Corn' patterns of genetic improvement for grazing animals, but the use of systems of crossing involving the rotational use of males of several defined genotypes (breeds?) may become widespread in beef production, and perhaps equally useful in sheep production.

The performance of sire-producing genotypes may be improved by genetic selection for their performance in crossbred combinations rather than, or at least in addition to, their performance as purebreds. The 'breeds' will have to be sufficiently large to avoid excessive costs from inbreeding and to provide sufficient males for their role in the industry.

If crossbreeding systems of production prevail, much of the emphasis on breed 'type' will probably disappear. The folklore of the judging ring will almost certainly be replaced by the output of computer-based programs of evaluation, using criteria such as fecundity, growth rates, feed conversion, and objective estimates of the attributes for which the average, perhaps unenlightened, consumer is prepared to pay. One hopes the criteria will also include the efficiency of use of resources such as land, labour, power and finance. Fat depth, fleshing, fibre diameter, fold score and fertility will increasingly replace backline, bearing and breed type. Animal improvement will thus be applying consumer evaluation in combination with Mendelian genetics.

Group or co-operative breeding schemes, in which several breeders, co-operatives, communes, or institutions combine their genetic resources, seem likely to replace, or perhaps to extend the scope of numerous individually and indivualistically managed flocks or herds. Provided problems of organization, communication, and finance do not inhibit management, the larger genetic pools generated in such schemes should confer worthwhile advantages. These may not be great in breeds such as the Merino in Australia, South Africa or U.S.S.R., where a nucleus flock may comprise tens of thousands of individuals under one management. The advantages of population size are likely to be especially important in beef cattle where herds are often small, as in the United Kingdom or France, but less so in Argentina. Already group schemes are operating with dairy cattle, and the Milk Marketing Board of the United Kingdom and the New Zealand Dairy Board schemes seem sufficiently successful to suggest the world-wide adoption in the near future of similar models.

The application of genetic improvement must depend heavily on definition of objectives. Two major components must be catered for. The grazier wants disease resistance, fecundity and stamina. The consumer will want a defined product at an acceptable price. The integration of the two requires manipulation of genotypes to satisfy both. The prospects for doing this seem brighter with rotational crossing systems than with pure breeding since heterosis for fertility in the dam, as well as growth rate in the offspring, have been demonstrated in cattle, and some problems, such as over-fatness, can be reduced by suitable systems of crossbreeding. Whatever the system, the definition of consumer preferences, not so much in absolute and discrete classes but in terms of prices for a wide range of products, must be subject to persistent study and the development of systems of communication which feed back information to the breeders. Consumer preferences can and do change and must be evaluated by breeders as they occur. There are signs that markets are being recognized more by breeders today than they were a decade ago (e.g. objective measurement of wool quality; fatness measurements in carcases) but widespread rationalization of marketing may not be with us for several decades.

4.3. Control of reproduction

Recent decades have seen spectacular developments in the control of ovulation and in embryo transfer. Indeed the technology of multiple ovulation and embryo transfer has probably developed already to a stage at which application is limited more by the cost of procedures than by the technical information required.

The future of such techniques seems at best uncertain. Although technically feasible, their contribution to selection of superior genotypes is doubtful, to a large extent because of the difficulties of identifying which females have genotypes sufficiently superior to justify the cost. In addition, the use of progeny testing as an aid to selection of both males and females could accelerate inbreeding to an unacceptable degree, ultimately limiting genetic progress through reduction in genetic variance, in addition to imposing the extra costs of maintaining inbred populations. This disadvantage may be minimal in rotational crossing systems. The value of embryo transfer may be greatest for the introduction of exotic genotypes for use in such systems, provided the risks of virus introduction can be kept to levels which do not offset the anticipated advantages. At present the advantages are not clearly recognizable, although extravagant claims are often accepted uncritically. The risks of disease are widely publicized, and seem to expand as the techniques of virology and serology become more sophisticated. No clear formulae for compromises between advantages and risks have yet emerged.

The research which has made these techniques of reproduction possible must command admiration. Although their application may seem of limited value at present, they may present unexpected opportunities for breed improvement or modification in the next few decades. At present, however, such opportunities do not seem obvious if the whole system of genetic improvement by selection, or by introduction, is examined critically in terms of gains and costs on a whole farm, or a whole continent.

4.4. Techniques of feeding

The problems of improving the nutrition of grazing animals are overwhelming. They embrace not only the problems of supplementation, but those of the selection of pasture plants, the management of pastures, and the techniques of livestock husbandry (B1: 6, 8, 13, 15, 21).

The recent explosion of information on the interactions between the source of protein in the diet, the fermentation of protein and non-protein in the rumen, and the performance of animals, has not yet been widely applied to grazing animals. The major limitations have probably been difficulties of supplementation coupled with uncertainty of economic benefit. Even with non-grazing ruminants applications are yet sporadic.

With grazing animals the use of this information is intimately bound up with the need to appreciate what the animal is actually eating as it grazes. Pastures are usually highly heterogeneous in respect to pasture species, the physical components such as leaves, petioles and stems, and their chemical composition. Moreover, the fate of the components in this diet, and the dynamics of animal responses to additional energy or nitrogen are not always clear, especially when diets consist of a variety of components with a range of susceptibilities to fermentation in the rumen, or eventual digestion and absorption in the abomasum and small intestine (B1: 5, 14, 15).

A full analysis and understanding of the processes is a tall order. At present there are some indications that supplementation of pastures with proteins which resist degradation in the rumen could improve liveweight gains in grazing animals, especially if they are being maintained on low quality roughages. The biology is being slowly elucidated; the economics will depend on local financial parameters as well as on the biology.

4.5. Pasture use

The extent of land which is of little value except for grazing, its vulnerability to mismanagement, and the low cost of the plant material consumed by grazing animals, which is not directly usable by man, underline the importance of systems of production from pasture relative to production from concentrated rations (B1: 7, 9, 13). Intensive systems of production from pasture have been developed for favourable mesophytic environments such as those of Europe and North America, but the applicability and relevance of this work for the extensive grazing enterprises of the Southern Hemisphere must be doubted. Production from these cannot depend on subsidies or tariff protection, and costs must be kept to a minimum. Such restrictions are unlikely to be lifted in the next few decades; indeed the use of subsidies and protection seems to be increasing.

The relatively recent development of technologies for tropical pastures has highlighted the tremendous potential for livestock production in the wet tropics and subtropics (B1: 8, 14). This may be of particular significance to Brazil and some parts of wet tropical areas of Africa and Asia, which are not yet heavily populated and on which pasture improvement is technically feasible. The perennially humid tropics of the Americas, Africa and Asia are sparsely populated but the seasonally wet/dry tropics contain some of the densest human populations, which almost invariably rely on the production of crops such as rice, cassava and pulses for their food supplies. Animal products from the wet and the wet/dry tropics may provide useful additional food, but not necessarily to those who need it most. Substantial efforts to improve the technology of tropical pasture development and production could be economically justifiable to provide an intermediate technology for developing such areas.

There are, however, considerable dangers in over-enthusiastic or precipitate development of areas such as the jungles of Brazil or Borneo, and some African forests. In these the effect of removal of the forest on soil fertility and lateritization, and the resulting need for substantial inputs of fertilizers to maintain the growth of forest or pastures may introduce technical and financial limits to productivity. It therefore seems questionable whether animal production from tropical pastures will develop to be a resource which is more important than crops. Where population pressures on limited land are severe, they seem unlikely to be relieved by clearing the jungle and establishing pastures, but the encouragement of the enlarged populations of people to support themselves, largely by growing crops for their own consumption, could relieve problems of under-employment and over-population. Combinations of tree crops, such as cocoanuts and oil palms, with seeded pastures offer worthwhile prospects for the perennially humid tropics, but the extent to which such systems can be extensively developed must be uncertain.

Pasture improvement inevitably poses problems of fertilizer use (B1: 3, 14). The anticipated costs of energy will almost certainly mean that nitrogenous fertilizers will become too expensive for the production of most pastures, unless such costs are balanced by unprecedented rises in prices obtained for animal products. Milk production may prove to be an exception, and some affluent people will almost certainly be prepared to pay prices for animal products which would make animal protein prohibitively expensive for most people. Nevertheless, legumes seem more likely than fertilizers to provide most of the nitrogen for grazed pastures. This may mean lower carrying capacities but not necessarily lowered animal production.

The availability and cost of phosphatic fertilizers may start to limit agricultural production in the next few decades, and even now higher costs are influencing fertilizer policies for pastures. The higher returns per unit of fertilizer from crops means that cropping is, and is likely to be, less vulnerable to increased fertilizer costs. Nevertheless, the use of phosphate in pasture improvement, and for the maintenance of improved pastures, could continue to expand because of expansion in the areas under improved pastures. The amount of phosphate used per unit area of improved pasture will come under careful scrutiny, perhaps resulting in decreased applications. The role of soil tests, plant tests, pasture history, and trial applications (strips or blocks) in relation to costs of applications, pasture management, cropping and fertilizer history, and the prices obtained for agricultural products will be examined closely (B1: 3, 14). The results may be incorporated into appropriate decision functions, with predictions generated by computer programs based on a variety of models of appropriate production systems.

The role of management procedures such as rotational grazing, deferred grazing, and variations in stocking rates has been worked over rather fully in temperate regions, mostly on improved pastures (B1: 2, 21). Understanding is still fragmentary in many areas, and extrapolations of experimental results to arid and semi-arid areas is certainly not justified at present. Nor is it clear whether tropical legumes require careful management if they are to persist and to contribute the nitrogen required for pasture production.

In the immediate future, as in the past, refinements of management, such as rotational grazing and fodder conservation, seem likely to be less useful than the expansion of pasture improvement and the adoption of higher (optimal) stocking rates. Until the consequences of high stocking rates have been examined over a range of environments, especially in the tropics, it will not be clear what management refinements, if any, may be necessary for high and stable productivity.

4.6. Disease control

Diseases of grazing animals were discussed briefly in Volume B1: 20 and little need be added here. The next few decades could see rinderpest and contagious bovine pleuropneumonia eradicated from the world. Brucellosis, tuberculosis, perhaps foot-and-mouth disease, and most external parasites will probably be eliminated, or brought under control, in countries with advanced technologies. In other countries they are likely to persist, perhaps at low prevalence in pockets of infection. Occasional epidemics may endanger wide regions so that quarantine between countries, and regions within continents, will probably continue to be necessary. Epidemics of diseases caused by viruses associated with insect vectors will probably continue indefinitely to plague our livestock industries. The eradication of most internal parasites seems as yet a very distant dream, but reduction to unimportant levels may be possible if problems of drug resistance can be overcome. Since this seems unlikely, internal parasitism will probably continue to make inroads into the health, and the production, of grazing animals.

In countries with advanced technologies, diseases of high production, associated with metabolic stress, nutritional deficiencies, or social stress (A1: 8 and B1: 19) will probably become more important, and more frequently diagnosed. The reported incidence of genetic defects may increase as aids to diagnosis improve, especially if coefficients of inbreeding rise as selection becomes more intense.

Disease control will probably hinge increasingly on livestock management through control of nutrition, especially at critical periods. Selection for genetic resistance to environmental stress will probably achieve useful dividends, but selection for resistance to infectious diseases and parasites must be considered with caution. The more rapid turnover of generations in the disease agent than in the host means that the agent can probably evolve more quickly than can the host. Nevertheless, the use of host resistance against babesiosis and other diseases, provides encouragement.

4.7. Conservation of genetic resources

Conservation of genetic resources is considered in detail in Chapter 5 of this volume. Some additional viewpoints are worth including. I am indebted to Sir Otto Frankel for discussion of these, and for access to drafts of a forthcoming publication.

The survival of some strains or breeds, even species, of grazing animal may be endangered by the ability of man to modify the environment, to block animal movements, and to penetrate to any part of the earth with lethal weapons. More subtle, but perhaps no less effective, may be the concentration of commercial breeding around a limited array of genotypes. For example, the pre-eminence of Friesian cattle in Europe, and increasingly elsewhere, could endanger the survival of many breeds of dairy cattle. The merits of the zho, and other yak—cattle hybrids, for transport and domestic purposes, could endanger the survival of the yak as a distinct entity.

Rare genes, and perhaps rare genotypes or gene-complexes, of agricultural animals may well be rare because they are of little present value, or because their value has not been recognized. That is not to say they will not be valuable in the future as a result of changes in the environment (clearing of forests, improvements in pastures), new food preferences (venison, game birds), new systems of production (multiple births in Finnish landrace sheep, fast-growing breeds of cattle), and new economic circumstances (sheep without wool). In plants conservation of genetic resources has emphasized sources of resistance to diseases such as rusts of cereals and fungal infections of trees. In livestock the sexual breeding system aids the preservation, at low frequencies, of resistance genes which could be important in the future. Indeed the efficiency of modern drugs in killing pathogens could well generate needs for genetic sources of resistance to complement, if not replace, chemicals rendered largely useless by resistance of pathogens. The use of genetic resistance to cattle ticks, and the consequent reduced reliance on acaracides by selection, provides a good example. Helminth parasites are already challenging drug companies by developing genetic resistance to new 'wide-spectrum' anthelmintics.

The cultural justification for preservation of threatened genotypes lies in the interest and delight, for town and country people alike, of the unusual, the bizarre, and the colourful. If these are part of the cultural background of man, as many rare breeds are, enjoyment is intensified. The fact that such genotypes may some day be commercially useful is almost irrelevant to the argument for conservation.

The agricultural needs for conservation can probably be met adequately by the maintenance of gene pools. Gene pools of threatened European dairy breeds, Zebu dairy and draught types, meat- and wool-producing sheep, and so on, could be maintained at a few centres around the world. A population of 4000, or even 1000 individuals would probably be sufficient to preserve even unrecognized genes having frequencies of about 0.01, provided care were taken to avoid extremes of natural, and especially artificial, selection. Perhaps it is questionable whether serious attempts to preserve genes which are found at frequencies lower than 0.01 can be justified. Rather the use of natural or artificial mutation may be a cheaper, and perhaps more effective, way of grasping the unusual. Entities such as the yak could be reconstructed, although modifications might be difficult to avoid, by appropriate crossing and selection.

The preservation of semen, ova and embryos by freezing, or other means, could provide a relatively inexpensive aid to conservation. The hazards inherent in such techniques limit their use to some extent, but they could well be a critical component in an effective scheme.

5. CONCLUSIONS

Animal production through grazing animals seems likely to continue to be an important activity for the foreseeable future. It may be expected to be, at least partly, phased out of areas which have a potential for intensive production of food or fibre from arable land, especially where irrigated. It may become enhanced in areas where crop production is not feasible. The food resources for grazing animals will tend to be largely pastures and roughages of various kinds rather than concentrates. The adaptation of animals to, as yet, relatively unexploited environments, and to developing technologies, will become increasingly important.

Draught animals will continue to play a critical role in crop production where they do so at present. They are unlikely to displace machine power in highly mechanized agriculture. If it is necessary for them to do so living standards could fall markedly in such regions.

The domestication of wild species is unlikely to be important to world agriculture, although it could be financially significant in certain areas. Genetic improvement of domestic animals will probably swing towards systems of rotational crossbreeding, and the objectives of selection will probably become more oriented to the requirements of consumers and commercial producers than to the views and interests of individual breeders. The technologies of control of reproduction, other than artificial insemination, seem unlikely to be important except in special circumstances, such as the penetration of quarantine barriers.

Improved nutrition of grazing animals will continue to be beset with practical and economic limitations. The use of protein supplements in which protein fermentation in the rumen is partly avoided could be a valuable aid to the use of crop by-products, and may be essential if economically viable systems of using such material are to be developed.

Pasture use will probably play a more significant role in animal production as competition with crops for food production resources intensifies. The potential of tropical and subtropical pastures for livestock production may not be realized because the areas in question may be already, or will become, so densely populated that only food produced directly from crops will be economically within the grasp of most people. Crops are expected to compete successfully with pastures for priorities in the use of fertilizers and fuel, but perhaps not for capital.

Disease control is likely to focus increasingly on diseases associated with metabolic, nutritional or social stress. Genetic selection for adaptation to new environments, and to chronic disease and parasitism, could be a major development in future.

In brief, grazing animals will continue to play several important roles. The development and maintenance of appropriate and efficient production systems will accentuate technical, and especially financial and social problems. The solution of these problems will continue to depend on the imaginative and purposeful application of appropriate research.

List of Contributors

Rolf G. BEILHARZ, B.Agr.Sc., M.Agr.Sc., Ph.D. Chapter 5. Born in 1936, he studied agricultural science (major animal production) at Sydney University and gained a Ph.D. in animal breeding, at Iowa State University, Ames, Iowa, U.S.A. Since 1965 he has been Lecturer, Senior Lecturer and now Reader in animal breeding at Melbourne University, where he has been responsible for breeding and behaviour of domestic animals. He is on the editorial board of *Applied Animal Ethology* and is Australian delegate to the International Ethological Committee.

Jane BELYEA, B.Sc. (Agric.) (Hons.), (Mrs. N. Elith), Chapter 11, is a graduate of the University of Melbourne where, after graduation, she served on the staff of the Animal Production Department as a Research Fellow and as Tutor in animal production. Her research interests centred on an analysis of the energetics of Australian agriculture and she has also published papers dealing with the inter-relationships between energy, agriculture and the built environment.

Petrus Leonardus Maria BERENDE, Chapter 10. Born in 1937 in The Netherlands, he studied agricultural sciences at the Agricultural University, Wageningen, The Netherlands. He graduated in 1969 in animal husbandry, with a specialisation in animal nutrition and physiology. From 1969 onwards he has been scientific staff member at the Research Institute for Animal Nutrition and Toxicology, in Wageningen, The Netherlands.

John L. BLACK, B.Agr.Sc., Dip.Ed., Ph.D. Chapter 6. Born in 1942, he studied agriculture at the University of Melbourne, Australia and gained his Ph.D in animal nutrition at the same university. For the last 10 years he has been a research scientist with CSIRO Division of Animal Production at Prospect, Sydney. His major research studies are concerned with amino acid and energy utilization in sheep, but he has also published papers comparing nutrient utilization between species. In 1979 he was co-editor of a book entitled *Physiological and Environmental Limitations to Wool Growth*. He was recently Secretary of the Nutrition Society of Australia and President of the N.S.W. branch of the Australian Society of Animal Production.

Sándor BÖKÖNYI, M.V.D., D.Sc. Chapter 1. Born in 1926, he studied veterinary medicine at the Hungarian Agrarian University, Budapest. After finishing his studies he focused his main interest on the origin and history of domestic animals and animal husbandry. He founded the Archaeozoological Collection of the Hungarian National Museum, Budapest, and was its curator through 1973. From 1973 through mid-1981 he served at the Department of Interdisciplinary Studies (in the last two years as its head) in the Archaeological Institute of the Hungarian Academy of Sciences, Budapest. Since mid-1981 he has been director of the above institute. Since 1952 he has also been lecturer at the Eötvös Loránd University, Budapest. He has authored seven books and over a hundred articles on archaeozoology, published in Hungary and abroad. He is a member of the executive committee of the International Council for Archaeozoology, member of the Archaeological Committee of the Hungarian Academy of Sciences, the Deutsche Gesellschaft für Säugetierkunde, the société d'Ethnozootechnie, corresponding member of the Deutsches Archäologisches Institut and serves on the scientific board of the interdisciplinary review *Paléorient* (Paris).

Richard CASSELS, B.A., M.A. Chapter 3. Born in 1947, he studied archaeology and anthropology at Cambridge University, Great Britain. From 1971 to 1982 he was Lecturer in Prehistory in the Anthropology Department of Auckland University, and is currently Curator of the Manawatu Museum, Palmerston North, New Zealand. He has published articles on New Zealand archaeology and was Secretary of the New Zealand Archaeological Association from 1978 to 1980.

H. EPSTEIN. Chapter 4. Born in 1903, he studied agriculture at Berlin Agricultural University and there gained his Agr.D. degree in agricultural economics. Since 1972, he has been Professor Emeritus of Zootechnics at the Hebrew University of Jerusalem. He has published the following books: *Domestic Animals of China*, Commonwealth Agricultural Bureaux, Farnham Royal, Bucks. (1969); *Fettschwanz- und Fettsteisschafe*, A. Ziemsen, Wittenberg-Lutherstadt (1970); *The Origin of the Domestic Animals of Africa*, 2 vols., Edition Leipzig and Africana Publishing Corporation, New York (1971); *Observations on Frozen Beef from Various Sources*, Beth Dagan, Israel (1971); *Domestic Animals of Nepal*, Holmes & Meier, New York (1977); together with D.E. Faulkner, *The Indigenous Cattle of the British Dependent Territories in Africa*, Colonial Office, London (1957); and in the press, *Humped Cattle*, International Publishing Enterprises, Rome; and *The Awassi Sheep*.

John HOLMES, M.V.Sc., Ph.D. Chapter 12. Born in 1939, he graduated in veterinary science from the University of Sydney in 1962. He received the M.V.Sc. from the University of Queensland and Ph.D. in nutrition from the University of California at Davis. For six years he was Senior Veterinary Officer (Cattle Production and Research) with the Department of Primary Industries, Papua New Guinea. Since mid-1979, he has been Senior Lecturer in Overseas Agriculture in the University of Melbourne School of Agriculture and Forestry, engaged mainly in postgraduate training of students from S.E. Asia and consultancies with Universities in Papua New Guinea, Indonesia and Malaysia.

Hans E. JAHNKE, Dr. oec. Chapter 14. Born in 1944, he studied agriculture and agricultural economics at the Technical University of Munich and the University of Queensland and gained his doctorate in social and economic sciences at the University of Hohenheim, Federal Republic of Germany. He has published books entitled *Tsetse Control and Livestock Development in East Africa*, Weltforum Publishing House, Munich, 1976, and *Livestock Production and Development in Tropical Africa*, Vauk Scientific Publisher, Kiel, 1982. From 1975 to 1980 he was principal economist at the International Livestock Center for Africa. He is presently lecturer at the University of Kiel, Federal Republic of Germany.

Ted LEDGER, N.D.D., C.D.D., Chapter 9, studied at the East Anglian Institute of Agriculture. Following war service in the Royal Navy he joined the National Agricultural Advisory Service in England and subsequently transferred to the Colonial Service in 1947 when he was posted to the Department of Veterinary Services, Uganda. In 1949 he joined the staff of the newly established East African Agriculture and Forestry Research Organisation at their headquarters at Mugeega, Kenya. There he was for many years responsible for research into the relative body composition of East African ungulates vis-à-vis their domesticated counterparts. Following retirement from E.A.A.F.R.O. in 1977 he joined the staff of the School of Agriculture, Melbourne University, as lecturer in animal production with special reference to the possibilities and problems of wildlife utilisation.

Fred H.W. MORLEY, D.Agr.Sc., H.D.A., B.V.Sc., Ph.D., F.T.S., F.A.S.A.P., F.A.C.V.Sc. Chapter 15. Born in 1918, he studied agriculture at Hawkesbury Agricultural College and veterinary science at the University of Sydney. After 5 years in the New South Wales Department of Agriculture he studied animal breeding and statistics at Iowa State College, gaining his Ph.D. in 1950. After further research in sheep breeding he joined the CSIRO Division of Plant Industry as a geneticist and plant breeder working on pasture legumes, and subsequently on the management of improved pastures. In 1977 he joined the Faculty of Veterinary Science, University of Melbourne, where his current interests include teaching and research in epidemiology and preventive medicine of livestock, and the development of computer models of pasture-animal-parasite systems. He is a Federal Past-President of the Australian Society of Animal Production, and formerly A.C.T. Branch President of the Australian Institute of Agricultural Science. He has edited volume B1, *Grazing Animals*, in the *World Animal Science* series.

Lynnette J. PEEL, B.Agr.Sc., M.Agr.Sc., Ph.D. Chapter 7. Born in 1938, she studied agricultural science as an undergraduate at the University of Melbourne, undertook postgraduate research in both animal production and history at the University of Melbourne and obtained her doctorate in history from Monash University. Since 1971 she has been at the University of Reading, England. Her book *Rural Industry in the Port Phillip Region, 1835—1880* was published by Melbourne University Press in 1974.

Elis Joost RUITENBERG, Chapter 10. Born in 1937 in The Netherlands, he studied veterinary medicine at the State University in Utrecht, The Netherlands. He graduated in 1962, specialized in pathology, and obtained his Ph.D. at the same university in 1970. From 1962 onwards he has been scientific staff member and, from 1980, Director at the Rijksinstituut voor de Volksgezondheid (National Institute for Public Health) in Bilthoven, The Netherlands.

Peter SINGER, M.A. (Melb.), B.Phil. (Oxon.), Chapter 8, was born in 1946 in Melbourne. He studied philosophy at the University of Melbourne and at the University of Oxford. He has taught at Oxford, New York University, La Trobe University, and is at present Professor of Philosophy at Monash University. His books include *Animal Liberation*, New York Review, 1975; *Animal Rights and Human Obligations* (co-edited with Tom Regan), Prentice-Hall, 1976; *Practical Ethics*, Cambridge University Press, 1979; *Animal Factories* (co-authored with James Mason), Crown, 1980; and *The Expanding Circle*, Farrar, Straus & Giroux, 1981.

Michael TRACEY, AO, MA, FTS. Born in 1918, he studied biochemistry at Cambridge. For the last 3 years he has been Director of the CSIRO Institute of Biological Resources after having been for 11 years Chief of the CSIRO Division of Food Research. In 1948 his *Proteins and Life* was published and in 1954 *Principles of Biochemistry: A Biological Approach*. He was joint editor of *Modern Methods of Plant Analysis*, published in seven volumes by Springer-Verlag between 1955 and 1964. He was President of the Australian Biochemical Society 1970—72 and is now President of the Australian Nutrition Society and a Foundation Fellow of the Australian Academy of Technological Sciences.

Derek E. TRIBE, O.B.E., B.Sc. (Agric.), Ph.D., D.Agr.Sc., F.T.S., F.A.I.A.S., F.A.S.A.P. Chapter 11. Born in 1926, he is a graduate of Reading, Aberdeen and Melbourne Universities and has been a staff member of the Rowett Research Institute in Scotland and the University of Bristol. For some 25 years he was on the staff of the University of Melbourne as Reader in Animal Physiology (1956—1966), Professor of Animal Nutrition (1966—1980), and Dean of the Faculty of Agriculture (1969—1973). He is presently Director of the Australian Universities' International Development Program. On many occasions he has worked as a consultant to international bodies such as F.A.O., U.N.E.S.C.O., I.B.R.D., I.D.R.C. (Canada), and the Rockefeller Foundation. He was a member of the U.N.D.P./F.A.O. East African Livestock Development Survey and for seven years was on the Board of Trustees of the International Livestock Centre for Africa. He is a Foundation Fellow of the Australian Academy of Technological Sciences. Together with A. Neimann-Sørensen he is Editor-in-Chief of the *World Animal Science* series.

Elizabeth S. WING, B.A., M.S., Ph.D. Chapter 2. Born in 1932, her undergraduate education was in biology at Mount Holyoke College and graduate work was also in biology at the University of Florida. For the last five years, she has been the curator in charge of the Zooarcheology Collection, Department of Natural Sciences, Florida State Museum, and has taught Zooarchaeology in the Anthropology Department at the Florida State Museum. She is senior author of the book entitled *'Paleonutrition: Method and Theory in Prehistoric Foodways*, published by Academic Press, New York in 1979. She is a U.S. delegate member of the International Council for Archaeozoology (ICAZ).

Subject Index